The Theory of Integer Classification Distribution And Factorization

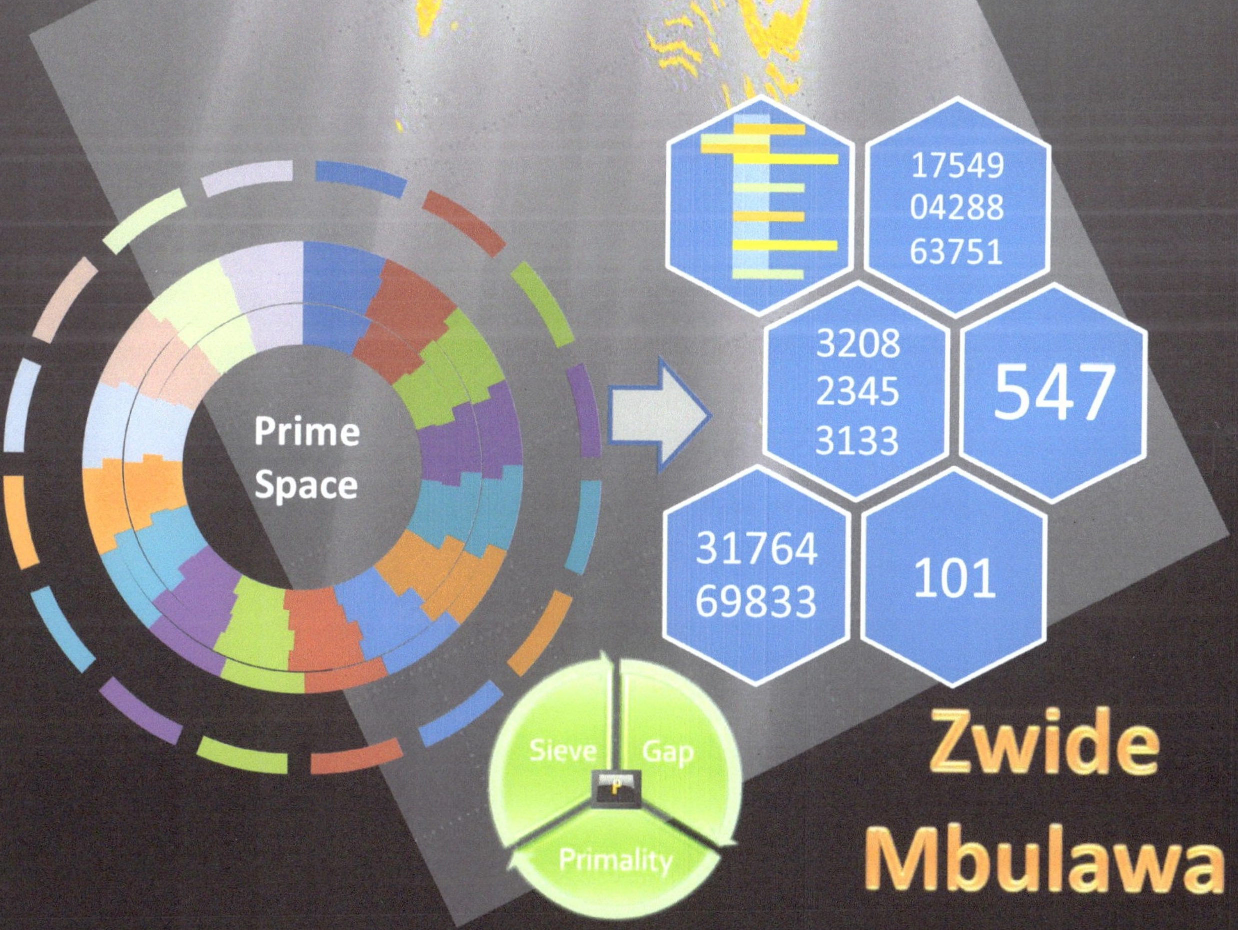

Zwide Mbulawa

Library of Congress Control Number: 2013907502

ISBN: Softcover 978-1-4836-3086-1
Hardcover 978-1-4836-3087-8
Ebook 978-1-4836-3088-5

Rev. date: 08/23/2013

To order additional copies of this book, contact:
Xlibris Corporation
1-888-795-4274
www.Xlibris.com
Orders@Xlibris.com

The circular conceptual model of the algebraic prime space that leads to structured factorization methods.

Key Words:
Number theory, integer distribution, factorization, prime number classification, number classification, primality testing, prime pattern, expository mathematics, fundamental theorem of arithmetic.

Private Bag 00184, Gaborone. Botswana
Telephone 00267-71845286
e-mail: zwide@hotmail.com
site: www.zwideprimes.com;
site: www.zwidembulawa.com

Contents

The Plumb Line

Mathematical fallacies arise because the constructs that have been formed and their manipulations are correct but the conclusions drawn thereon are inaccurate.

-And-

Hidden mathematical fallacies arise because the constructs that have been formed are incorrect but their manipulations are correct in as far as the framework in which they exist, and the conclusions drawn thereon are nevertheless accurate and applicable.

-Yet still-

The consistency of the system of axioms and axiomatic constructions defines the elements of a successful proof, yet effecting an axiom or an axiomatic construction in isolation results in a fallacy.

Yet one must always dare to the challenge. Something interesting may be behind the fallacy!

Zwide Mbulawa

22 January 2013

Dedication

To God the Father, the God of the Compressed Factorization Space[1].

To the development of expository mathematics, this is responsible for the context of the human face in mathematics.

To Batswana, who are serious, innovative and hardworking!

Zwide Mbulawa
April 2, 2013.

[1] God reveals His character through either an assignment or a task. To Moses He said " (Exodus 6:3 [NET]) *I appeared to Abraham, to Isaac, and to Jacob as God Almighty, but by my name 'the LORD' I was not known to them*." In Jesus Christ, He revealed Himself as the Savior, as the God of Reconciliation. As the Source of Knowledge, my testimony is that I met Him as the God of the Compressed Factorization Space in the quest for faster factorization. I also met Him as God of the Amazing. This was in the discovery of the Merge operation in prime numbers, it has amazing arithmetic properties! And I assume I have done my best to uplift His name by presenting this theory as accurately as possible.

Preface

The core of mathematics is the thought process, the fundamental axiom is that without a thought there is no mathematics. This book is an explorative mathematical thought experiment based on "*what you think*", "*how you think*", in the context of "*the art of thinking that results in discovery*". The formulas used here are not new, they are all over, what is new is the thinking and result derived from them. This book is a sequel to another book The Theory of Prime Number Classification, which focused on looking for a model for prime number behavior and distribution rather than a predictive formula. As more findings were made, it became necessary to have a different approach and focus rather than emphasize on prime numbers alone. Hence the first book is designed as a discovery book while this book is meant to consolidate and bring a more structured approach by presenting a model for all integer numbers. The book presents the only work in mathematics so far that has a concept of three things; classification, distribution and factorization of integers in an integrated and related manner. Generally in number theory, the prime numbers have sort of been the lone actors, and they generate the most research and attention. One may say that the composites are the supporting actors on the mathematical stage, and their role has been that of a shadow more than anything else. This book intends to demonstrate that this is not the case, that composites and pseudoprimes have a specific role to play in classification and distribution of integers. Infact, pseudoprimes generate prime numbers, and the context of this theory puts them in order as pseudoprimes, primes, and composites.

Indeed, it is a great personal challenge to deviate from normal and accepted thinking. It is even more discouraging to follow a path that ignores existing proofs or conjectures that have stood the test of time, though there was really no intention of disproving or disregarding any existing theorems and proofs. When in such a position, you question your motives, your radical approaches and conclusions, and wonder why you are swimming in a different direction. Yet at the same time you are confronted with an overwhelming conviction that the alternative methodology offers greater insights and conclusions, and pushes the frontiers of mathematical thinking. You then realize that application of innovation and doctrinal frameworks are based on personal decisions.

Two paradigms of recognition exist, the personal and community recognition systems. That is, there is a way you can recognize who you are and what you can achieve at a personal level, the same being true at group or society level. The community recognition plays a more important role than the personal recognition system because of the context of universality. Furthermore, the challenge is that the two paradigms are mutually exclusive and deterministic. There exists an Axiom of Recognition in mathematics that can be stated thus: *Any credible mathematics derives cumulatively and consistently from already proved constructs*. Loosely interpreted the axiom means conform or don't go out of the box, subsequently new and fresh ideas outside an existing paradigm attract suspicion. Credibility and acceptance is available as long as you have an identifiable community that defines the box, and secondly, mathematics is immutable. Once a proof, always a proof!

However I have not written this book nor developed this classification theory with the aim of getting recognition from the mathematical community, extremely important though it is. The reason is I became aware of the two types of recognition systems, and decided to ignore the community one. That would have killed all the creativity I had in mind as a community enforces conformity and tradition, which is the price I would have to pay. I wrote the book merely as an experience of my thinking, and to participate in the joy of creating new constructs that yield interesting insights. Consequently, fruitfulness rather than recognition became my major yardstick for validity and personal acceptance. Secondly, innovation and creativity force you to recognize yourself by trusting your intuition and believing in your thoughts and ideas. As a result, the element of human context in mathematical thinking and discovery has been preserved rather than being distilled and hidden in loftiness of deduction, neither has the religious experience been made to be inconsequential. It is only by taking such a position that I could courageously go out of the box to see and express things differently.

The big lesson learnt in the reality of life is that discovery starts with a painful but determined departure from the norm, or the support and credibility it provides. The other lesson learnt is that a previous discovery is a stepping stone to the next discovery. Some discoveries leave me baffled and amazed, and make it so difficult to imagine that someone else somewhere has not walked a similar path of reasoning and adventure. The discoveries have occurred through a combination of the following.

a) Personal consistent analytic efforts based on knowledge, expertise and the scientific investigative framework.
b) Through inspiration from persistent focus and desire to get an answer.
c) Through pure doodling that bursts into creativity in an environment of knowledge, experience and foresight.
d) Through God's inspiration and revelation. As the creator, God has the power to give you new and fresh ideas from His throne.

For me through my Christian faith, the last one is the cornerstone of my research activity, and I have no hesitation in giving my Father the glory for that. God's presence and guidance in the research made the big difference.

When I began this project, the only thing I knew about prime numbers was the college definition. So 2008 was spent doing research and developing the first classification framework, and 2009 and 2010 in producing the first book and computer program. The first book is an introductory emphasis on the concept of prime number classification. Then from October 2011 and the rest of 2012, these have been dedicated to producing the second book on the subject using a general context for integer classification. Effort to keep record of the discovery process was made because of the following questions I was trying to answer;

a) Do doctrinal revolutions occur at thought level only, or can there be a revolutionary change in the paradigm of interpretation and application. That is, can new mathematics come about dramatically?
b) If so, what are the personal experiences, and the inhibitions?

The two questions were inspired by Crowe's law that revolutions do not occur *in* mathematics. The article on The Fruitfulness of Doctrine In Mathematics is a short record on my findings and thoughts on the subject. It is actually a paper to me, writing to myself as a record of what I found out in the experiment I was conducting. I was in school. I also had to choose the language to use, the ontological and epistemological context. I chose and spoke the language of imagination and creativity, the language of ideas as a technique of expressing expository mathematics. I could have chosen the language of purely logic structures deprived of intuition, of strict rigor, analysis and proof methodology.

The ideas, concepts, structures, and theories developed here have been tested for consistency and applicability through computer programming.

Zwide Mbulawa

March 2013

PART ONE

As the book is both an academic exercise and an experiment,

Part One is actually a record of some of the major happenings and processes of the experiment.

I thought....and thought...

....thinking is an amazing thing. Where does it come from?

The Power of Classification

This number below has 141 digits. It took 20 minutes from 00:17am to 00:37am (20th March 2013) to show that this number is not a prime number using the rudimentary natural primality testing method. (See page 205 and merge classification in page 406). No super computer used! Infact the number of digits is not material using the merge operation.

7899005412897531889750091232154
6798076857297139752575031022984
5601221980081261201243217863908
7036578435895329811783290794738
29175386732109611

This is what this book or the thought is all about!

You will be surprised that even the product between any two numbers follows an exact distributive pattern (Integer Product Law) besides the ordinary odd-even framework.

The Framework

From my perspective, the discovery of the factorization primality test is the most important result. It was much unexpected because I spent a lot of time and energy trying to get a method for finding prime numbers using the developed framework. I had developed an algebraic test and had almost assumed there is no methodical approach because of the fact that prime numbers are random, and that the same applied to pseudoprimes.

The graphic below summarizes the journey of discovery. One can say that there were different phases, so these are just highlights. It all began with an innocent doodling of my Barclays ATM number code 1346 using the numeric keypad structure to get numbers I called prime pairs.

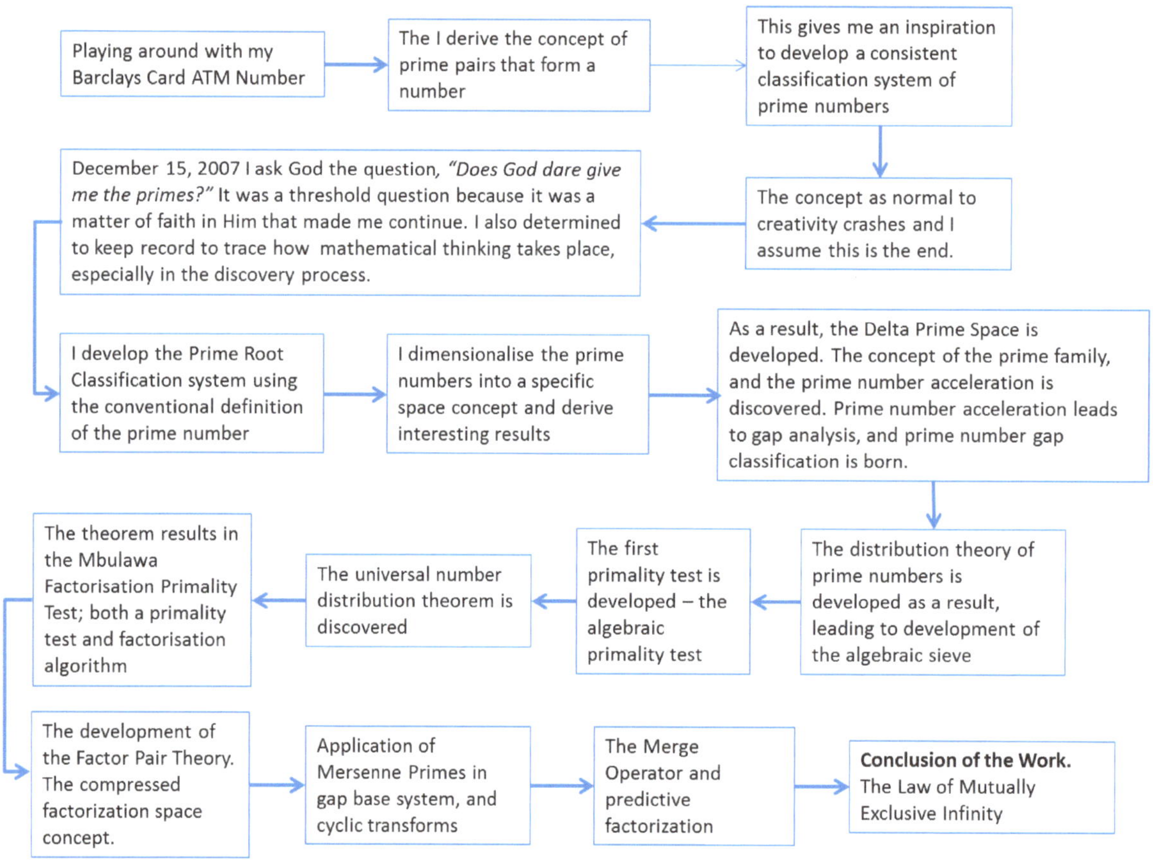

Figure 1

Some of the results are as follows.

The Delta Classification. This classification is based on prime gaps and derives the important concept of a prime family. This classification is based on an algebraic definition of a prime number, and is a foundation for other classification techniques that follow.

The Gap Theory Classification. Using prime families and prime gap axioms, one can demonstrate that there is a gap base system to describe all prime gaps. This becomes important for the UNDT framework.

End-Digit Classification. The occurrence of primes and pseudoprimes is defined by a specific pattern and cycle of ending digits to derive specific set behavior.

Polynomial Classification. Primes and pseudoprimes can be classified into consistent and well behaved sets having the same properties through the expression $a^n \pm h$.

The Primality Condition. This is an algorithm using inverse operations in the algebraic sieve in order to define a condition that establishes whether a number is prime or not. It consists of the P1-test (exclusion) that is also a composite test, the P1-test (inclusion) and the P2-test that test for primality.

The Universal Number Distribution Theorem (UNDT). This is a general approach to how numbers are classified and distributed with respect to each other in the context of primes, pseudoprimes and composites. The theorem justifies most of the approaches proposed by the classification theory. According to the theorem, all numbers arranged in a structured manner algebraically. This becomes a powerful framework in regard to development of primality, distribution patterns, behavior and factorization and other and mathematical structures.

The Mbulawa Primality Test condition. This is an algorithm for deriving primality and prime factors at the same time within the algebraic sieve.

The Mbulawa Factor Pair Primality Test. This is based on searching for primality using factor pairs. It is the fastest test of the four methods that have been devised.

The Structured Factorization Method. Making use of the gap base 2 and 4 structure, a construct is developed that becomes a method for carrying out factorization in a quick and efficient manner. This is done by introducing the concept of a bounded search that produces prime factor pairs.

Factor Pair Theory. This is the development of a comprehensive framework for the process of factorizing numbers in the UNDT environment.

The Compressed Factorization Space. The factor pair theory is then applied to optimize the factorization process.

Merge Operation. This is a form of addition-multiplication operation that results in a pattern called predictive factorization. No matter how large a number is one can predict what prime factors it will have, at least one or two. This means other factors can then be determined. The operation is also a powerful tool for integer classification, and therefore can be used to determine a number DNA.

The Law of Mutually Exclusive Infinity. This is the conclusion of the theory that seeks to demonstrate that integers follow a specific distribution pattern that has impact on how we conceptualize infinity.

Probability of a prime number. The probability of a prime number occurring is a challenge to calculate because of the random appearance of such numbers. However, using the UNDT framework, the probability is calculated to be a maximum of 0.78125%. Practically it decreases to zero towards infinity.

Prime Number Population Theory. This states that there exists a specific fixed prime space $[a, b]$ which defines the distribution and population dynamics of prime numbers. The theory has a fundamental consequence for the conception of infinity, and distribution of prime numbers at infinity.

Key Research Findings

This is a summary of some of the key findings of the research. The foundation of the research is based on UNDT expressed as follows.

A. UNDT

Universal Number Distribution Theorem (UNDT)

Let $-\infty \leq x \leq \infty$ *be an integer, then all integer numbers are universally distributed as composite, prime and pseudoprime by* $f(x) = (y_1, y_2, y_3, y_4, y_5, y_6)$*, where* $y_i = 6x + n$ *, for* $n = \{-3, -2, -1, 0, 1, 2\}$.

This lays the foundation for a unified approach for understanding integers in terms of classification, distribution and factorization. The distribution is in terms of gap base 2, gap base 4 and gap base 6, where gap base 6 is a subset of either gap base 2 or 4 for a given value of x. That is for $n = \pm 1$ this defines a unique structure of the set of primes and pseudoprimes called the ditation set D_t. The gap base defines a gap distribution system of prime numbers, and the probability of occurrence of a gap base 6 prime number is higher than for gap base 2 or 4.

- *Definition of Gap Base 2*. Let $n = +1$ then this forms a particular group of primes and pseudoprimes.
- *Definition of Gap Base 4*. Let $n = -1$ then this forms a particular group of primes and pseudoprimes.
- *Definition of Gap Base Zero*. Let $n \neq \pm 1$ then this forms the set of all composites.

In particular, the ditation set is given by $f(x) = (6x - 1, 6x + 1)$, $-\infty \leq Int(x) \leq \infty$, or described as the multi-valued function $f(x) = (x, y_1, y_2)$. The set defines the algebraic prime number. The definition is independent of the standard prime number notion, such that 2 and 3 are not elements of D_t, and 1 is an algebraic prime number.

An algebraic prime number is any number p such that;

a) $p \in D_t$.
b) *it is divisible by one and minus one; and*
c) *it is divisible by itself, and by its magnitude.*

B. Mersenne Theory

Sieve conditions are defined to operate on the set D_t to derive specific sets, for example Mersenne numbers. Similarly $M_p = 2^p - 1$ defines $h = 1$ in particular, other Mersenne numbers in general exist for $h \in D_t$.

Mersenne Theory	Comment
Mersenne Numbers – conventional definition	$M_n = 2^n - 1$
Mersenne Primes – conventional definition	$M_p = 2^p - 1$
Mersenne Primes **Gap Base 2**	$M_p = 2^p - 1$
Mersenne Numbers **Gap Base 4**	$M_n = 2^n + 1$
Mersenne Primes **Gap Base 4**	$M_p = 2^p + 1$
General Mersenne Primes generation	$M_p = 2^p \pm h$, where $h \in D_t$
There exists a sieve structure, the M(b) set $M_b = 6^n \pm h$	*For* n *an integer,* $M_b = 6^n \pm h$*, where* $h \in D_t$*, and* $0 < h \leq \infty$.
Primes for Gap Base 2 M(b) set	$f(p) = 6^p + h$
Primes for Gap Base 4 M(b) set	$f(p) = 6^p - h$
Exclusion Theorem *Let* p *be an element of* M_b*, then* $p \notin M_p$	A prime number cannot belong to the Mersenne numbers and at the same time to the set M_b.

Table 1

C. UNDT Laws

Law	Consequence and attribute of law
The Prime Gap Law	*a) The gap base structure has a probabilistic influence on the distribution of gap formation between prime numbers.* *b) The average natural prime gap increases in proportion to $n\{f(x_a) \cdot f(x_b)\}$ where n is the number of pseudoprimes in range $[a, b]$ at position x.* **Consequence and attribute:** The gap base structure creates a sieve that determines position in terms of 2, 4, 6. A prime or pseudoprime must fall into this structure without fail. This law works in conjunction with the Gap Population Law.
The Law of Pseudoprimes	*Let $W = f(x_a) \cdot f(x_b)$, where $f(x_a) = 6x_a \pm 1$ and $f(x_b) = 6x_b \pm 1$, then W is a pseudoprime.* **Consequence and attribute:** That pseudoprimes owe their initial existence to the prime numbers. But once established, pseudoprimes dominate the primes and whatever position they miss is occupied by a prime number. Because of the attribute of product, they self-generate.
Fixed Space Distribution Law	*There exists a self-determining fixed space [a,b] consisting of prime and pseudoprime numbers.* **Consequence and attribute**: This makes the prime distribution to be non-linear and set-based. !t helps to derive a more accurate behavioral theory of prime numbers.
Distributive Autonomy Law	*The distributive behavior of prime numbers and pseudoprimes is independent of composite numbers.* **Consequence and attribute**. The primes and pseudoprimes do not involve the composites in their fight for position in the space [a,b]. The law discriminates the behavior of integers, and makes composites a distinct autonomous class. The law has direct impact on the idea and concept of infinity in regard to prime numbers.
Gap Population Law	*For a given prime space $[a, b]$ in D_t the population of smaller gaps diminishes towards zero as x increases to infinity.* **Consequence and attribute:** The consequence is that primes are not infinite, and that the smaller gaps are the first casualties to go out of the scene. It also implies that the twin prime conjecture does not hold under this law. The law is observed practicality as prime number gap acceleration which can be measured for each gap.
Pseudoprime Gap Law	*The average natural gap to the next pseudoprime decreases tending towards 3 as $f(x) = 6x \pm 1$ increases.* **Consequence and attribute:** All the above laws make this law to exist. The exact pseudoprime difference is 2 between gap bases, 4 diagonally and 6 within a gap base. But once 2 is established, 4 and 6 are automatic. The behavior is unlike prime gaps.
Prime Gap Structure Law	*The natural prime number gap increases in multiples of six that are randomized by the inter-gap base relationship.* **Consequence and attribute:** That is, prime numbers are generated in an orderly manner, but are randomized by the gap base structure. This explains "weed" effect.
The Co-existence Theorem	*For every (x, y_1, y_2), then (y_1, y_2) are either a pseudoprime or a prime number.* **Consequence and attribute**: The theorem defines the algebraic sieve. This theorem is a consequence of the Distributive Autonomy Law.
Law of Integer Distribution	*For the range $[a, b]$, then $f(b) - f(a) = n_1 + n_2 + n_3$ where n_1 is the number of primes, n_2 the number of pseudoprimes, n_3 the number of composites and $f(b) - f(a)$is a constant.* **Consequence and attribute:**_There is a deterministic distribution of numbers such that this defines a space $[a, b]$. This is a result of the Fixed Space Distribution Law. Combined with the Distributive Autonomy Law, this law gives us a clear picture in regard to the distributive behavior of prime numbers towards infinity.

Table 2

D. Integer Product Law

The UNDT framework derives the above law and number categories. There are three types of odd numbers; the gap base 2 odd numbers, the gap base 4 odd numbers, and the composite odd numbers, called Class 1A. There are three types of even numbers in the UNDT framework; the Class 1B even numbers, the Class 2B even numbers, and the Class 2A even numbers, where zero is a class 2A number. This results in the integer product law stated thus.

> *For any two integers $p_1$and p_2, then the product p_1p_2 is either a composite in gap base zero, or a pseudoprime in gap base 2 or 4.*

Hence this can be elaborated further in the following manner.

1. [composite]x[composite] = [composite], where this is a pure composite or a general composite
2. [composite]x[prime] = [composite], where this defines a general composite
3. [composite]x[pseudoprime] = [composite], where this defines a general composite
4. [pseudoprime]x[pseudoprime] = [pseudoprime]
5. [prime]x[pseudoprime] = [pseudoprime]

The law is the basis for the factorization framework.

E. Factorization

The factorization space is defined in the context of the algebraic sieve, and the count function based on x_2 that seeks prime factors of a given number W through verification levels. The algebraic sieve results in four types of factor pairs that define the factor pair theory responsible for creating transient localized databases. An example is made using $W = 3{,}480{,}992{,}563$ to illustrate the factorization space and compression, and corresponding verification levels.

$W = ax_3 + b$	Primitive Factorization Space	Standard Factorization Space	Compressed Factorization Space	
	Trail pairs	**Standard Pairs**	**Efficient Pairs**	**Turbo Charged Pairs**
	$x_2 < x_3$	$x_2 = Int\left\{\frac{W+7}{30}\right\}$	$x_2 = Int\left(\frac{\sqrt{W}+1}{6}\right)$	$0.25\sqrt{x_3}$
$W = 3{,}480{,}992{,}563$	$x_2 =$ 290,082,714	$x_2 = 116{,}033{,}085$	$x_2 = 9{,}833$	$x_2 = 4{,}257$

Table 3

F. Primality Testing

The UNDT framework results in six primality tests, where these are mainly based on the factorization space (failure test approach). There is also structured primality testing not reflected here. The general condition for primality testing is $x_1 < 1$, but this may not be observed under compression conditions.

Primality Testing Procedures	Description
Algebraic sieve primality testing	Defines an algebraic sieve to derive prime numbers in a given range.
Algebraic primality test	Uses the algebraic sieve to do primality testing
Mbulawa primality test	Uses standard pairs in factor pair theory to do primality testing
Natural primality testing	For extremely large numbers, where testing is limited to classification techniques
Compressed primality testing	Two types of primality testing algorithms based on efficient pairs and turbo charged pairs

Table 4

G. Self-Testing Primality

This is an unusual concept hypothesized as follows. It uses programmable conditions for primality testing.

Self-testing primality derives prime factors and verifies primality structurally.

It is a structural determination of primality using the factor pair theory based upon specific conditions. For example;

Let $W \in GB2$, then if n = 2 and $a_1 = b_2$, then a_1 and b_1 are prime numbers.

It does not matter how large the number is, the above condition is true. This is for two prime factors. Another example is given for three prime factors, where n is the number of factor pairs.

If $n = 6$, and $W \in GB2$, then b_1, b_2 are prime, and if;

i. $b_1 \cdot b_2 = b_3$, *then* b_4 *is prime,*
ii. $b_1 \cdot b_3 = b_4$, *then* b_3 *is prime,*

The prime factors of W are either b_1, b_2, b_4 or b_1, b_2, b_3 .The CRAWL function is a self-testing operation designed for up to three prime factors in cycles using self-testing. The conditions define the self-testing structure.

H. Predictive Factorization

An operator $\oplus$ such that $\alpha_1 \oplus \alpha_2 = \beta$ denotes a merge operation, where α_1, α_2 are prime, pseudoprime or composite. The operator is like a mixture of addition and multiplication. It results in predictive factorization. For example the number 718997 has the factors 29 and 24793, and performing a merge operation gives the number 718997718997= 718997$\oplus$718997 which has factors 29, 101, 9901 and 24793. This also derives root factorization, for example 101 and 9901 are roots. The merge operator is also used for classification of large integer numbers.

I. Revised Concept of Infinity

The UNDT environment takes a second look at Euclid's proof of infinite number of primes and results in a revised concept of the type of numbers that exists at infinity through the infinity number theorem stated thus:

The infinite number space consists of composites and pseudoprimes only.

The theorem is a result of the Law of Mutually Exclusive Infinity which is a direct consequence of the application of the Law of Integer Distribution, the Law of Pseudoprimes, and the population theory of prime numbers.

J. Philosophical Position

From the book as an experiment, an experience, and a research exposition then the following is noted.

Axiom of Thought Type

There exists a specific type of thought process that results in discovery, as well as a creative mathematical thought, and a non-mathematical creative thought.

The Discussion

This section is a discussion on the experience of thinking and discovery. It shares the experience of the author, and the insights gained and learned.

1 The Fruitfulness of Doctrine In Mathematics

My greatest interest while carrying out this research was to establish;

a) how do I think mathematically?
b) what makes a thought to be a mathematical thought?
c) how do I discover ideas?
d) what are the requirements and demands upon my mind and personality to enhance the discovery process?

Indeed I had that great interest in the experiential part of the mathematical process rather than the clinical presentation of the result. My aim was to compliment my book on the science of mathematical thinking and also to contribute towards the development of expository mathematics as a powerful tool for theoretical framework in contrast to cumulative or organized mathematics. Cumulative mathematics is understood as a format of mathematics organized around axioms, theorem and proofs on the basis of strict rules of rigor and objectivity. It is cumulative on the basis that the founding assumption expects no revolutionary occurrence to take place such that it can nullify an accepted and tested proof. Consequently proofs tend to build on each other to form a solid unshakable foundation.

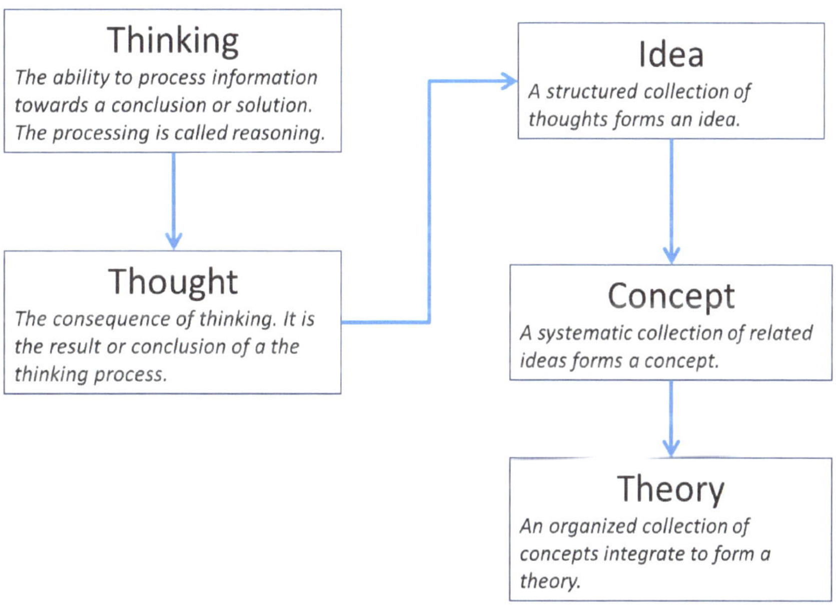

Figure 2

Mathematics is a thought process, without a thought there is no mathematics. When thinking occurs, it can happen as an image flash, unprocessed and without a distinct structure, or just as an observation. Whatever causes one to start thinking about something happens at an ordinary level of thinking, and must be converted into a mathematical thought. The ordinary level of thinking is defined by what is already common in terms of thinking about a particular subject, and the mind is not applied exhaustively in any particular manner. However, the occurrence of thinking as a stimulus from what has been observed internally or externally of the mind must rise above the ordinary level, because mathematical thought is structured reasoning that obeys certain rules.

The greatest challenge in the development of this theory of integer distribution has been the conversion of what I am thinking into a mathematical thought to form a concrete idea that has an element of consistency. I may take hours trying to release and arrange the thinking in order for it to produce a thought that is now mathematical in nature. For example, to discover predictive factorization, I used color coding to identify pattern. Before the color coding there was no thinking about something called predictive factorization, because the thinking was at an explorative stage. Color coding is a good instrument as a technique to aid the thinking process and to create an opportunity for conversion to take place, so as a matter of practice I followed this procedure. I already had a pattern from which the ideas were beginning to develop, but the color coding produced sub-patterns that indicated other relationships between prime factors and number structure. An analysis then took place whose aim was to elevate and structure the thinking towards a solid observation. During the analysis, the mind then plays around with various possibilities as it tries to find "the solution", which is the conclusion of the thinking – the mathematical thought. From the pattern I then discern the structure and by discerning a particular structure this assisted me to precipitate a mathematical thought, where this led to the formation of an idea I called predictive factorization. This was followed up by further development into a concept for systematic application. Thinking therefore is an escalating process beginning from the fuzzy to the structured thinking called a thought. The basis of a collection of thoughts then results in the formation of the idea.

It is not entirely mathematical to use the word "theory" to describe mathematical ideas, because a theory is a speculation. However, expository mathematics is an explanatory subject, and from that point of view it can consist of theoretical exposition of mathematical phenomena. Prime numbers and their general relationship to integers are phenomena. It can be noted that theoretical frameworks are normally preceded by well-developed doctrinal positions. Mathematicians admit doctrine as a mathematical framework for thinking out ideas, but are never completely comfortable with doctrine as an emphasis since it may admit too much intuitive methodology and possible thought error. Consequently the word theory is worse as it takes out the wind from the power of what mathematics stands for, since the structure of the mathematical system is exact through the axiom, postulate, theorem and proof. That's where the beauty of mathematics lies, and it is this approach that makes it to be a lofty and an objective subject. A proof can be improved, but the fact that it should be true for all time is the basic axiom for developing mathematics further. However in the application here, the collection of ideas and concepts has led to the use of the word theory in order to capture the essence that this is a different ideological approach to solving certain mathematical problems.

It is acceptable not to use the word theory provided you are dealing with one or two ideas that demand little doctrinal content and no theoretical formulation. Once the idea is established, then you simply derive the axiom, postulate and theorem with its proof. That's it, neat and to the point! Probably it is also acceptable as a format for cumulative mathematics because it allows for rigor and builds itself upon a system of axioms, postulates, theorems and proofs. As stated before, when an approach has a collection of ideas to define the mathematical framework, the use of the word theory becomes appropriate, especially in the context of expository mathematics whose emphasis is idea and concept building, meaning and interpretation, methodology and result.

1.1 Of ideas and consistency

The starting point of expository mathematics is the idea because its emphasis is to explain or discover things. The idea is then developed to mature into a concept. Therefore a concept is an idea with a definite identifiable structure. When an idea is born, it is fuzzy and not well defined. It may have several gaps that need to be filled with extra research and thinking out thoroughly. Therefore the law of discovery can be described in this manner:

> First let the idea flow with all its errors and short-comings, and then contextualize the idea for application. Once an idea or concept has been fully structured and developed, establish it through developing a property of consistency and clarity in the relationships it defines. As a feedback process, educate yourself on the idea or concept in order to mature it. Educating yourself may lead to further discovery, confirmation and

conceptual consolidation. Finally follow it up with thorough revision for setting, flow, clarity and expression. At the mature stage, establish validity through balancing consistency of result and fruitfulness of idea application to promote the process of acceptability.

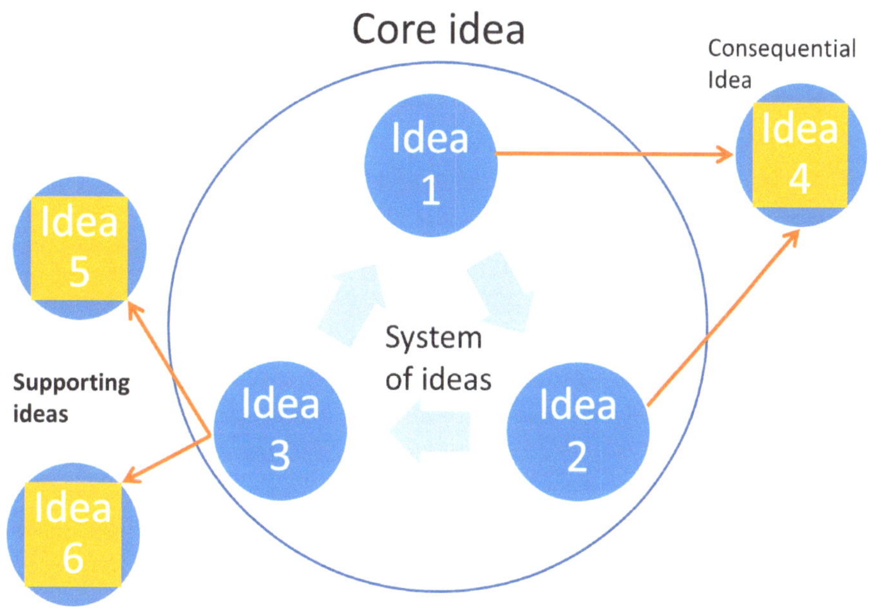

Figure 3

Setting of an idea is important for defining the type of structure it attains. It involves looking at issues around the idea, how they relate to the idea and conversely. Educating yourself on what you have created is a major challenge as you have to develop the ability to be detached and self-critical. When all the above steps have been observed, the last step is then to take note on the point of departure to already existing ideas, to find out what causes the difference and what explanations can be offered. This gives you the background for establishing contextual validity and realizing the extent of consistent result.

A concept is a structure that consists of several ideas put together such that they demonstrate a particular system observable as a reproducible construction. This normally consists of a core idea and supporting ideas, where consistency is critical within the core idea. For example three ideas can come together to form a system of ideas. A particular idea may generate one or more ideas, or two ideas may contribute to the formation of one particular idea, the consequential idea. The turning point in development of a theory is the formation of the structure of the idea or concept. In my case, this was the recognition and acceptance of the definition of the multi-valued mapping function $f(x) = (y_1, y_2)$ as representing one mathematical structure, and as a solution to describing prime number behavior and distribution. The fact that it rejected 2 and 3 as prime numbers converted it into a doctrine, since I now had to find reasons why this should be justified. The two numbers have been recognized as prime numbers for over two thousand years, this is really unacceptable, I would reason. However, the power of doctrine lies in its fruitfulness, hence I could not reject the "evidence" before my intellectual eyes presented by the doctrine and consistency of the ideas as a system.

There could be different ways in which an idea matures such that it results in a theorem or mathematical model. The above diagram is actually experiential; this is the general pattern that I find my mind works in order to conceptualize an idea.

a. *Axiomatic Development Stage*. The mind makes an observation, a problem is identified, and assumptions are made. Consequently an idea is born.
b. *Experimental Stage*. No one ever knows the power of an idea unless you experiment with it to establish its context and determine its viability. Some ideas die at this stage, while in other cases the substance of the idea begins to take significance.
c. *Research Stage*. The axiomatic and experiment stage will demonstrate that the idea works, but this stage investigates relationships to find out why it works. Consequently knowledge about the idea or on the idea develops. It now transforms the idea to a concept.
d. *Breaking idea barriers*. Sometimes after working on an idea for some time, it is difficult to continue with the idea, it's like one has come across a wall. Also some mistakes and thoughts in the idea are hidden, and the mind is unable to penetrate for a solution. The best way is to leave the current work, and try a new angle so that some other conclusion will explain some of the challenges of the first idea. That is, the two ideas interrogate each other within a system of thinking.
e. *Doctrinal Development Stage*. This is not a necessary stage for all ideas, but some ideas that are unique and question a current paradigm will require a justification for departure. This implies doctrinal development, where this leads to a revision of the idea, and testing of its basic tenets, and an explanatory development of the idea.
f. *Distillation Stage*. The idea begins to have implications, consequence and application. It is now in a position to see how it fits into the overall picture of its paradigm. It may even give birth to other new ideas and applications. A lot of error and shortcomings may be ironed out at this stage. Distillation is almost like the opposite of brainstorming as it seeks conceptual purity.
g. *Maturity Stage*. The idea comes to this stage through demonstration of consistency and also the extent of generality. Consistency is the proof an idea, and generality is its power of application. Consistency results in the establishment of theorems which may precipitate a particular mathematical model.

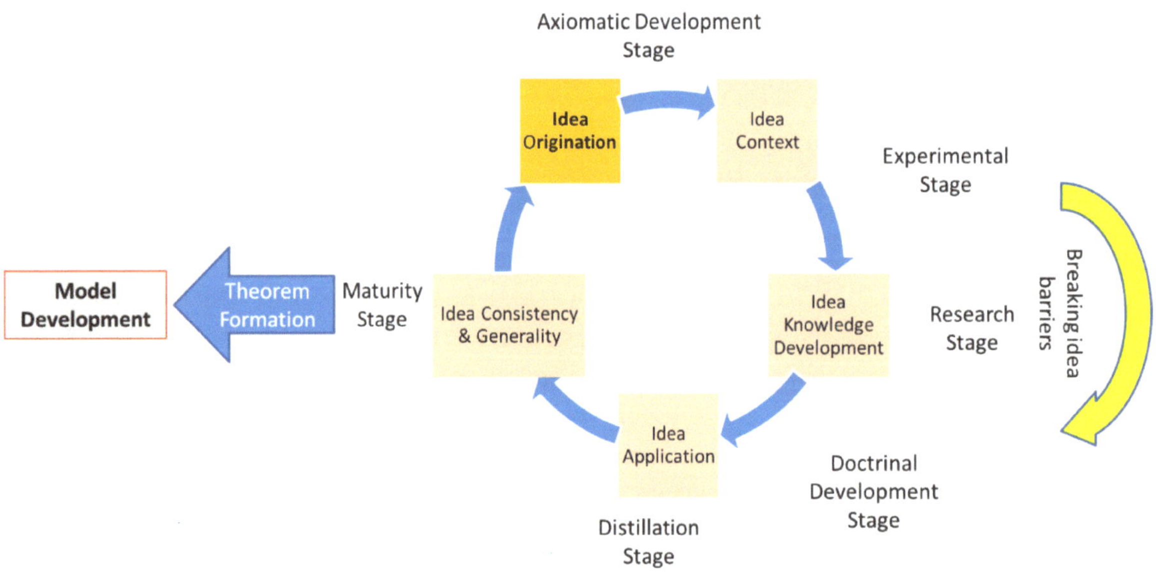

Figure 4

The development of the factor pair theory or the Euclidean factorization space underwent all of these stages. The distillation stage is probably the most intellectually painful because it answers the "why" question. Doctrine normally finds a lot of support at this stage because of the process of verification that is involved. However, the emphasis has shifted from the fruitfulness of the doctrine towards a validation framework. Having established the cause, it looks at the cause-effect relationship between ideas in order to understand why something is happening the way it is happening. Consequently it locates erroneous thinking and assists the idea to reach maturity. For the factor pair theory, looking at how k_1, k_2 and k_3 affect distribution of factor pairs helped me penetrate the theory or carry out the process of distillation.

The final process on dealing with ideas and concepts is that of polishing them. This may be regarded as an aspect of the maturity stage. This consists of two activities;

a) That of looking for gaps
b) That of expanding on implicit and implied statements

The process of polishing should occur when the idea or concept is already finished, and existing in a particular framework. Actually there should be a reasonable flow by this time in how the idea is expressed and developed so that polishing indeed brings the shine out of the idea. Often a paragraph will have gaps because some of the thinking was completed in the mind of the author, and the author understands where he or she is coming from. The reader may not have this advantage, and as mathematicians we normally don't have the patience to fill in the gaps. However, for expository mathematics it is an important activity because of the nature of expounding on new ideas and concepts.

Implicit statements are very natural, and they can also contribute to a paragraph having gaps. These are statements that express a fact but the meaning and the interpretation does not completely come out, and one has to read the sentence once or twice before comprehension begins to dawn. Such statements need to be unpacked by giving them an introductory sentence as well as a wrapping up sentence. Implicit statements mostly arise out of observations that are made as a result of imagination or are a picture in the mind of the author. Consequently they may not be so strong in their logical presentation. A lot of such statements are made when the idea or concept is still developing, hence the need to revisit the paragraphs as a polishing process.

The process of polishing is very good as it also adds the dimension of self-education on the idea that has been developed. This increases the accuracy of presentation and comprehension of the idea. Polishing also offers another opportunity to discover new relationships to other Ideas, and this is what has helped me to move from discovery to discovery.

1.2 Of doctrine

Doctrine is key in expository mathematics. It offers the explanation and the context of ideas and concepts. What then is significant about doctrine, or to put it in another way, what is the role of doctrine in mathematics? Doctrine may be regarded as providing four things towards mathematical development.

a. *Principle*. It gives us the reasoning of the principle on which the idea, the theorem is based. The principle helps to give us understanding of why the idea works the way it works in the context of principle.
b. *Rules*. The rules are useful for the application of an idea, and they are the ones that ensure that there is consistency observed and applied. Rules give credence to the logical process contained in the idea. Doctrine gives justification for the rules and explains how they work.
c. *Guideline*. Doctrine creates arguments that tend to guide us as to how to execute a method, they show expected direction in the application of the idea. This is the explanatory nature of doctrine.

d. *Code*. Doctrine gives an explanation of the pattern of the idea. The code is like a combination of principle and rules, the type of explanatory outline that develops the structure of ideas and concepts.

The development of rules tends to be a significant factor in doctrinal development such that sometimes the arguments and principles expressed in the doctrine are overshadowed. This is due to the fact that rules enhance the application and understanding of the logical processes, which is more useful in proof methodology. The explanatory nature of doctrine is important for expository mathematics because it creates the opportunity for discovery or mathematical growth to take place.

Doctrine development occurs in four ways:

a. *Doctrine as truth*. This is doctrine that is independent of events and activities. It is doctrine that uses or results in a set of principles, and tries to relate cause and effect. It can be identified by particular reasoning structures and outcomes of its consequent practice. Its power is universality, and the essence of truth. Proof methodology would have its emphasis here.
b. *Doctrine as experience*. This is doctrine that results from events, activities, interpretation and meaning of events and activities. It is not universal, but may be very strong in similarities between two or more people, or between two situations. Its power and justification is fruitfulness, and its limitation is that it is contextual. This would be the starting point for expository mathematics.
c. *Doctrine as a combination of truth and experience*. This occurs when doctrine as experience is subject to doctrine as truth. It represents the ideal condition where the objective guides the subjective. The advantage is that it combines the strengths of each; that is universality, essence of truth, the contextual construct and fruitfulness. Expository mathematics would have its emphasis here.
d. *Doctrine through ideas.* The way to change any doctrine is through new ideas, new methods, and new thinking. Unlike logic which is rigid once it has been formed, doctrine is more elastic in giving an opportunity to develop further through revision of existing assumptions. This is foundational for expository mathematics.

In order to bring about dramatic change in mathematics, one has to focus on the doctrine behind the logical structures. However in reality, the explanatory nature of doctrine always has a weakness of foundational expression and conceptualization. Consider the following argument between two students.

L: *You say 11 is a counting number, but what do you mean by a "counting number".*

M: *Yes its counting, not all numbers express the counting concept, some express a measure concept, numbers like π, others express a relationship concept such as the exponential e.*

L: *So if there is nothing to count, then 11 does not exist, it's an abstract thing?*

M: *It does not have to exist because you are counting!*

L: *Which means it exists before counting, I want that deeper definition that describes 11 before counting.*

M: *That's not a mathematical thought! That is why a definition or concept can be axiomatic. The number eleven has other properties besides counting that define its characteristics as number. For example, 11 is a prime number, which is a special attribute that says it divides itself and no other number divides into it to derive a counting number. So counting can be linked to other concepts that show a relation that identifies the number. The number also exists before 12 but after 10, which means it has a sequence attribute referred to as position. That implies 11 is a construct or symbol that is precisely identifiable because of a framework of order. Secondly, you don't have to count to get 11, you can use an arithmetic process to derive it.*

L: So that is what you mean by 11?

This argument illustrates the gestation that takes place in trying to develop the doctrine around a given mathematical activity. It also illustrates the power of fruitfulness, the fact that it allows us to understand something without contradiction in terms of the way we apply it, though there may be contradiction in the way we explain it. The basic question that "O" is asking really is, "What is mathematics?". The key is the desire to gain understanding about what makes it work, and this precipitates the doctrinal:- that reasoning that tries to tell us why it works, and deeper still, why it works the way it works? But we can still do a lot of things with 11 without that deeper understanding. It's like knowing how to drive the car, but you have no knowledge of the intricacies of the engine. In this case, "O" wants to know about the engine, not the car! Yet deeper still, the argument between the two student is captured by the following statement.

Conjecture

One cannot prove doctrine of a concept, but only the results thereof.

The conjecture demonstrates the dilemma of doctrine, because doctrine is not a proof of why something works, but it is necessary as a setting towards that proof. The only way for doctrine to have credibility is to link itself to results or application. The results, which are also an expression of the fruitfulness, are a consequence of the structure of the concept, and hence if correct they demonstrate the validity of the concept. We may then say by implication, the concept is proved. But there is no mathematical mechanism so far that prescribes rules for proving that a concept is true without reference to result and the accuracy of its application. So it is not possible to prove "11" as a concept, but we can use results of application to demonstrate the usefulness of 11. Consequently one cannot say, "here is a proof about this thing we call counting". It is for this reason that the emphasis of expository mathematics is conceptual consistency and fruitfulness of application rather than proof as this comes as a latter process of justification.

As a result, emphasis in doctrine is always avoided in mathematical result or proof methodology because of the above conjecture. You create a better impression if you walk without a using a stick, because the stick demonstrates a weakness of the body. In a similar manner, doctrinal emphasis is like using a walking stick in cumulative mathematics, but it is the fuel necessary for new thinking in expository mathematics.

1.3 Of fruitfulness

Besides the significant power of a proof, the power of fruitfulness in mathematical applications and in solving problems is equally appealing. Therefore fruitfulness has a history in mathematics, and its role has been recognized as being necessary for development of the subject. It is the lack of reliance on logic and expectation that makes it create controversy, as it also promotes freedom and the power of creative approaches in problem solving. We note the following from the quote provided.

> The Axiom of Solvability expressed his (Hilbert) determination to overcome such arbitrariness and to claim full measure of his freedom as a mathematician. The freedom of the mathematician permits him to solve by creation. This freedom is limited only by the need to be consistent and fruitful. For Peacock and other formalists of the early nineteenth century, fruitfulness was a matter of preserving computational utility, and preserving computational utility was essentially a matter of preserving the basic algebraic laws of number. In

other words, it was adherence to the Principle of Permanence. Fruitfulness in other words required consistency with the basic laws of arithmetic[2].

Certainly, the basic tenet of expository mathematics is the freedom to create provided the parameters of consistency and fruitfulness are strictly observed. This freedom allows alternative thinking and opposite way thinking as mechanisms for finding new solutions and precepts in mathematics, it does not emphasize preservation. Cumulative mathematics tends to be restrictive and mechanical in its discipline and development, and is ever suspicious of whatever does not build on a current paradigm. This is because it seeks to establish mathematics as a subject whose conclusions through proofs are immutable, hence emphasizing preservation. This then creates a limitation in terms of freedom to bring on board new concepts to solve anomalies or create new purposeful developments. It is for this reason that there is a fundamental difference of approach and methodology in the two frameworks.

Indeed it is difficult to justify fruitfulness as a doctrine unless we consider it under expository mathematics. The pursuit and exaltation of the mathematical result significantly dims the things that caused the result to be achieved because of the clinical approach that removes the emotion, experience and human face of mathematics. Similarly, the role of doctrine for development of mathematics is not that firm either as a practice, but it is a fundamental issue in expository mathematics. In practice mathematicians are taught how to prove rather than to develop doctrine to support ideas and concepts. The two key ingredients of doctrine is meaning and interpretation, and these are exactly the two tools employed in thinking and development of expository mathematics. Therefore the ability to develop doctrine is taken as a key practice for successful advance of expository mathematics.

We can consider fruitfulness under three categories.

a) *Fruitfulness of result.* This is the most common emphasis of the concept of fruitfulness. Here fruitfulness is in the solution of a problem to get a certain result. It is indeed difficult to argue against consistent result. This is how complex numbers gained their foothold in mathematics.
b) *Fruitfulness of method.* This is also a common understanding of the application of fruitfulness. A method may not have the standard of rigor required in the subject but provides a way of handling and deriving solutions to problems. This is how the derivative in calculus was born or accepted.
c) *Fruitfulness of doctrine.* If a doctrine leads to certain useful concepts, methods and results being derived, it demonstrates fruitfulness. Cantor's handling of infinity probably illustrates this point.

In any of these three categories, fruitfulness is as important as having a proof, and sometimes it may even overshadow the proof. The nightmare of any mathematician is to have a powerful argument that seems well constructed and sparkles with truth only to find that it is redundant due to some overlooked point. In that respect, fruitfulness provides a platform for demonstrating consistency between theory and application, whilst a proof would deal with issues of validity and generality. Whereas an idea is housed by doctrine in the way it defines its context, the development of an idea begins with three activities;

a) an observation, where this considers attributes and behaviors in the individual object of study
b) an association, where this considers attributes and behaviors through a collective pattern in a group of the objects of study
c) and a relationship, where this seeks an exact expression that describes the structures of the individual object or connects the group of objects.

---

[2] *Shapiro S., Vainwright W.J. The Oxford Handbook of Philosophy of Mathematics and Logic. 2005. Oxford University Press.*

The development of observation and association are the likely activities that contribute to fruitfulness of doctrine. Normally the observation should start by trying to answer a research question or problem definition. The association activity then tries to put structure into the idea or concept by looking for trends, consistent occurrences, patterns that form a link and connectedness between the components. The activity of observation and association together are the premise for discovery to occur in the thought processes of the mathematician.

1.4 Of pattern methodology

The tool that is used to develop an idea into a concept in expository mathematics is called pattern methodology. A lot of mathematical energy is spent on proofs and proof methodology. There is nothing wrong with that, but like anything in this world it has its disadvantages. The focus of expository mathematics is to spend the same amount or more of that energy upon pattern methodology – that is the ability and technique of looking for patterns with a systematic mathematical mind so as to be able to create idea constructs from a reasoning structure. The methodology rests upon;

a) developing and applying pattern search methodology
b) deriving an idea from the pattern
c) determining the context of the pattern, and hence defining its conceptual limitations
d) the consistency of the pattern in its power to develop the concept
e) looking for meaning within pattern as well as new interpretations arising from the pattern
f) the strength of the consistency in terms deriving generality for the concept.

The most challenging aspect is probably developing techniques and methods that assist you to identify patterns in the subject of interest. For example, in the development of the concept of predictive factorization, the ability to accurately state prime factors of a given number because of structure, I had to use a particular technique to extract the logic behind the pattern. And this technique I have developed over time, but it varies how I use it each time there is a challenge. First of all I noticed a particular pattern in numbers as well as their factorization result; that is observation. This was the pattern search methodology stage. To study the observation I then colored and categorized the table according to classification techniques. The effort was to derive an idea from the pattern. I then got a deeper pattern result that expressed relationships between the numbers, hence the context, which I then assigned the name predictive factorization. Consistency was established by finding out the conditions that make the idea to work. The big challenge was looking for generality, it was really a headache. I just could not penetrate the issues of generality. Then I colored and separated the table into odd and even patterns, and did further classifications using the UNDT environment, and the answers arrived that delivered the context of generality. All of this was happening in a back and forth manner of conceptual development.

Focus and concentration, with a desire to get a result are important techniques to develop. Such techniques can come as a result of experience, imagination, insight, inspiration or idea construction. In my case the extensive use of imagination has been the main driver as a pattern search methodology technique. To derive an idea from a given pattern is probably the next most important step. The idea develops from the type of information and solutions that the pattern seems to provide. These should also give an indication as to the context that the pattern raises in terms of the area it addresses, and the type of information and solution it provides. After that consistency becomes the "proof" that there is a finding, and once this is established, meaning and interpretation become major factors of consideration in the pattern methodology.

The above technique defines an approach in pattern methodology in the line of discovery. I call it thinking by PI-methodology (or $\pi\_methodology$ if you want to be mathematical); the manner of thinking that I apply in my reasoning process in order to get a given result. The "P" stands for presentation, the "I" for imagination. The method combines presentation and the use of imagination in such a presentation so as to derive patterns of thought that precipitate structure in ideas. It involves an approach where all facts or information are put together to form a picture and generate the structure of an idea. At times as mathematicians when we have worked on an idea and derive a structure, we then place our argument in such a manner as to disguise its origination to give the impression that this was purely a logical deduction. I have done the same thing myself, distancing the structure so derived so as to give the impression that its arrival was clinical and pure until I realized that I am actually under-developing myself. Yet many of the ideas and structures we derive are a consequence of a particular type of process necessary for mathematical discovery, and the imperative is to develop this as a mathematical tool rather than let it suffer an obscure position.

Input a number:		☐ Show repeated factors as p^k	Run time:
990078557990078557	Factors:	7*11*13*13*13*19*52579*5858453	00:00:00.013
990078559990078559	Factors:	7*11*13*19*1453*52579*681403	00:00:00.010
170093761170093761	Factors:	7*11*13*19*52579*170093761	00:00:00.010
888877775888877775	Factors:	5*5*7*11*13*19*59*113*5333*52579	00:00:00.004
537902193537902193	Factors:	3*7*11*13*19*379*52579*473089	00:00:00.006
67431068967431053689	Factors:	7*11*13*13*19*257*52579*201829	00:00:00.009
410986745410986745	Factors:	5*7*11*13*13*19*181*181*193*52579	00:00:00.004
390009593390009593	Factors:	7*11*13*19*52579*390009593	00:00:00.010

Figure 5: Using www.javascripter.net as structure for doodling and derving factorization root law.

The $\pi\_methodology$ can also be regarded as an advanced way of doodling as in it you arrange things in a particular order of presentation to look for a possible picture. I discovered the factorization root law (FRL) because I was doodling with numbers in a matrix structure, having a row of numbers on the left and a corresponding row on the right consisting of the factors of those numbers. If there structure of a matrix was not present, the possibility of the discovery is very small, but because I was doodling within a structure, a pattern emerged. Figure 5 shows 7, 11, 13, 19 and 52579 as the factorization root of the numbers, which were chosen at random.

Therefore the matrix structure was able to form a picture, and the reasoning was able to spot the pattern. There is no logical reason that could have derived this pattern, but perhaps after such an observation, then the mind can apply itself to look for the logical reason behind the pattern. It was by imagination that I said to myself, "*what if I arrange the numbers in this manner, what result will I get*", but at first it began in a random manner. Then I looked for pattern, discovered a general pattern, then I discovered a specific pattern that can state a law. And since the mind is in a process of thinking, I will then wonder if it is possible to create numbers with roots as a way of generating predictive factorization. What is important to note is that this is also a skill to think in this manner, just like it is a skill to think in a logical or creative manner. However, though it is a skill, anybody could have stumbled across what I did, because discovery from doodling really is an unexpected blessing. This was a blessing because it is so simple, though it is not accurate to attribute things to being a blessing in mathematics, but under pattern methodology it is a possibility. At the end of the day, the fundamental axiom is that every mathematician is a human being; as such we can be gifted in particular ways and blessings do occur.

Two other examples are given to illustrate the technique, this is for $W = 71 \cdot 71 \cdot 71 \cdot 103$ and $W = 17 \cdot 59 \cdot 43 \cdot 43$. The first is classified as a gap base 2 number and the second as a gap base 4 number. By presenting the information as a picture, it is then easier to look for a pattern that explains how factorization takes place. It is just like drawing a graph and using its shape to understand something. The picture formed by the presentation can relate complex entities together to form relationships for consistency, interpretation and meaning. Here the idea was to look for a pattern that describes a concept I was trying to build on compressed factorization procedures. What then develops is that the consistency of the pattern proves a point or observation, but it does not prove a theorem. For example for the gap base 4 number, the factorization level keeps on going down by 1 whereas it remains constant at level 2 for gap base 2. That is, P4-3, P3-2 and P2-1 for the first number and P4-2, P3-2, P2-2 for the second number. Furthermore, that the number that is factor-2 will always be prime, and this is called a "system observation". Laying out information in this manner of factor-1 and factor-2 makes it easier to study and comprehend behavior. The same applies to the observation on levels, it also represents a system behavior and differentiates types of numbers. Once a system observation is picked up, then it is tested for consistency. Once consistency is established, then knowledge development takes place and a theory is constructed.

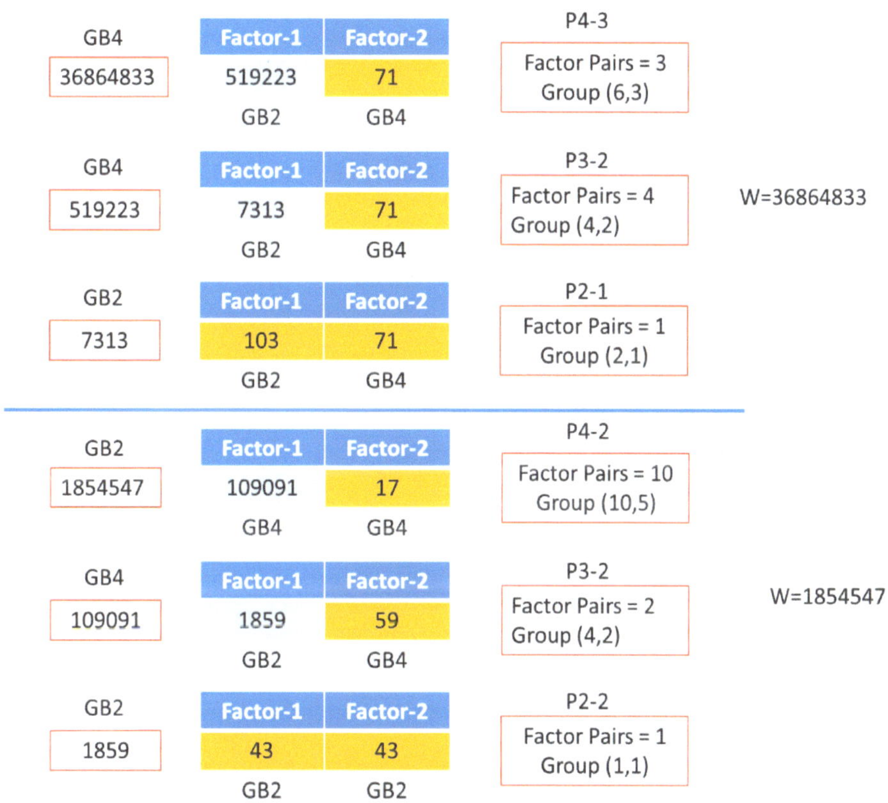

Figure 6. . The PI-Methodology in compressed factorization space concept

The application of this method is evident from the plentiful graphics and tables used in my books. Even though I have been using it for some time now, but I noticed it as a tool when I was developing the concept of compressed factorization. This is a factorization technique based on self-testing primality. I was using previous knowledge to develop this concept, but it was proving difficult to formulate until I changed the way I think about it through imaginative presentation. The presentation then helped me to develop a pattern, and the pattern helped me to develop the concept. So how I present information to my mind aids the thinking process tremendously, and is especially useful for discovery.

The PI-methodology is then used along with thinking by elimination. This is a common approach, and constitutes the logical method to deal with idea building. Thinking by elimination involves;

a. identifying the core idea, and the key assumptions that lead to a standard context of the idea
b. identifying anomalies that arise from the application of the idea; what cuases the anomalies and how to deal with them in the context of the overall idea
c. identifying areas of error or deviation from the standard context
d. identifying peripheral ideas that support the core ideas.

Pattern methodology as a system therefore provides proof of relationships through consistency, and in so doing contributes in defining structure for concepts. Proof of relationship can then lead to formation of key assumptions called axioms and the construction of a theorem. Further knowledge development around the concept can provide an environment for proving the theorem.

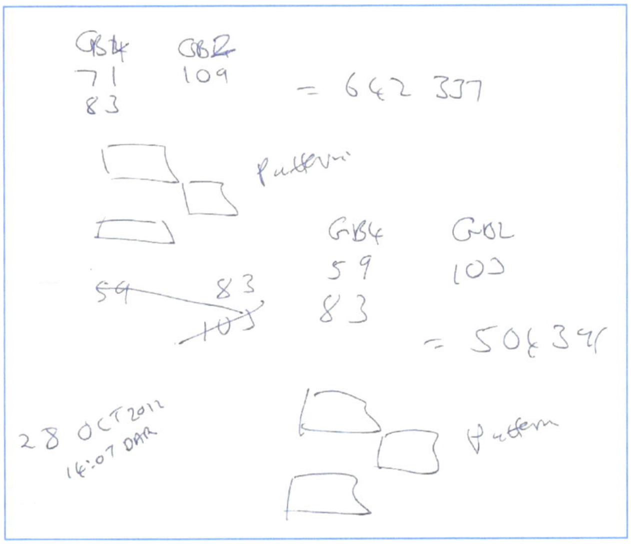

Figure 7. "Thinking-in-patterns" in distributive behavior of prime factors

Another very good example of pattern methodology in this book is the derivation of the structured factorization approach, and the development of classification techniques. The graphic example of a "thinking-in-pattern" is given illustrating pattern methodology in action, where the boxes indicate how factor pairs distribute to derive prime factors. Here I was thinking in terms of patterns, not in terms of numbers. I was looking and investigating behavior, and my mind had to arrange understanding and perception in terms of patterns rather than logic. Generally this approach here and elsewhere led to the derivation of the basic rule for pattern methodology expressed by the following rule.

$$Pattern = pattern + 1$$

The rule implies the observation, associations and relationships that we establish for a given pattern are consistent with the next pattern. Therefore "$pattern + 1$" conceptually denotes subsystems that successfully come together to form a system, where this is a principle of idea to concept development through the technique of pattern search. Consistency of pattern has the same power as the proof method under this rule, since consistency provides circumstantial or demonstrative proof in contrast to logical proof. Circumstantial proof depends on interpretation and context while logical proof depends on the power of inference and structure that is already established. That is, demonstrative proof uses concepts and results to provide the proof so as to build the required construct as a valid mathematical structure. As noted already, once the structure is defined, a doctrinal environment is created for logical proofs to be derived. The rule does not mean the patterns are exactly the same, rather they are a subject to the same observations, associations and relationships. For example prime numbers and pseudoprimes are different types of numbers, yet they subscribe to the same observations, associations and relationships.

Though powerful in its own right, pattern methodology tends to be despised in the mathematical tool set because it does not appear rigorous but rather is explanatory. Secondly, it does not necessarily provide a proof and therefore appears to be of no consequence. However, its power lies in the ability to observe, create and innovate mathematical ideas to generate new thinking and fresh insights. Pattern methodology depends on the following.

1. *Recognition of pattern trends within a group and between groups.* This is an evidence searching mode of mathematical thinking, and this is where innovative observation can play a role. This type of observation is in contrast to contextual observation that seeks to establish an observation within an already defined and existing framework.
2. *The basic idea framework that forms the concept.* Looking for an idea framework is an important activity, it is almost like developing a proof. An idea can fail to produce a structure that is consistent, where this is a possible indication of failure.
3. *The arguments of justification.* These are arguments that sound like a proof but are not strong enough to define one. They are responsible for developing doctrine around the idea and strengthening the idea into a concept.
4. *Consistency of pattern.* This is defined by the rule above. In expository mathematics, consistency of pattern has the same role as a proof in cumulative mathematics.
5. *Integration of pattern outcomes to develop theory.* A theory is a context that defines and expounds a relationship between a set of doctrines. The UNDT is a typical example.
6. *The interpretation maxim* stated thus: meaning emerges from pattern to create context, and pattern emerges from consistency, and consistency creates an interpretation paradigm. These four elements are critical as tools for creating, forming and testing new ideas and concepts.

All of the above make a conceptual structure or setting for defining fruitfulness derived from application in mathematical concepts. In particular, arguments of justification, definition of meaning of created constructs, and interpretation of application are responsible for making the framework for the development of doctrine that goes along with the idea or concept. If one studies the work of Rowan Hamilton on real numbers and imaginary numbers, it is clear that he was following such a methodology of thought.

One may argue and say the emphasis on pattern methodology is overboard and not really practical. For example consider the following theorem:

> **General Factor Pair Limit Theorem:** *Let $W \in D_t$ have m prime factors, then in a range* $[1, Int(\sqrt{W} + 1)/6)]$ *the number of unique factor pairs n;* $n \leq 2^{m-1} - 1$, *where* $1 \leq m$.

If $m = 1$, then $n = 0$, and the number is a prime number, it is its own prime factor. Where then is pattern methodology here, since theorems are the essence of mathematics? The stark reality is that one will not see any evidence of pattern methodology. If we assume that idea development is an incubation process, then pattern methodology is in the pre-incubation phase as well as the incubation phase, while the production and exposition of a theorem is at post-incubation phase. The metamorphosis between these stages may be so great that there is indeed no visible evidence. The graphics show the thinking that led to the above theorem. The intent of the theorem is to relate the number of prime factors a number has got to the number of factor pairs of the given number – the transient database. This is a limit defining theorem because it states the range in which such pairs are found. If $n = 0$, then the number is prime, and no factor pairs exist. The power of the theorem lies in the fact that the bigness of a number does not necessarily change the number of factor pairs, which is what we are inclined to think naturally. A big number may have three factor pairs which you use to determine the prime factors, so finding such prime numbers is speedy, and the same being true for a smaller number.

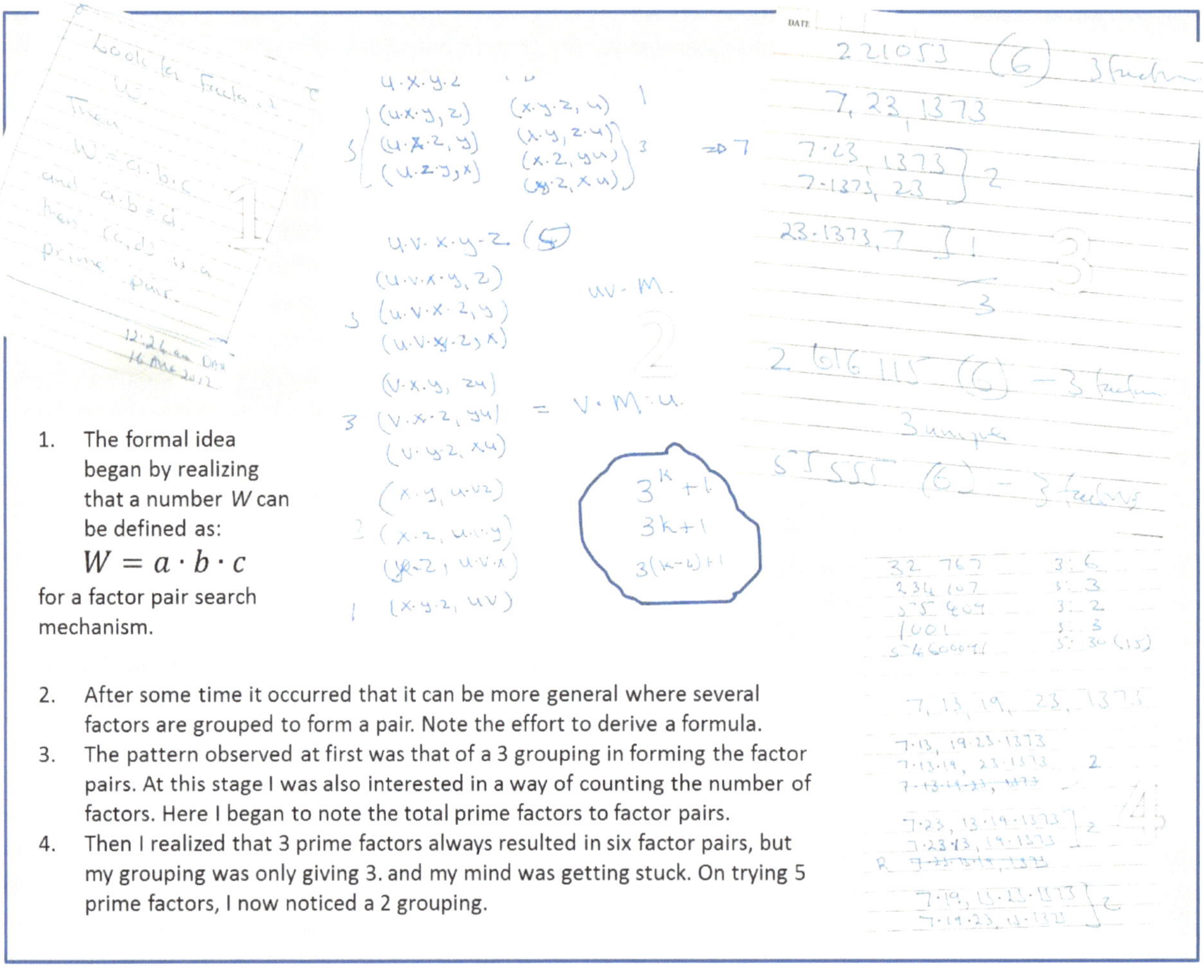

Figure 8

However, the theorem did not just drop into the mind no matter how clinical it looks; it came as an end-product of a process of pattern methodology. That is, one cannot have post incubation if there was no pre-incubation, and furthermore what you get at the post stage depends on the scope and scale at the foundational stage. Even then, in dissecting the theorem;

a) $W \in D_t$ is a doctrinal statement. The doctrine arises from the construction of the set D_t.
b) $n \neq 0$ is a conceptual statement based on the validity of the doctrinal statement, and defines the simplest primality test criteria.
c) $[1, Int\left\{\frac{W+7}{30}\right\}]$ is a direct consequence of the UNDT[3] framework. The non-UNDT[4] framework does not have the power to produce such a result. Hence this is a structural parameter in the theorem.
d) "Factor pair" is a terminology that is directly based on and is representative of pattern methodology. Factor pairs cannot be developed as a concept independent of the UNDT framework.

That is, a forensic audit will bring out the footprint of pattern methodology as shown by the above graphics.

[3] UNDT: Universal Number Distribution Theorem. This is a unified theory on number distribution and behaviour.
[4] Non-UNDT refers to the traditional approach concerning prime number definition, prime number theory and integer theory.

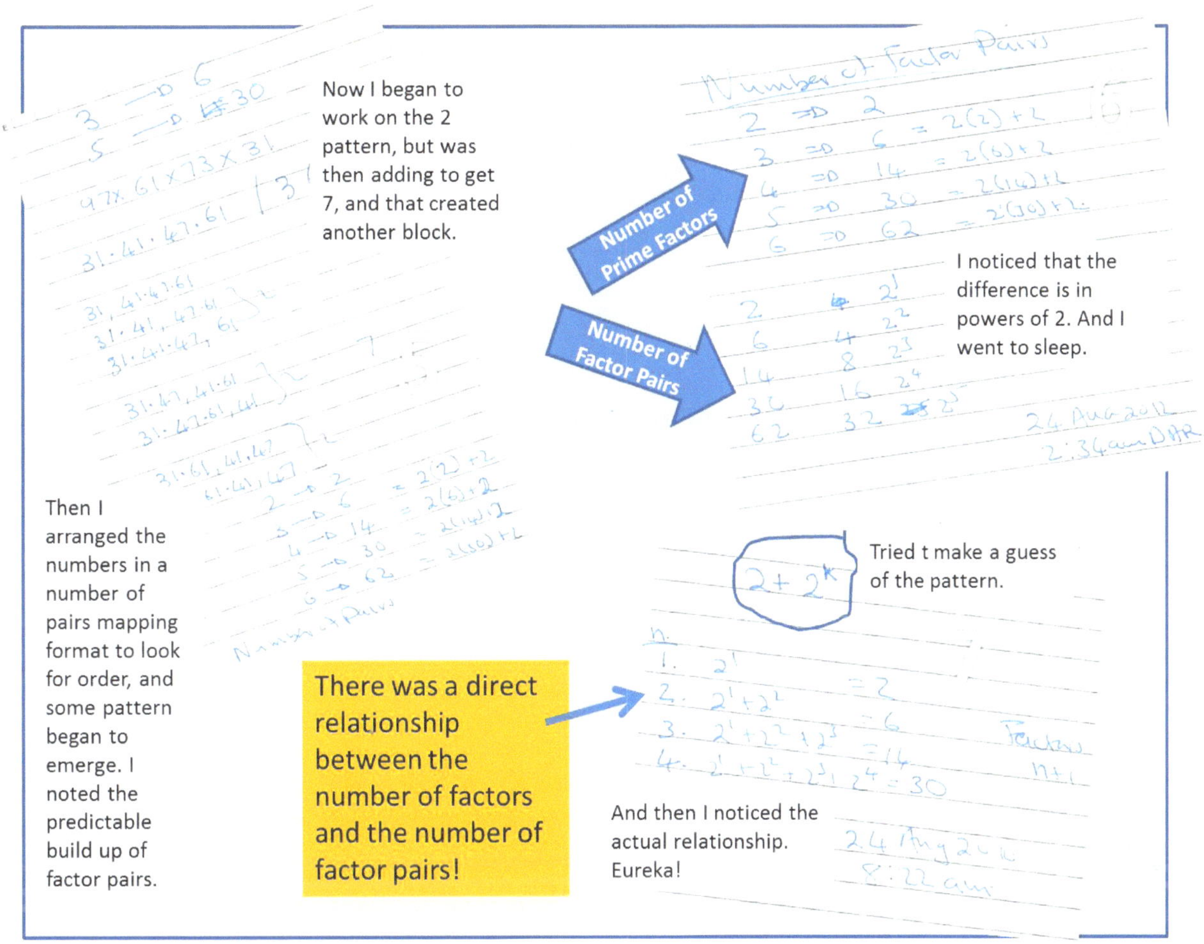

Figure 9

Besides consistency and fruitfulness, one of the important aspects of pattern methodology is the development of ideas or concepts that are significant on the basis of resulting theorems or problem solution. This is also a critical parameter for expository mathematics. The significance parameter defines the impact that the concept results in, and helps to propel it towards acceptance much faster. A good idea may fail to be significant because of limited relevance in application, or because it achieves a poor perception. For example, deriving factor pairs alone may not be a significant result, but using them to get prime factors makes the idea significant. Significance can also be established through a resulting paradigm shift in thinking. For example, the theorem creates a paradigm shift of making us not to look for single numbers in prime number factorization, but rather focuses on looking for related numbers called pairs. Through factor pairs, the theorem introduces the concept of an instant database with an upper bound from which to search for prime factors. This is conceptually significant for application. Consequently, significance is regarded as a psychological parameter that may or may not be material depending on fruitfulness in application.

> *Now how do you establish consistency in pattern methodology that has the same level of acceptability and practical implication as a logical proof?*

Probably the answer here lies in the strong combination of significance and fruitfulness of the idea or concept. But all of this will also be a function of the current mathematical paradigm. Expository mathematics depends a lot on fruitfulness of the concepts developed as it does not emphasize on the power of proof. It is therefore perfectly at home with pattern methodology. The assumption is that proof methodology will take over in due course to prove the concepts so defined. The difference between proof methodology and pattern methodology is that the latter is a concept building technique, and an evidence searching and gathering tool while the former is a truth establishing technique from an objective position.

In the context of concept building, evidence searching and gathering, consistency is measured and studied in two ways. The first is fruitfulness of application, and the second is prediction, where these two can frame the level of significance. If these two are coherent in structure, then consistency exists that demonstrates proof of relationship. As noted already, proof of relationship differs from proof of a theorem because we can demonstrate that structures form a relationship without necessarily proving theorems about them. Consequently foundationally, proof of a relationship a step towards proof of a theorem.

An illustration of this is made where pattern methodology was used to find the exact relationship in line with the given theorem above.

Factor Pairs	Factor-1	Factor-2
1	7313	71
2	5041	103
3	103	5041
4	71	7313

W= 519 223
Group (4,2)

Prime Factors
71, 71, 103

Table 5

The number 519 223 belongs to a group defined as (4,2) meaning 4 factor pairs in gap base 2, and two in gap base 4. The maximum under 3 prime factors is (6,3) shown at level 1. Numbers that have 3 prime factors will fall in any of the three groups (6,3), (4,2) and (2,1) according to the graphic below.

					Prediction		Actual	
No. of Prime Factors	2	3	4	5	6	7	6	7
Level 1	(2,1)	(6,3)	(14,7)	(30,15)	(62,31)	(126, 63)	(62,31)	(126,63)
	1	2	4	8	16	32		
Level 2	(1,1)	(4,2)	(10,5)	(22,11)	(46,23)	(94,47)	(46,23)	(94,47)
		2	4	8	16	32		
Level 3		(2,1)	(6,3)	(14,7)	(30,15)	(62,31)	(30,15)	(62,31)
			3	6	12	24		
Level 4			(3,3)	(8,4)	(18,9)	(38,19)	(18,9)	(38,19)
				4	8	16		
Level 5				(4,2)	(10,5)	(22,11)	(10,5)	(22,11)
					5	10		
Level 6					(5,3)	(12,6)	(5,5)	(12,6)
						6		
Level 7						(6,4)		(6,3)

Table 6. Relationship between the number of factor pairs and number of prime factors.

The number of prime factors row indicates the prime factors involved, for example 3 implies 3 prime factors. Then Level 1 is given as (6,3) meaning that the maximum number of factor pairs is 6 or 3 depending on the gap base of the number. Infact, for any number of prime factors m in level 1 then $n = 2^{m-1} - 1$, and we have $(2n, n)$ according to the theorem. Hence each level defines the number of factor pairs we expected in the transient database, but the theorem defines the maximum. Now this was done practically up to five prime factors. Then a pattern was investigated by using the above creative presentation. The colored boxes indicated the differences, for example for (6,3) and (4,2) the difference is (6-2)=2. Using the pattern observed, and then the number of factor pairs was predicted for a number with 6 and 7 prime factors. This is where the power of pattern methodology comes in as pattern is the basis for consistency. Having consistency meant a prediction could be made, and this was done successfully because only (5,3) and (6,4) were wrong in the prediction.

However, the power of pattern methodology at the end of the day depends on observations made, where such observations can be quantified as a system through equations and other mathematical tools. The aspect of observation though powerful introduces the human element in discovery. As already stated, observation really depends on what we see and do not see. Whilst you can be certain of what you see, you may not be aware of what you do not see. This is the basic premise for refereeing and the strength of cumulative mathematics – the consistent reference to work already done ensures that all that has to be seen has been seen. It is quite common even for the genius to make a huge blunder on what has not been seen. Even the power of prediction may miss out what is not seen; the prediction is accurate but is based on what has been seen. Hence the mathematics is fruitful and consistent but does not present the full picture because of what is referred to as systematic error. This type of error is difficult to detect because it is part of a given structure and conforms to the structure in behavior. The above example is actually a case in point.

This is my experience and observation. Systematic error is not a factor of consistency or fruitfulness rather is an issue of two things.

a. *Accuracy*; this is achieved provided all possible permutations of a problem or behavior have been considered. This is a factor of computing power, experience, further insight and observation, and also sheer luck. Additional changes normally are able to correct accuracy issues without affecting the first observed result. They may extended and broaden the understanding or pattern observed without necessarily changing the underlying structure.
b. *Anomaly conditions*. These are observations that do not affect the overall behavior of a system, but when they occur will reflect a deviation from the expected. What is important about anomalies is to understand when they occur or why they occur. This gives them their place in the overall structure of the concept, and in this manner they are "controlled".

Accuracy may be a function of determination and rigorous approach in handling a problem, as well as looking for an opinion to cover what you may have missed out. Anomaly conditions may be much more difficult to pick up, or they may be easy, but they are not predictable. The process of determining accuracy and establishing whether there are anomaly conditions is normally the last stage of pattern methodology. It is carried out easier when the concept is fully developed and the areas of consistency are clearly defined.

One of the mathematical theories that I developed, called the Theory of Ditation Mathematics, is based on the discovery of systematic error that actually precipitates the Hamiltonian Triplet problem.

1.5 Of fruitfulness in Mathematics

Fruitfulness in mathematics implies the ability of an idea, construct or method to provide a consistent and productive result in a profitable and effective manner inspite of limitations in rigor. The result becomes the "proof" of the method or idea, and this is ideal for the context of expository mathematics. What is then important in the fruitfulness of a doctrine is the evidence that is used to build it, the context the doctrine establishes through its ideas, tools and methods, and the results or fruits that come out of it. Consistency of both application and results then lends credibility to the usefulness of the doctrine, and consequently helping it to find a place in cumulative mathematics.

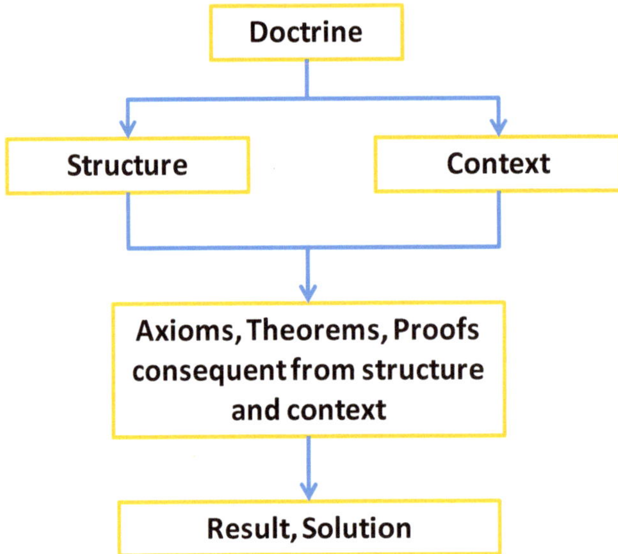

Figure 10

Fruitfulness of doctrine implies that the doctrine is able to generate ideas, tools, methods and results that can be used to solve problems, and shed light to particular mathematical findings. Consequently the ideology developed here contradicts the established format of a prime number, where it now attains the structure of an algebraic definition. However the gap base classification of primes leads to a new sieve methodology, two new primality tests, and a theory of number distribution that allows us to develop factorization algorithms. These are the fruits we cannot argue against inspite of the strength of our mathematical tradition.

It is only in the studying of the history of mathematics that we begin to see and understand the reasoning, and the doctrine behind some of the great mathematical conclusions. Yet this reason or doctrine was actually responsible for the discovery or result. Expository mathematics is one area where doctrinal context as a mathematical tool should have a legitimate place as a form of discipline to expand mathematical research.

In conclusion, doctrinal fruitfulness as a mathematical tool or method is key for expository mathematics. The work of Sir Rowen Hamilton was done as an expository work, however cumulative mathematics only focused on the desired result of $(a,b)(c,d) = (ac - bd, ad + bc)$ and its application. His work illustrates the fact that in order to create a change in paradigm in values, tradition and methodology, there is need to have a strong doctrinal context to bring about the transformation. In other words, doctrinal fruitfulness is a necessity for a mathematical revolution to occur in mathematics – this is almost a law. This must be accompanied by a strong character willing to sacrifice the benefits of following the current paradigm and standard expectations.

1.6 Of logic and immutability of proofs

Logic consists of a series of steps that are repeatable as a pattern and have three types of arguments:

- *an initial argument*. This argument can be expressed as a set of assumptions. The assumptions are axiomatic, and consequently they give the logical structure that results a rigid nature.
- *a substantive argument*. This steps is formed from a deterministic arrangement and reasoning of thoughts to form a struture that can be named and assigned attributes. The deterministc arrangement discovers or expresses relationships within the consutruct or to other constructs. Theorems and consequent proofs are

tools that are used to develop the deterministc arrangment and reasoning to form specific structures that become points of reference.

- *and an end argument*. This is the outcome of the substantive argument expressed in the form of observations and a conclusion. Normally a proof is a substantive conclusion in mathematics, it settles arguments and issues with regard to the fundamental assumptions made.

While a logical structure can be discovered, once it is set, changing it is generally regarded as an impossible thing, otherwise we would not have something called mathematics. Hence the focus of mathematics is the production of a logical structure, where a combination of such structures acts as a referential system of knowledge.

In the context of proof methodology, we may conclude that doctrine plays an important role in defining the context in which a proof exists. Doctrine creates the structure that determines the validity conditions necessary in a given proof. Since the focus comes on the structure, not much attention may be paid to the doctrine that created the structure. Mention of this has already been made where Hamilton's product represented a structure that resulted from a doctrinal position that he took pains to develop and expound on. This brings us to the following conclusion; that a proof is true for ever provided the doctrinal context is valid for ever. That is, as a principle:

> *A change in the doctrinal context of mathematical thinking will bring about a change in the conditions for validity of a given proof, given that structure follows doctrine.*

This is referred to as a contextual relativity principle in mathematics. Notice that the change is not in the structure, but rather in the doctrine, because the structure will not change unless the doctrine changes.

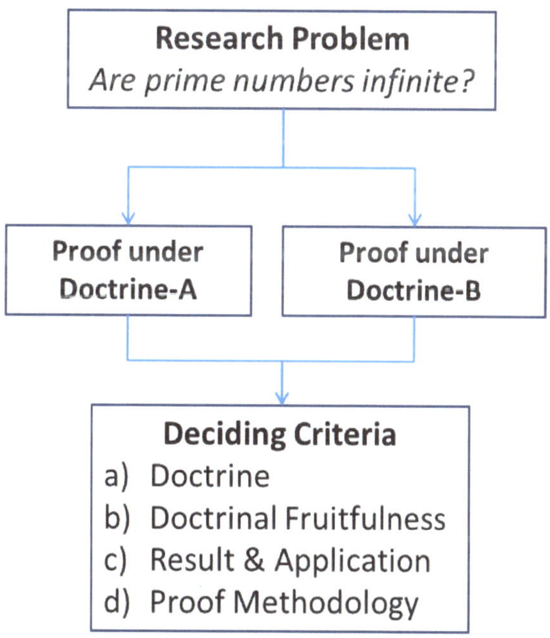

Figure 11

The principle makes the strong implication that proofs can indeed change! Mostly they are known to change in terms of improvement or use of different techniques, but not in terms of validity. The principle therefore establishes contextual relativity in mathematics rather than immutability. The reason why mathematicians minimize the dominance of doctrine in mathematical development is precisely because of this principle; the more clinical, mechanical and rule based mathematics is, the less likely that there will be error in inference, logic and result. The assumption being made is that this will affirm conditions of validity of proofs, and hence immutability of the mathematical result. But the fruitfulness of doctrine tends to disturb this equilibrium.

Given two doctrines that determine the validity of a proof, the question then becomes how do you determine which proof is true, Proof A or Proof B? The author came across such a situation when answering the question as to whether prime numbers are infinite or not. Current thinking proves that prime numbers are indeed infinite.

POOF AND DOCTRINE

- *Structure is a consequence of doctrine.*
- *Doctrine is a position of interpretation from a collection of consistent observations.*
- *A proof is correct because it is consistent with structure. If it is consistent with structure, it will be consistent with doctrine. If the proof is consistent with doctrine, then the arguments so constructed are then considered valid.*

Zwide Mbulawa

29 December 2011

It is by no means an easy task, but doctrine and doctrinal fruitfulness can assist in creating the framework for weighing arguments of justification, while result and application, and proof methodology will contribute to substantiating issues of validity, credibility, and mathematical commonsense. Validity will consider construction of arguments, the logic of such arguments, and the context of the logic in determining that validity. Credibility looks into the overall picture starting with doctrine, result, application and the proof to determine the integrity of the whole presentation. Mathematical common sense is the acumen that is applied by a mathematician using experience, depth of knowledge, and general expertise to discern and make a reasoned judgment. The whole process is not an event, and it eventually helps to transform the expository nature of the discussion to the more standard traditional expression of subject.

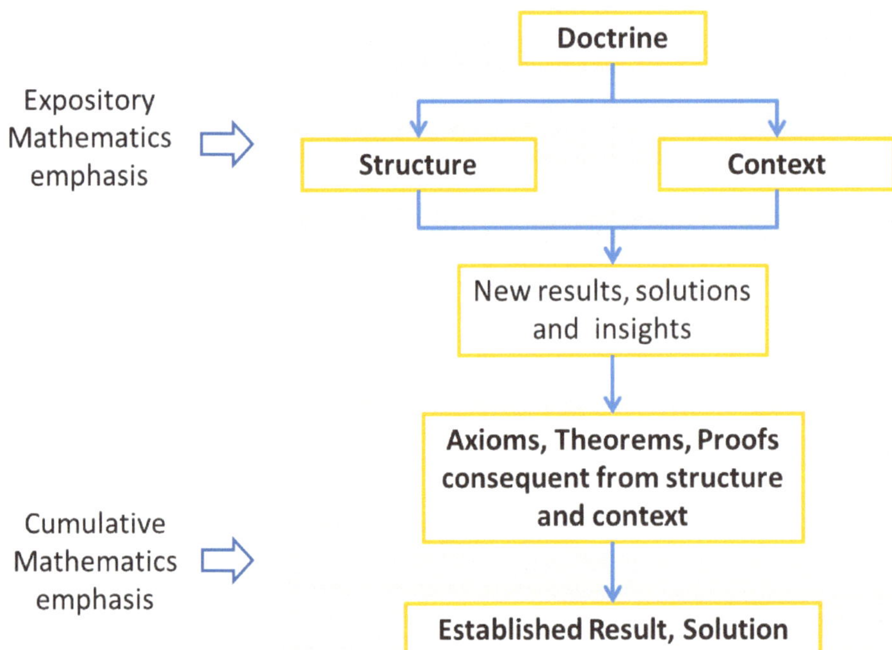

Figure 12

In presenting a case before a judge, one is presumed innocent until proven guilty. One can also plead guilty but proof still has to be provided in order to substantiate judgment. Mathematical common sense dictates the same observation in regard to proof on the following three conditions.

a) *Sufficiency Condition*. A proof is true because there is no valid argument against it.
b) *Consistency Condition*. A proof is true if the validity of the arguments is irrefutable in a given context and structure.
c) *Contextual Condition*. A proof is true provided certain conditions exist to define its limit of application.

The conditions are not necessarily mutually exclusive, they can apply in an inter-related manner. The sufficiency condition is like being innocent until proven guilty, and the consistency condition is like substantiating the guilt or lack of guilt with evidence of proof. The contextual condition is like having an alibi inspite of some points indicating a contrary observation (for example defining $i = \sqrt{-1}$ is a contextual condition). The basis of the sufficiency condition is logic and inference. The basis for the consistency condition is structure, as well as the model and doctrine presented by the structure. The basis of the contextual condition is fruitfulness of result. The three conditions highlight the fact that writing a proof is not just about creating systematic arguments. It is about creating arguments that provide evidence to substantiate the truth position. Arguments are a construction that demonstrates that evidence is systematic, related and consistent. The construction of an argument presents a truth, and evidence is also a truth that must be substantiated. Arguments therefore provide a constructed validity while evidence is a substantive validity. Therefore one may have evidence but fail to prove through argument due to limiting factors such as advanced development of syntax and context. The basis of fruitfulness in application is on evidence of something that works or provides a solution without strong backing of constructed validity.

Consequently, all work done at an expository level eventually becomes part of cumulative mathematics as the structure matures and the results get extensive application and acceptability. This then positions expository mathematics as having a foundational role in the development of cumulative or organized mathematics.

The two subjects are of the same mathematical coin, and should be viewed as such inspite of their apparent differences of approach and emphasis.

1.7 Of intelligence

One of the intriguing questions that I asked myself is a consequence of a research into factor pairs. These are a transient localized database used to determine prime factors and primality of numbers. This is paraphrased as follows:

Can a computer go beyond thinking comparatively into thinking as an expression of intelligence?

Infact the question led me to an unusual hypothesis.

Hypothesis

Mathematics is an expression of intelligence.

The reason for this assumption is simple: mathematics involves a specific type of reasoning that we can identify as mathematical thinking, a reasoning that is not only logical but also creative and constructive both in the physical and abstract realm. The theory of factor pairs has a part that deals with efficient pairs that are responsible for higher speeds of factorization and determination of primality. The thinking here focused on how to create the transient database much faster. Now two variables are used to derive factor pairs for particular numbers, X2 and K3. The question arose because of the recognition that it is not thinking that provided a solution as to how efficient pairs could be developed, rather it was intelligence. Consequently, thinking and intelligence differ – thinking rationalizes while intelligence is an aptitude that causes a creative decision. Therefore intelligence did three things, it;

- processed the information
- perceived a method
- ...and made a decision

that the roles of X2 and K3 can be interchanged in order to shorten the processing of factor pairs. The critical aspect is decision making, implying lack of appropriate decision making at a certain level is an indication of lack of intelligence. That is, two components are necessary; the thinking process and the aptitude in the thinking process that we call intelligence whose tool is the decision. This aptitude in thinking is not just based on comparative decisions but in the ability to solve, in the ability to decide how to solve, in the ability to make a precise creative decision, and to consequently create a system that expresses the content of the whole process. In process terms, intelligence is a particular type of decision making involving discovery, comparison, association and creativity.

These factor pairs will be produced by X2

X2	K3	Factor-1	Factor-2
2	396381	2378287	11
7	106346	638077	41
12	61411	368467	71
35	20862	125173	209
79	9218	55309	473
130	5597	33583	779
225	3232	19393	1349
294	2473	14839	1763
509	1428	8569	3053
1498	485	2911	8987
5337	136	817	32021
5583	130	781	33497
9668	75	451	58007
101400	7	43	608399
229484	3	19	1376903

These factor pairs will be produced by K3

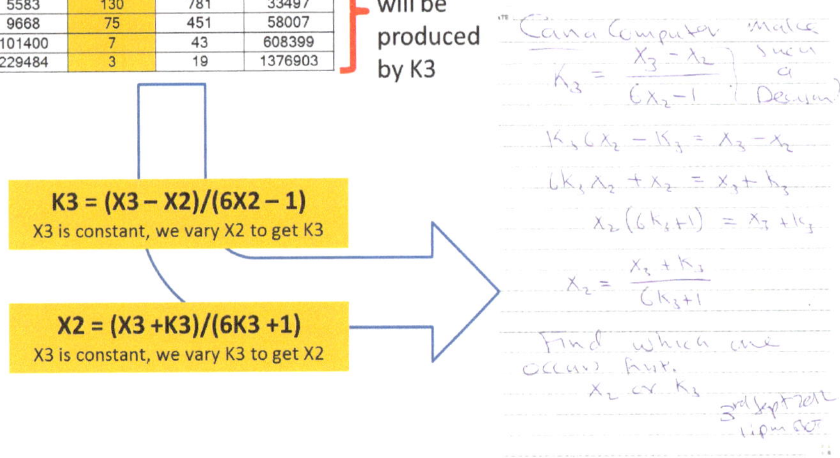

Figure 13

The question is what led me to think in this particular direction? I could not comprehend how my mind could connect points to trace a specific path that determined the solution of the problem that faced me. It was in my experience beyond just figuring out things logically, but my mind was applied intensely to seek a solution in order to capture speed and efficiency in developing the transient database. What I noted from the experience is that the type of decisions that I made in my thinking resulted in the type of solutions I obtained. This is not the only example, but I now began to observe it as a pattern that enhances the process of discovery.

The follow-up question then became:

How does this definition of intelligence affect expository mathematics?

It affects it in the context of the ability to make decisions which reflect intelligence that brings about discovery and creativity. The implication is that not necessarily all types of intelligence have the ability to bring about discovery or creativity. In plain language, discovery and creativity can be a form of intelligence or expression of intelligence. This can probably be developed as a skill and aptitude, but it can also come as a natural talent or gift to a person. And most probably, it is mathematical, and therefore human.

There is part of the expression of intelligence that is a little recognized but forms a vitally important process. This is the function of naming. A construct cannot be effectively developed unless it is given a name. That is, a name gives an idea or constructs the power to be identified in a unique manner. The name normally tries to reflect the attributes of the construct as well as forming its unique identity. For example, in seeking an algorithm that efficiently develops factor pairs, I called this the double barrel approach, symbolizing the fact that we are using two variables K3 and X2 in an interchangeable manner simultaneously. Can a computer create a variable and give it a name as a concept in order to create an identity? This was the essence of the question that I asked myself.

Infact there are so many ideas that are compressed into this construct, the meaning of X2, K3, the framework and definition of Factor-1 and Factor-2, and the fact that for a number W, then we can look for prime numbers as far as $\sqrt{W}$. It is probably this observation that led to the conjecture that mathematics is an expression of intelligence. This is not an exclusive definition, but it is particular in identifying a particular characteristic that we expect from the practice of mathematics. Descriptively, naming is a process of assignation, and results from making observations that lead to a particular decision in regard to that which is named. Without naming there would be no unique identity of a given object, and without that unique identity mathematics would become a confusing subject without any expression of intelligence. Hence naming becomes an effective tool for conceptual communication, as well as a direct reflection of the process or expression on intelligence involved in the creative activity. Naming achieves three things;

a) Assignation of meaning. The named object is geiven a unique meanining.
b) Assignation of location. The object can be stored by name, and retrieved in a simialr manner.
c) Assignation of relationship. The objects so named can be related though common attributes.

The three result in the formation of a thought map by which we can find our way to build and develop thoughts, and thoughts into ideas. Hence one of the skills that one must develop in expository mathematics is the ability to be build and name a construct to accurately reflect its attributes such that a concept can be clearly identified.

1.8 Of discovery

In response to the last two questions as to how ideas are discovered, and the type of requirements and demands upon the personality to enhance the process, the fundamental statement is that one needs a mind that queries the norm and is persistent in believing in an alternative explanation or framework. What has been established by my experience is that doctrine is imperative as an area to create the space necessary for the discovery activity to take place. That is, discovery is as deep as doctrine will allow, and the capacity for doctrine is a determinant of the mindset. On the other hand, the mindset is driven by attitudes and values, many of which may result from cultural and group dynamics, including personal inclinations and the sense of purpose. In my case, the sense of purpose has been the strongest driver, and my personal values and attitudes has been based on this.

In terms of psyche, discovery is a more a function of what you see rather than what you think. What you see is in turn is influenced by your belief system, your personal values and the mindset. If thinking was the main driver behind discovery, there would be numerous discoveries, and it would not have much influence from personality. The value of persistence for example is a well-known attribute that contributes to discovery, so personality counts.

What you see can be in two ways; a physical form or it can be in abstract form. Discovery can be initiated in two ways, by what you see or by what you think. Discovery that is initiated by "to see" is referred to as primary discovery. Here seeing is followed by thinking. This is the main discovery or initial discovery that leads to other discoveries. Sometimes the practice of "to see" is referred to as having insight, intuition or awareness. The discovery that is initiated by what you think is referred to as secondary discovery. Here thinking is followed by seeing (if you did not think like that you would not see). This is because thinking as well can lead to discovery since it makes you to look for certain things which you begin to see. What you see in the abstract may depend more on what you think rather than what you see, that is the emphasis is the reasoning as a consequence of the discovery. Hence it is deductive and inductive in nature. Pattern methodology depends a lot on the physical construct, that is the pattern that you physically see which then leads you to think in a certain way. Your mind sees something in a pattern, and then begins to think. This is then supported by deductive thinking to create ideas and structures.

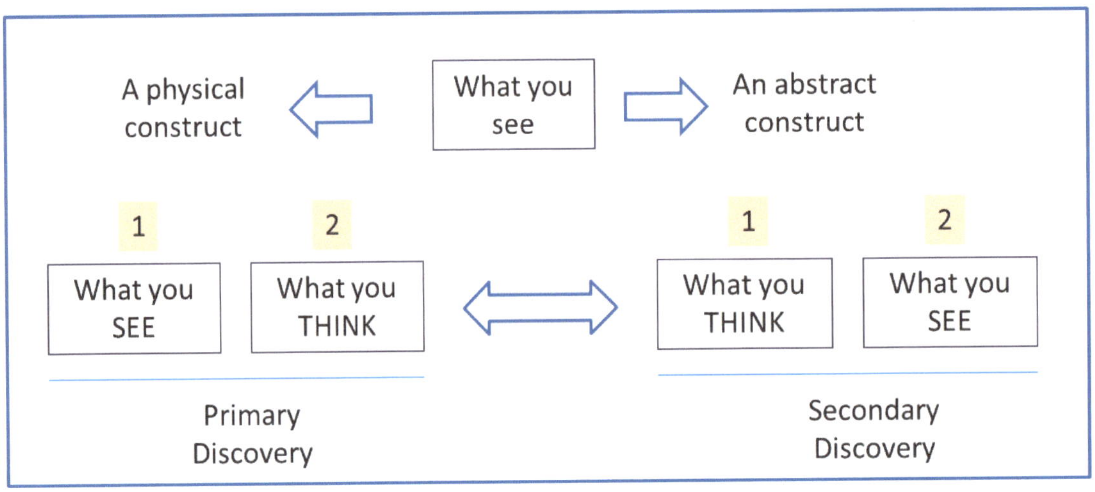

Figure 14

In developing the theory of integer distribution, classification and factorization, the experience has been a combination of both types of discoveries processes occurring almost in a cyclic manner. Note that it does not imply that primary discovery is necessarily more significant than secondary discovery; it is primary because it was mostly premised on insight followed by deep thinking. On the other hand, more people are likely to make the same discovery under secondary thinking than under primary discovery because the latter has a higher factor of personality influence than the former. This is the major difference between the two.

The interesting thing about discovery based on what you see is that if you note in a limited manner what you have discovered, and try and explain it later; your mind may draw a blank. It's like you have to re-think and re-learn what you were trying to do. An example is the development of the structured factoring method (SFM) coordinate structure. I developed this structure as an observation, but did not write-up much explanation on how I derived it. The structure forms the basis for the factor pair theory that is used in factorization. I could not even understand why in particular I had two instead on one structure, and why x_1 was set to zero! These are some of the practical experiences in the road to discovery.

Base	x2	x1	Factor-1	Factor-2
1	0	0	1	-1
2	0	0	1	1
1	1	0	1	5
2	1	0	1	7
1	2	0	1	11
2	2	0	1	13
1	3	0	1	17
2	3	0	1	19
1	4	0	1	23
2	4	0	1	25

Base	x2	x1	Factor-1	Factor-2
1	0	0	1	-1
2	0	0	1	1
1	1	1	5	5
2	1	1	7	7
1	2	2	11	11
2	2	2	13	13
1	3	3	17	17
2	3	3	19	19
1	4	4	23	23
2	4	4	25	25

Figure 15

Indeed it became extremely difficult to understand the thinking behind the development of the structure due to limited explanation. But the structure works, that is what I saw! However, the same thing can happen for discovery based on thinking; you come to a conclusion but if you have left some gaps, then they become difficult to fill in unless you attend to the situation while fresh on your mind. This implies there are techniques or some form of discipline that is necessary to keep discovery accurate and flowing, and this is to record as much as possible the thinking and the seeing. I did eventually work out the basis of the structure!

As a process, discovery is context driven. The figure below shows a discovery that was made but was difficult to substantiate because of lack of context.

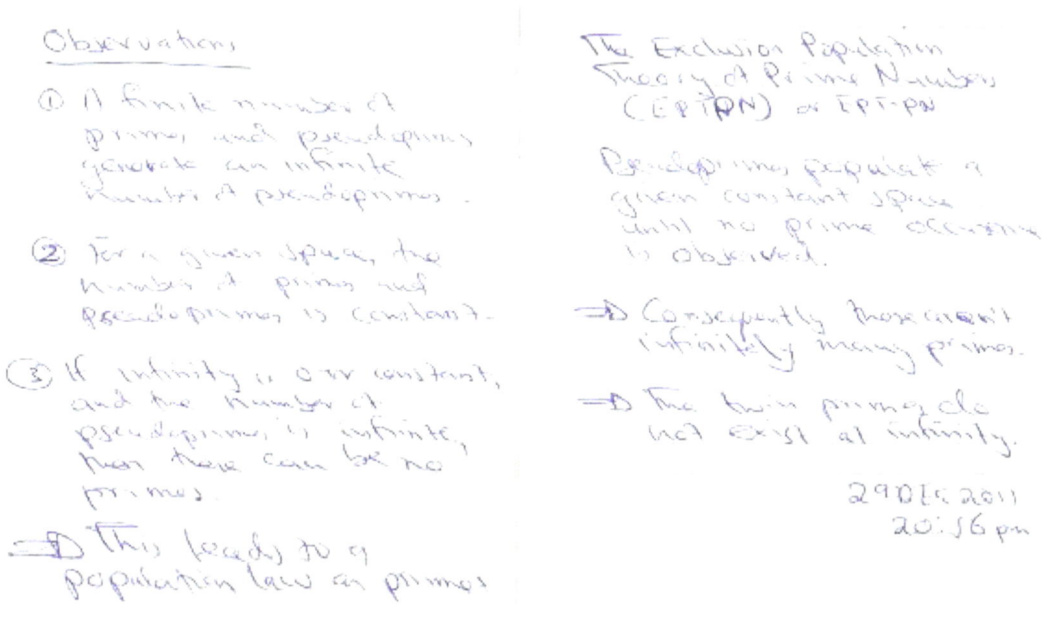

Figure 16

The observations were used to work on the concept of infinity in regard to prime number distribution, but the population theory still lacked context. Then on 21 August 2013, the concept of finding the probability of prime number was used to provide the context. This emphasizes the way expository mathematics works with idea building. The notion of *"pseudoprimes populate a given constant space until no prime occurrence is observed"* could not be developed until a context was provided, and this was done through the use of probability. It is possible that many ideas get thrown away because of lack of a strong contextual framework, as was the case with this idea. Infact, without context discovery is a major hurdle. For example, some of the observations made in this book are not new, but it is the context that has given the new dimension. Also note the conclusions that drawn from the observations that contradict Euclid's proof that there are infinitely many primes. This strongly illustrates the effect of context and proof validity. Euclid's proof is accurate in the structure that it exists in, which implies that to some extent, mathematics also has a relativity provided by contextual frameworks.

Discovery may initially be non-linear intellectually until it settles on a specific path of mental activity. It is fuelled spiritually by inspiration and revelation, as well as being developed technically through knowledge, experience and expertise. The non-linear process is referred to as brainstorming, where the mind considers any possibility in order to create a new thought or trend of exposition. One may consider inspiration and revelation as gifts within the personality that allow one to perceive beyond the norm that is an academic standard, and recognize a meaning and interpretation that lends a new dimension to a problem or solution. Just like a conscience may guide a person towards a certain moralistic behavior, these two gifts guide the user to where the discovery is hidden. In my life, these two gifts have operated as a blessing from God whom I refer to as my Father because of our relationship. When the discovery is born, logic then becomes the broom that sweeps the clutter and dross through the use of knowledge and expertise. Indeed, it is unusual for logic to give birth to discovery until and unless discovery has already made its mark.

Discovery cannot be expressed as a mathematical algorithm; it is an attitude of the mind, a function of the spiritual, and a practical divergent position taken from the norm. Therefore it is human, and will remain human. This means there will always be a relationship between mathematical thought and the person that conducts it. Because the thinking as shown before depends on what the mind is trained to see, and trained to think in a discovery context, this leads to the axiom of thought.

There exists

a) a specific type of thought process that results in discovery,
b) a creative mathematical thought, and a non-mathematical creative thought.

We can think creatively in different areas and discover things as a result, but a mathematical thought is unique. This is the personal experience I have noted. The greatest fundamental discovery in my opinion in this work is the algebraic set D_t and the UNDT framework for distribution of integers. A lot rests on these two conceptual constructs.

1.9 Of personality

Personality counts in terms of patience with oneself. This includes lack of bitterness after the rejection of ones ideas since it often becomes a stumbling block for new discoveries taking place. This results from the emotional upset where the mind tends to spin around the rejected idea glorifying how good it is, and how bad others are in not giving due recognition. Emotions of disappointment can also put a full-stop to a discovery process unless persistence and self-belief is part of the character. If these are well managed in an environment of intense focus, then discovery becomes a product of personal discipline, inspiration and revelation.

Consider the above response to the concept of classification of prime numbers. There is indeed a great ideological divide between what the author of the paper sees and what the referee understands. The author thought the idea is great, here was a novel way of classifying prime numbers with a consistent result in terms of the general pattern. The curve derived was always the same irrespective of the range chosen, and this was not significant! On the other hand, the referee saw nothing unique to be excited about, and was rather annoyed by the simplicity of the idea as a mathematical thought.

Indeed, the paradigms were different; having chosen a path of expository mathematics the author was creating theorems, these were a consequence of the ideas that produced axioms and definitions. The referee had his focus on cumulative mathematics where everything has to conform to existing theorems and ideas, as well as being connected to other important mathematical figures for credibility. That is, rejecting the paper was actually natural and understandable. But one may conclude that the referee failed to understand the basic concept since it was not about decimal presentations, rather it was about identifying prime numbers using a root theory in order to understand their distributive properties. The paper clearly said it was looking for an underlying behavior in the distribution of primes, and according to him this beautiful curve had no meaning. The referee saw an *"idea of decomposing a stream of digits"* when the author saw a root classification system that can be used to understand distributive behavior of primes. The author wanted to avoid *"number theoretic properties of prime numbers"* which was also a well-trodden path clearly defined by cumulative mathematics. Infact the author wanted to be adventurous in thought, discover new ideas, and find ways of solving things with fresh insights. These differences and perceptions then precipitated the decision by the author to do the research for personal satisfaction and as a fulfillment of purpose rather than recognition and praise.

> Referee's report on
> "The classification of prime numbers using prime roots"
> by Zwide Mbulawa
> for the Journal of Integer Sequences
> November 16 2009

In this paper the author classifies prime numbers according to their base-10 representation. In my opinion the paper does not have any interesting mathematics in it. For one thing, the classification is based on a particular base (base 10). Why not base 2 or some arbitrary base? For another, the author does not prove any interesting theorems; there are many more axioms, definitions, and hypotheses than there are results. A general rule is that if there are more definitions than theorems then the paper is in trouble.

The idea of decomposing a string of digits into concatenated decimal representations of prime numbers could apply to any number, not just primes, so the idea seems poorly focused.
Finally, the author does not use his ideas to actually accomplish anything or relate his work to known work. I would encourage the author to focus on the arithmetic and number-theoretic properties of prime numbers, not their decimal representations.

I recommend rejection of this paper.

Figure 17. Letter from referee

The phrase *"not their decimal representations"* shows that ideas do have the chance of being severely despised primarily because of mathematical values and expectations, and the dictates of tradition in methodology. The referee severely despised the fact that the tool for the classification amounted to trivial decimal presentation, and therefore the author should desist from using such techniques. Yet this very same method of decimal presentation comes out as a powerful tool for predictive factorization, showing how intellectual attitudes can block areas of creativity and discovery.

However, the referee made a point that was very valid nevertheless, on the matter of significance. This can be inferred from the comment *"the paper does not have any interesting mathematics in it."* He was right that the theory was not significant, the classification may give a distinct pattern but may not distinguish between odd numbers. You could apply the same approach to odd numbers that are not prime. Hence you would probably get the same curve using odd numbers that were not prime. Significance is a major consideration for publishing in a journal though it is relative because it depends on what one is looking for to define the significance. The aim of the paper was to show that through classification there is an underlying structure to prime numbers even though they were random, and in the authors mind that was the basis of the significance. Furthermore prime numbers do have a decimal property that defines their generation pattern referred to as cyclic transforms. So the fundamental question is; how many interesting questions, ideas and mathematical proposal are rejected because of differences of perception between a referee and an author?

The author was also of the opinion that mathematics was about discovery of new ideas, and that what mattered was the truth established by the mathematical finding rather than who you are (credibility issues). But this was a naïve approach as the interest is also strong to defend existing ideas and paradigms, to exalt and have pride over certain work, as well as extolling certain ideologies and personalities. This is besides the fact that most journals have no interest in a developmental platform or sharing of interesting ideas; the focus is result, proofs and conformity to the standards and expectations of the journal. It is in this respect that I found that discovery is a lonely road, and sensed some anger and frustration in the response to my work by the referee. The author felt like why should he have had the audacity to send something so elementary dealing with decimals presentation of numbers. At that point there was great need for self-motivation, determination to continue inspite of lack recognition. Instead of letting disappointment decide destiny, the author consciously embarked on a lonely road and decided to enjoy what he understood as his purpose – producing a classification theory and looking for applications. So really significance is a slippery issue, what you discover is not significant if you are not known, and it can be extremely significant to the extent of creating some high impact waves if you are some guru. This is inspite of the fact that mathematics is assumed to be an objective and scientific subject, since in practice it is very subjective and is a clan issue.

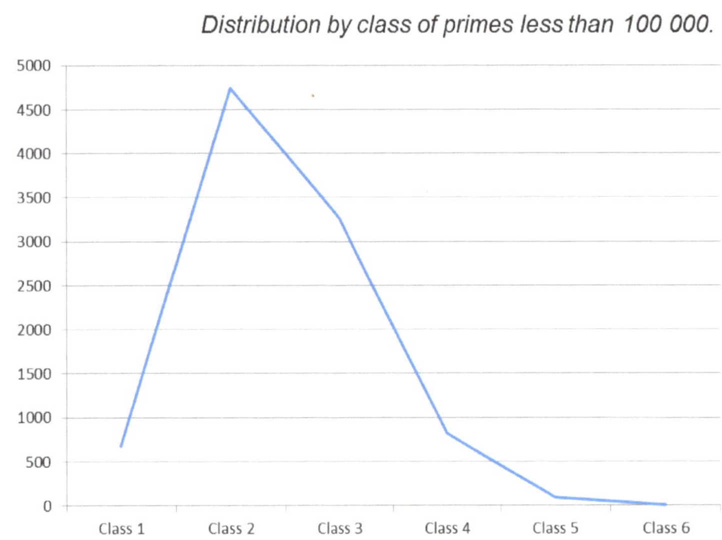

Class 1	673	7%
Class 2	4745	49%
Class 3	3260	34%
Class 4	818	9%
Class 5	93	1%
Class 6	3	0%
Total Primes	9592	100%

Figure 18. Distribution of prime numbers according to prime roots.

Note that the referee states that the author has proved no interesting theorems. And he is right because that was a major shortcoming. The author thrives in an environment where he constructs theorems, not to prove the theorem. The opinion of the author was that constructing a theorem is also a skill. Under the cumulative mathematical framework, one must have the idea, produce a theorem, and then prove it to complete the process. I realized that it is a different skill to construct a theorem, and quite another to prove it. I preferred to spend my energy on the former as a focus of activity focus because it worked well with idea and concept building. In practice you are not an accomplished mathematician unless you can show your mantle by proving a theorem, but I had to accept this limitation in exchange for constructing a theorem and demonstrating its fruitfulness. Again it's an issue of advantage and disadvantage; one has to make a choice.

These are some of the lessons that I picked up in my personal experiment to find out how I think mathematically, how my thoughts develop, and how do I harness the creative energy and use it to drive the process towards a particular fruitful conclusion. Ultimately, I give God the full glory for all the discoveries I have made, because it is after prayer to Him that my mind would open to a solution, a perception, a different interpretation, a possibility, and off course a discovery! That is the personal angle to expository mathematics, and this book is my evidence. It took years to develop my mind as a tool, to discipline myself how to handle a discovery, how to create ideas and make them have structure, how to remain in a given context to avoid error development, how to use different angles of approach to confirm validity or consistency, and to consistently practice the discipline of revising an idea and re-explaining it from different perspectives as a triangulation technique. The greatest difficulty was recording mathematical failure of ideas and the greatest challenge to record success as excitement would override practicality. But it is the recording that helped me many a time to transform failure into success, or to sharpen and consolidate success into fruitful application.

Finally, do not hurry to prove an idea because it takes time to turn it into a logical structure that can produce a proof. It takes time to build an idea into a concept, to understand what it is saying to you, and what you can get out of it. What is key about an idea is consistency and fruitfulness, and then its ability to solve a problem, generate knowledge and demonstrate explanatory power. Thereafter one may then seek to construct a theorem leading to a proof arising out of the application of the idea as a logical structure. Many ideas are killed because of the immediate desire to conform to standards by establishing a proof. And lack of development of doctrinal strength contributes to less significance of the idea.

Yes, personality does contribute in to the discovery process, and how far you go in the discovery you make. Because I did not have an opportunity to bask in recognition for each discovery I made, I was able to continue to develop the theory as deeply as I could. Basking in glory of an idea can lead to a shallow effort, and so I took advantage of the fact that nobody is interested in my ideas, even if they already on a website! My joy was based on what I produced, not the recognition that I received. Personal values like;

- *knowing your purpose and destiny*. Everybody God has given something to do here on earth, whether you believe in Him or not. If He has given you something to discover, as long as you stick in your purpose, you will discover it. It is up to you to choose to give Him the glory or not. There are probably many a genius who have never made significant discoveries, so discovering things is not logical, it is also a gift.
- *persistence and discipline*, over your mind, thoughts and ability to withstand criticism to a positive position.
- *keeping a consistent focus*. Emotions can easily distract your from your course.
- *mastery of what you are doing*. Mastery holds the final key, it brings quality and confidence to your work.

The practice of keeping a diary helps to trace your thoughts, and remember that to despise the simple is to reject the creative and the imaginative. It is the simple ideas that are the source of much inspiration and discovery rather than the lofty.

1.10 Of revolutions

In conclusion, having performed an experiment on myself, noting how I think, imagine, create and innovate, and consequently seeing the result thereof; I can only make one conclusion that contradicts Crowe's tenth law[5] which states that revolutions do not occur in mathematics. On the contrary, revolutions do occur in mathematics provided we recognize one fundamental principle – that recognition is the essence of a revolution.

A revolution is the fight for recognition of a new paradigm over an old paradigm, it is a fight because the older paradigm does not want to disappear or lose its hold. It becomes a revolution because of the fight and because of the replacement. Whilst a political revolution is normally violent, the same analogy cannot be used in mathematics. What is common to both is that a revolution has an impact, the result must make a significant impact in order for the criteria of revolution to apply.

Crowe argues that no replacement actually takes place because mathematics is cumulative in development. This could have been true if all mathematics is cumulative, however some mathematics is expository. In this case one can replace the fundamental theorem of arithmetic with the universal number distribution theorem. The former concerns prime numbers deriving natural numbers, while the latter concerns the precise relationship of all integer numbers.

It must also be noted that it is people who cause a revolution to occur, and it is also people who try by all means to stop a revolution from occurring.

The foundation of expository mathematics;

1. *is idea centered,* where ideas are necessary as a tool to challenge an exisiting paradigm.
2. *does not focus on consolidation of the logical and proof structure of exisiting theorems*, but rather emphasises doctrinal research and concepts supporting the doctrinal framework.
3. *focuses on understanding proofs*, their strength and weaknesses, and the ideas and arguments that build the proofs.
4. *searches for doctrinal explanation of anamalies,* as a possible way of contributing to new developments, especially anomalies that are an outcome of a system.
5. *creates an environment for theorem development*, this is an expected outcome. "Discovering theorems" and providing consistency of obersvation and argument is a greater emphasis than of providing proof.
6. *focusses on the why of construct relationships*, which makes it an explanatory subject
7. *generates new knowledge* as a consequence of the research activity.

The above make expository mathematics the crucible for mathematical evolution and revolution! The logical and proof structure is attended as a consequence of all of the above, and precipitates cumulative mathematics. Therefore the correct question is not whether revolutions occur in mathematics, rather it is as follows.

At what category and phase in mathematical development do revolutions occur in mathematics?

The answer is expository mathematics because of the above advanced reasons. Revolutions do not occur in mathematics when they have reached the cumulative structure of existence because of the consolidation and proof conclusions that have substantiated the results. However, changes brought by the expository development of mathematics can cause foundational changes in cumulative mathematics, and can eventually lead to some form of overthrow of cumulative mathematics. Therefore it depends on the level of interrelationship and connectedness between the expository mathematics and the cumulative mathematics in the given area of interest. If the two are

[5] Gillies,1992. Revolutions in Mathematics. Oxford Press.

mutually exclusive, then the expository mathematics will cause a revolution that does not create an overthrow condition for the cumulative mathematics.

Figure 19

Perhaps it is proper to conclude by noting the remarks by Kuhn[6] in regard to scientific revolutions on the simple assumption that indeed mathematics is also a science.

> *This need to change the meaning of established and familiar concepts is central to the revolutionary impact of Einstein's theory. Though subtler than the changes from geocentrism to heliocentrism, from phlogiston to oxygen, or from corpuscles to waves, the resulting conceptual transformation is no less decisively destructive of a previously established paradigm. We may even come to see it as a prototype for revolutionary reorientations in the sciences. Just because it did not involve the introduction of additional objects or concepts, the transition from Newtonian to Einsteinian mechanics illustrates with particular clarity the scientific revolution as a displacement of the conceptual network through which scientists view the world.*
>
> *These remarks should suffice to show what might, in another philosophical climate, have been taken for granted. At least for scientists, most of the apparent differences between a discarded scientific theory and its successor are real. Though an out-of-date theory can always be viewed as a special case of its up-to-date successor, it must be transformed for the purpose. And the transformation is one that can be undertaken only with the advantages of hindsight, the explicit guidance of the more recent theory. Furthermore, even if that transformation were a legitimate device to employ in interpreting the older theory, the result of its application would be a theory so restricted that it could only restate what was already known. Because of its economy, that restatement would have utility, but it could not suffice for the guidance of research.*

However Crowe's argument is that even though mathematics is a science, there is nothing that is discarded in mathematics on the basis that it is both systematic and cumulative. The exploration here then focuses on the non-cumulative nature of mathematics so as to identify this as a possible area for a revolutionary incident. Kuhn also points out that change of meaning is the key to discovery as well as revolutionary change. And change of meaning implies change of doctrine. He also notes that the differences between discarded scientific theory and its successor are real, where this becomes a form of progress and evolution of the subject. As humans indeed we seek continuity by trying to make the older version a special case of the new version so as to minimize the impact on what has already been gained.

At the end of the day, there must be something hidden to expose that is revolutionary in thinking, result and fruitfulness! It is probably the combination of these three that result in the significance that makes the ultimate difference in establishing a new paradigm of practice and thought.

[6] Kuhn Thomas. *The Structure of Scientific Revolutions*. 1962. University of Chicago Press.

PART TWO

This is a record of the outcomes of the gestation of the thought processes of the experiment which demonstrates work in progress in the conceptualization and development of expository mathematics.

.....that new ideas and concepts can indeed be born that revolutionize our current paradigm, if we are willing.

On the lighter side;

This is a story of the fight amongst the integers.

> It is a fight for space
>
> And it is a fight for infinity

There is only one winner!

(Hence the Law of Mutually Exclusive Infinity)

The military strategy used is theorems,

> ...a clash of theorems in the quest of occupation of space and of infinity.
>
> ...a race established using gaps for speed and size
>
> ...and subterfuge in multi-valued functions
>
> ...the defection of top spy Zero to even numbers
>
> ...where 2 and 3 are mortally wounded in the quest
>
> ...and the civilians, the Composites watch helplessly

This is the story...

And the unintended target is Factorization!

The Definition of Number Types

The section gives;

1. the algebraic definition of a prime number
2. the definition of the ditation set, and indicates how it relates to prime numbers

The following is noted in regard to the section.

a) Classification is a consequence of relational definitions.
b) Therefore a limitation in the context of definitions provided also provides a limitation in the context of conceptual development.
c) The broadness of mathematical scope depends on the use of number as a systematic tool for construct formation and application.
d) Therefore what we get out of a number and the constructs it forms depends on how we conceptualize its role and the attributes that we assign.

2 Distribution Theory of Integers

The major interest of the book is in studying integers, specifically the classification of integers and their consequent distribution. Consider the number 3500. As we know it, it is even, and we probably don't go beyond that in understanding its classification, far less its distribution determinants. As will be elaborated later, this number has a specific class that it belongs to, and this class determines exactly how factorization occurs. The class also determines exactly how it relates to the prime numbers and other integer numbers. For example, when we test the number, it belongs to Class 2B, under the gap 2 category. In terms of grouping it is family 2, and it is a composite.

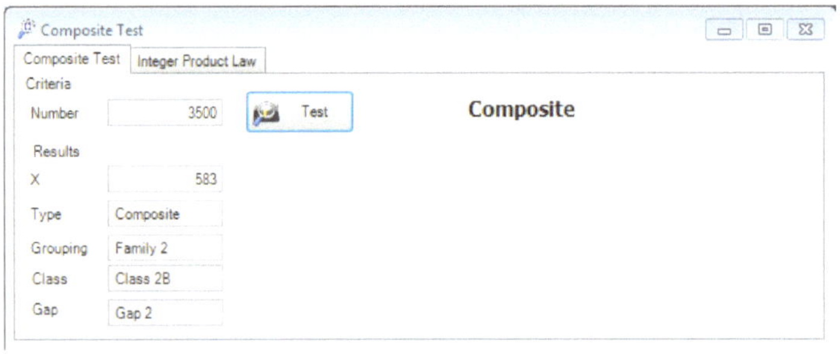

Figure 20

This tells us that the products that make up this number are likely to come from the following pattern. For example, it can be made up from a number in Class 1B and Class 2B, or Class 1B and Gap Base 4.

Class	Class	Resulting Product Class
Class 1B	Class 2B	Class 2B
Class 1B	GB-4	Class 2B

Table 7

This is the type of information that helps us to build an algorithm for factorization as it is a distribution theory on integers. Generally we note that there is such a strong pursuit for properties of prime numbers and their distribution that the relationship between numbers in terms of classification, distribution has not benefited much attention.

The origin of numbers can be said to be the concept of counting, which then led to arithmetic operations. Counting is done through the so called natural numbers denoted as $\mathbb{N}$. All counting numbers are positive integers, and they normally form the starting point of defining number classification. On a technical basis, the natural numbers are said to define whole numbers if they include zero. Therefore whole numbers are also positive integers.

Since there are positive integers, it also implies that there are negative integers. This set of integers is denoted $\mathbb{Z}$ and it is inclusive of the zero. The set $\mathbb{Z}$ can further be broken down into two sets, the set of even and odd numbers. All even numbers are divisible by two to give an integer that is either odd or even, and zero is defined to be an even number. Infact, all even integers that end with one of the following; 0, 2, 4, 6, or 8 are even. On the other hand, odd numbers are not divisible by two to yield another integer, they always leave a remainder of one. Both the even and odd numbers can be negative or positive. This is the basic staff.

Then we have the prime numbers. This is defined to be natural number with the exclusion of one, where such a number has exactly two distinct natural number divisors: 1 and itself. All prime numbers are odd. Just like even and odd numbers, if a number is not a prime, then it is said to be a composite number. Therefore a composite number is said to be an integer that has factors besides one and itself, and they can be odd or even factors.

The set of natural numbers and the set of all integers are a subset of the real set denoted $\mathbb{R}$. From this the general conventional classification for integers can then be shown in graphical form.

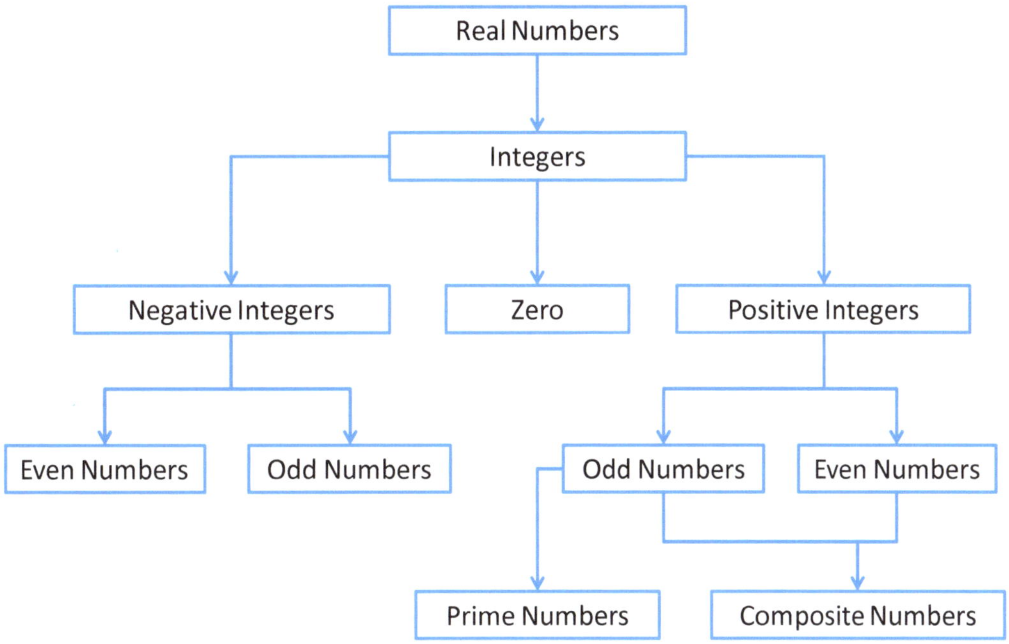

Figure 21

There is not much that can be said in terms of a distribution theory of integers. The significant fact is that we have even and odd numbers, and these are distributed along a real number line of the set $\mathbb{R}$. This is the simplest structural description of the distribution. However, when we consider the set of prime numbers, here there is an intense study of distribution patterns as a result of trying to break into the mystery of the prime numbers. There is no interest or concept of how composites are distributed as their existence is more or less trivial. Consequently one can say there is no significant theory that describes how integers are actually distributed since in a way, there is nothing complex in structure except in the occurrence of prime numbers.

This book defines two types of approaches that determine understanding of integers as shown.

Conventional Approach	UNDT Approach
Defines the set of integers Z	**Defines the set of integers Z**
There are negative and positive Integers	**There are negative and positive Integers**
Zero is between negative and positive numbers in terms of sequence	**Zero is between even numbers in terms of classification**
There are no negative prime numbers	**There are negative prime numbers**
A prime number is a natural number inclusive of 2 and 3, but 1 is not prime	**Prime numbers are algebraic by definition, where 2 and 3 are not prime numbers, but 1 is a prime**
Pseudoprimes have varying definitions depending on the type of prime numbers being studied. Generally it is a number that is not prime.	**Pseudoprimes are a specific set that is distinctly related spatially to the set of prime numbers. Only one type of pseudoprime exists.**
Composites are odd and even. A composite is	**Composites are odd and even. Any number**

any number that is not prime, that is, inclusive of pseudoprimes.	that is not prime or pseudoprime is composite.

Table 8

This is the conventional approach, which has existed for centuries and the UNDT methodology. The UNDT framework results in the following tree for defining the distribution of integers. The major difference is the treatment of prime numbers, and the introduction of a new distinct number into the number system; the pseudoprime. Also geometrically and in the concept of the real number space the zero is a number between positive and negative numbers, but in terms of classification it is between even numbers and is not related to the odd numbers!

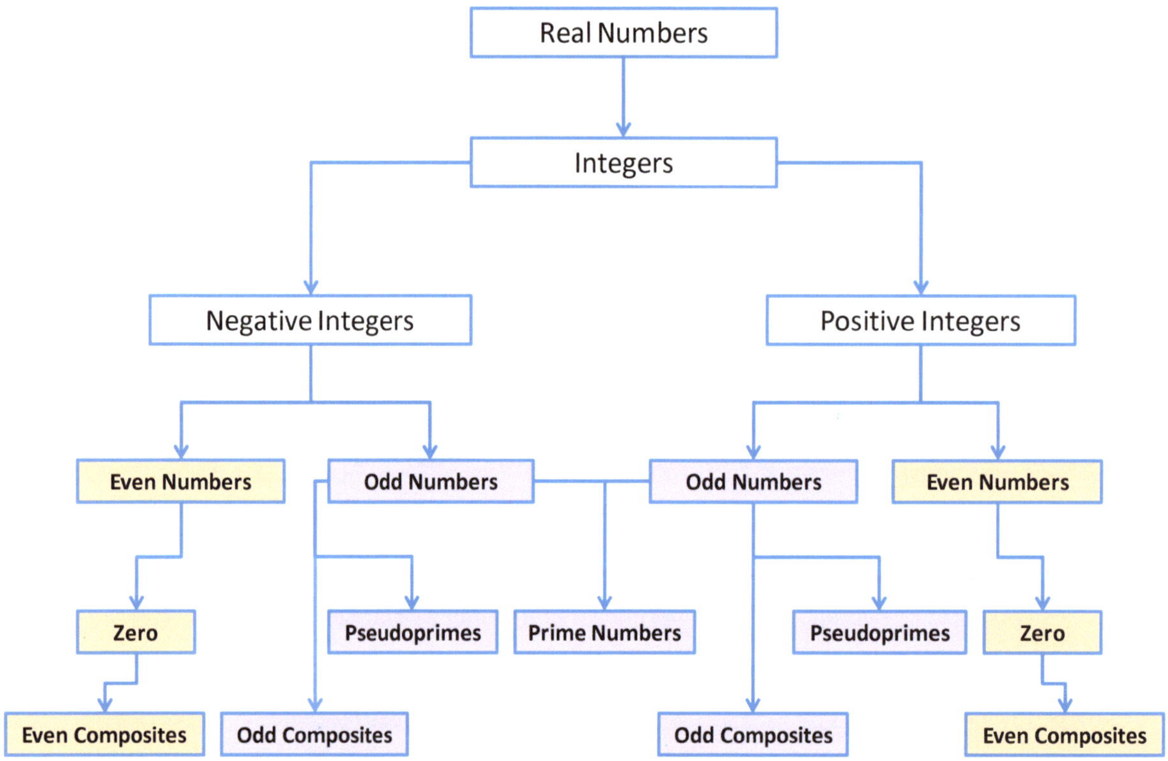

Figure 22

When it comes to prime numbers, the primary focus is on their randomness and gap occurrence. This is exemplified by the prime number theorem which approximates the prime count. The generation of prime numbers has been a major challenge due to the fact that there is no formula for predicting a prime number, and the behavior of such numbers is random. The book on The Theory of Prime Number Classification by the same author derives a model for prime number generation rather than a formula. The findings will be applied in this book.

2.1 The Challenge of Prime Numbers

The prime numbers have a long history of study spanning centuries. The ancient Greeks studied prime numbers extensively. The Greeks understood the idea of primality. The greatest interest that they generated is their capability of being used to describe any integer number by use of a product. That is, any integer can be expressed as a unique product of primes where this is called the fundamental theorem of arithmetic. Even though they are random in nature, prime numbers also exhibit particular patterns using modulo arithmetic as well as polynomial expressions.

In the Book IX of the Elements, Euclid proves that there are infinitely many prime numbers, and even today this proof is taken as standard. Euclid also gives a proof of the fundamental theorem of arithmetic, where this shows the extensive interest in the area of prime numbers. Around 200 BC Eratosthenes developed an algorithm for calculating primes that has been called after him as the Sieve of Eratosthenes.

However, prime numbers have posed quite a challenge in developing a classification methodology for them. Classification is a powerful tool for interpretation and building further models for prediction of behavior.

2.2 Definition of a Prime Number

One of the tough decisions that had to be made was the application of a different definition of a prime number. Historically a prime number is divisible by 1 and itself only, and must be positive. The assumption is that such a definition was the consequence of the mathematical paradigm at the time of its formation since only positive integers were known, and zero did not even exist. That is, before the advent of the negative numbers, the prime number definition was satisfactory in this manner:

A prime number is any number that is divisible by itself and one only.

During Euclid's time, the notion of a natural number did not exist, only a counting number. Many mathematicians still give this as a valid definition. With the advent of the natural number concept, the definition had to develop a little further as shown.

A prime number is any natural number that is divisible by itself and one only.

The natural number is positive, and it developed as a notion in relation to the negative numbers. In the above, we are mostly considering conceptual consistency of the idea of defining a prime number. However, when we relate this to other mathematical objects such as the negative numbers, we realize some inconsistencies. That is,

- a prime number p is divisible by minus one, and
- a prime number p is divisible by $-p$.

Hence, this makes us to try and improve the definition in order to manage the interpretation of relationship between other mathematical objects. That is, we can formulate the definition as;

a prime number is a natural number that has two distinct natural numbers as divisors: itself and one only.

The qualification is that the prime number is a natural number, and its divisors are also natural numbers. Relative to the negative numbers, we are saying we do not consider them as divisors of prime numbers. However, this still does not prevent us from dividing by a negative number because doing so is still a very valid mathematical process. This is where doctrine comes in as a developmental tool as it will question why we should confine ourselves to positive numbers only. So far there is no strong doctrinal context that requires or justifies this paradigm. The other shortcoming is the fact that an even number 2 is also regarded as a prime number, where this is not consistent with the observation that all prime numbers are odd. The number is an anomaly, but it is consistent with the definition of a prime number that is based on natural numbers and divisibility.

As stated before, it must be remembered that prime numbers were discovered before the existence of the concept of the real number, so they have been confined only to be natural numbers. That definition has probably made sense until now.

On the other hand, there must be a justifiable reason for demanding a change of perspective in a given definition. This comes from the following observations.

1. That prime numbers should also include the condition of being divisible by minus one. Actually, all natural numbers are divisible by minus one, including prime numbers. The omission of minus one reflects the fact that the definition was made when negative numbers were nonexistent in mathematics. Therefore, the prime number 19 is divisible by -19 to give -1. We overide all of this by making the assumption that prime numbers are positive, but this does not imply that primes cannot be divided by a negative number. The assumption therefore is that 19 is divisible by $\pm1, 19$ and -19.
2. Prime numbers can also be negative. For example, -19 is divisible by 1, by -19, and by 19 only. It shares the same characteristic as the positive integer!

Indeed we can then develop a different interpretation system or doctrine that allows us to recognize these valid arithmetic operations. The changes are made and accepted on the basis of conceptual consistency, fruitfulness and recognizing that at times it may be necessary to remove certain limitations or conditions in order to introduce a different perspective whilst at the same time retaining some aspects of the previous thinking. The concept of the algebraic sieve works because it assume the existence of negative prime numbers.

We can introduce the notion of systems in order to substantiate the definition framework for the prime number. The system can be devised on the following concept axioms.

a. *There exists a given set of elements in D_1 and D_2.*
b. *The only operation allowed in the sets is division.*
c. *The result of that division must be an element within the set D_1 only or within the set D_2 only.*
d. *The result must be an integer.*

These are called concept axioms because we want to form or justify a definition of a concept out of them. Now, let $D_1 = \{p, a\}$, where P and a are integers. The rule of constraint is that the result of the arithmetic operation must be a closed set result. Then we have the following outcomes describing the system of the set.

System	**A**	D_1	**System**	**B**	D_2
Divide	**By**	**Result**	**Divide**	**By**	**Result**
p	a	P	p	a	P
p	P	a	p	p	a
a	a	a	a	a	a
			p	-a	-p
			-p	-p	a
			p	-p	-a
			-p	p	-a
			a	-a	-a

Table 9

If, for example, $p = 10$ and $a = 5$, the the result is 2, and this does not satisfy the axioms. Therefore, we see that $a = 1$, and P must have no factors; hence, we have $D_1 = \{P, 1\}$ in order to satisfy the above system.

Similarly, let $D_2 = \{P, -P, a, -a\}$. Then we have the outcomes shown describing the system of the set. Hence, to satisfy the set conditions, then $a = 1$, and P must have no factors. Therefore, $D_2 = \{P, -P, 1, -1\}$. Then we note that $D_1 \subset D_2$.

The above approach uses the set concept to describe the justification for a systematic definition of a prime number based upon a set of concept axioms. This is in contrast to the definition of the prime number that is based on attributes of divisibility by one and itself only which may be described as intuitive since it is solely based on observation.

2.3 The Multi-valued Function for Prime Numbers

Generally, $y = f(x)$ is the standard approach in the concept of functions and defines a one-to-one mapping. If we want to extend the dimensionality of the function, then we consider $y = f(x_1, x_2, \ldots x_n)$ also as a one-to-one mapping as it represents one point in a multidimensional space. However, if we consider a circle whose general equation is $x^2 + y^2 = r^2$, then there are two possible results for y, that is, $\pm y$ for the every x, and this is a one-to-many mapping $f(x) = (y_1, y_2)$. The example of the circular function shows a multi-valued function in the same dimension, that is, having more than one point to represent the function but still being in the same dimension. Secondly $\pm y$ is an inverse function of $x^2 + y^2 = r^2$.

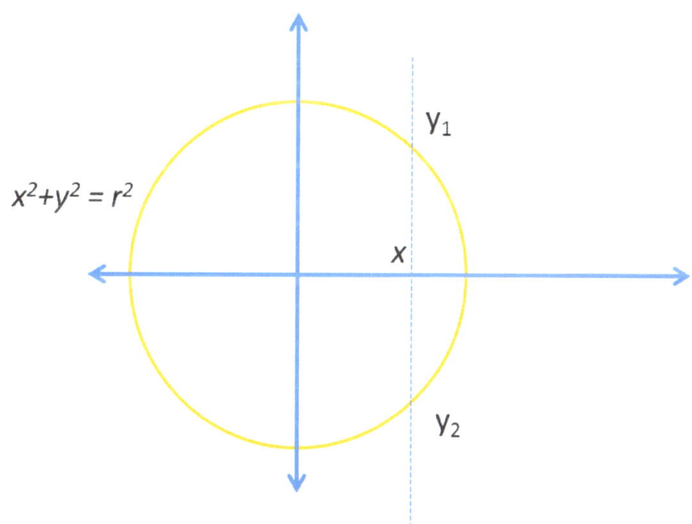

Figure 23

The development of uniqueness in regard to the one-to-one mapping of the function has meant that this technique may have been seen as representing real-life situations to problems or mathematical context, and defining a proper concept of space. It became the dominant technique for interpreting uniqueness, relationships, and reality for problem solving. The UNDT framework is based on a concept of a multi-valued function of the form $(x, y_{1,}y_2)$, one value of x giving in two functions at the same time. Therefore something like $f(x) = (y_1, y_2)$ would not be seen as presenting a proper function or framework for real life problem solving because of the lack of unique solutions in the linear space. It evidently influenced how research into prime numbers is done, which explains the reason that most attempts to describe the prime number in formulaic expressions have tended to look for an equation with one-to-one properties for prediction.

Definition - The Multi-valued Function

Let there be a set $f(x) = (y_1, y_2, \ldots y_n)$ *provided;*

a) $(y_1 \cap y_2 \cap \ldots \cap y_n) = \emptyset$.
b) y_i *is not an inverse function*

Note that the function $x^2 + y^2 = r^2$ is not defined as a multi-valued function in this context because of the second condition, since such a function must not be an inverse function. Therefore each function y_i are independent of each other and are not inverse functions. A simple example of such a function is $f(x) = (y_1 = 4, y_2 = 7) = (4,7)$. This can be a description of two parallel rays of light from negative infinity to positive infinity. Typically we would say $y_1 = f(x)$ and $y_2 = f(x)$.

An extensive use is made of the multi-valued function $f(x) = (y_1, y_2, \ldots y_n)$ as a problem solving technique to develop the model of the algebraic sieve. Note that when $n = 1$, then we have the standard function $y = f(x)$. In some cases $f(x)$ defines a special set called the ditation set when an additional condition is inserted.

2.4 The Ditation Set

The relationship between prime numbers provides evidence to conceptualize a linear space that is multi-valued and presents this as a problem-solving technique. This is done by introducing the concept of a ditation set. Even though this is a new term, but the basic idea is already there. The word "ditation" is a special word that arises from the definition of the ditation condition in a subject called Ditation mathematics. Further reading on the concept is well defined in the book called The Science of Mathematical Thinking. A simple introduction will be pursued here.

Consider for example the elements of the Cartesian space defined by two axes $(x,0)$ and $(0,y)$. There is an angular relationship between all these elements such that we may write them as $(x,0) = (x,0°)$ and $(0,y) = (y,90°)$, hence $(x,0)$ and $(0,y)$ are at 90 degrees with respect to each other. That is why it has been stated that the concept basically is already there as this is a polar notation. Hence we can relate these two sets by defining the notion $(x,0) + (0,y) = (x,y)$, and we know that y is at ninety degrees with respect to x, and that this angle is constant.

Hence we have elements in the set of all points $(x,0)$ and set $(0,y)$ such that the angle between these two sets is always 90 degrees. These two sets together form a space. In a similar manner, a ditation space can be defined simply as a space whereby elements are related by position that is measured in a linear or angular manner. For example consider a practical example of the calendar which consists of 12 months shown in the graphic.

The initial position is represented by zero, and therefore there will be twelve positions ending with position 11. However because we are defining a ditation space that is based on a constant angle base, this can also be expressed in circular terms. Since a complete circle is 360 degrees, each month is represented by 30 degrees. In the same manner, the first month is said to be at zero degrees. This gives us a degree measure until 330 degrees. Hence the 25th of May can be represented as (25,4) or (25,120). This is the basic idea of the concept of a ditation set.

MAGNITUDE	MON1	MON2	MON3	MON4	MON5	MON6	MON7	MON8	MON9	MON10	MON11	MON12
POSITION	0	1	2	3	4	5	6	7	8	9	10	11
ANGULAR	0	30	60	90	120	150	180	210	240	270	300	330

Table 10

As in the example of the Cartesian space, each month represents a set; hence the notion of the ditation set is that we have a number of particular sets consisting of elements which exist at a given angle with respect to each other. The constant angle is called the fundamental angle β that defines the relationship between such sets. This is the second notion of the ditation set, that of sets related to each other in angular terms.

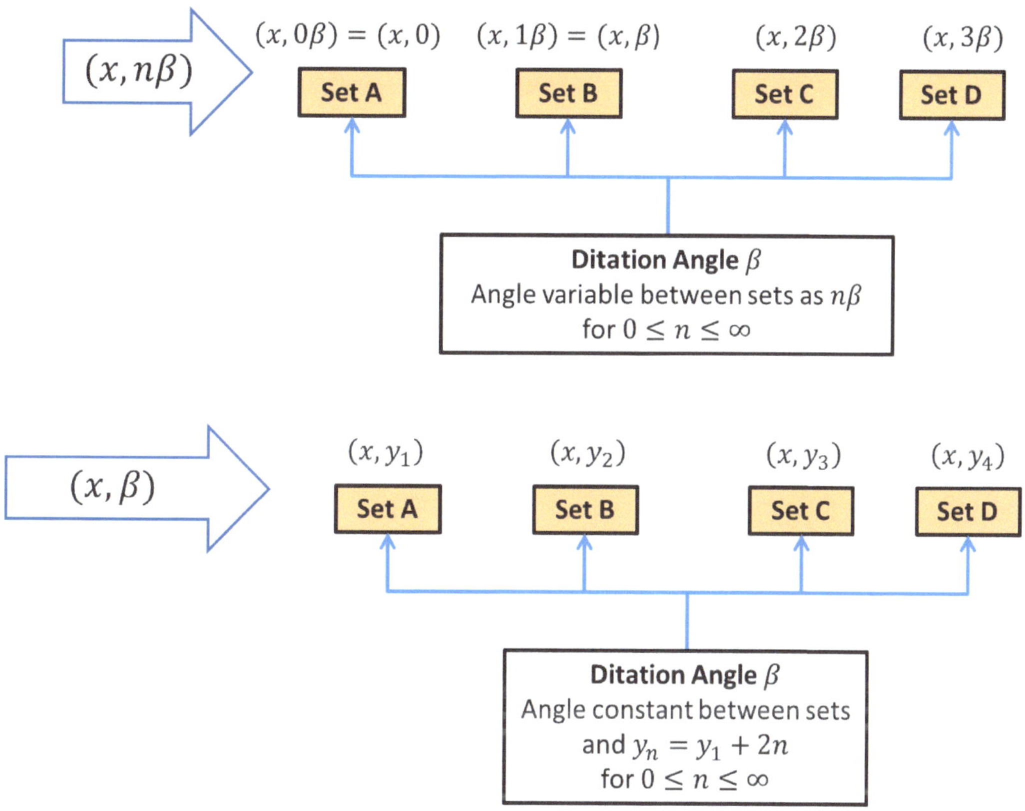

Figure 24

In defining a ditation space, we assume a ditation angle β, where this angle relates two or more sets. The angle can be variable or it can be constant as follows.

a) *A variable angle $n\beta$ for the same magnitude x defines a circular space*. For the same value of x, the position of x is defined through a constant angle. For example $(51{,}40^\circ)$, and $(51, 80^\circ)$ are elements of such a set, where $\beta = 40$. The notation is similar to the polar notation (r, θ), but differs in the meaning, interpretation and application of the angle β. The circular distance between $(51{,}40^\circ)$, and $(51, 80^\circ)$ is called the displacement. The circular ditation space is used to solve Hamilton's triplet problem, and to derive the commutative quaternion.
b) *A constant angle β for the same magnitude x defines a linear space*. For the same value of x, then $y_{i+1} - y_i = k$, where k is a constant and is referred to as the displacement. Since we have (x, y_i) and (x, y_{i+1}), this can be written as a single notation (x, y_i, y_{i+1}). The linear space is used to define and describe the characteristics of the prime number space, and integers.

That is, we now extend the same notion to $f(x) = (y_1, y_2)$, where we make the function have the same angle.

Definition - The Ditation Set Axiom

Let there be a ditation set $f(x) = (y_1, y_2, \dots y_n)$ provided;

a) $(y_1 \cap y_2 \cap \ldots \cap y_n) = \emptyset$.
b) y_i *is not an inverse function*
c) $u = y_{i+1}$ *is a reflection of* y_i *at an angle* β *about some axis where* $0 \le \beta \le 90$.
d) *For some* y_i *where* $v = y_{i+1}$ *, then* $y_{i+1} = -y_i$.

For a value x, this derives y_i where this relates to y_{i+1}, but y_{i+1} can have two values as $y_{i+1} = u$ and $y_{i+1} = v$ where the condition $y_{i+1} = -y_i$ is satisfied.

Consequently for the multi-valued function, not all $f(x) = (y_1, y_2, \ldots y_n)$ define a ditation set. In some functions the third condition is not satisfied and as such cannot be defined as a ditation set. The condition of y_i not being an inverse function implies that the set of functions $y_1, y_2, \ldots y_n$ cannot be derived from each other, they are distinct such that the union of the sets forms a specific space. The main issue is that all four conditions that define the set must be satisfied. The third condition defines the notion of the ditation set as a reflection space at a given angle β. It must be noted that β expresses a relationship between the two sets.

There are two conditions that apply here, for $u = v$ and $u \neq v$ in applying the ditation set axiom. Assuming we have an element y_i in set y_1, then we can determine a relationship to the elements of the second set y_2 using the angle β. Secondly, we may look for a relationship between the two sets where $y_{i+1} = -y_i$ for a given y_i. Then we consider the angle θ between y_i and y_{i+1}. Hence if;

1. $u = v$, then $y_{i+1} = -y_i$ and $\beta = \theta$ the angle of reflection; the angle of reflection between y_i and y_{i+1} is constant.
2. $u \neq v$, then $y_{i+1} = -y_i$ and $\beta \neq \theta$ the angle of reflection; the angle of reflection between y_i and $y_{i+1} = n$ is variable.

The set defined by angle β satisfying such conditions is called a ditation set, and $\theta = 90$ is the biggest angle of reflection between any two sets. In general, a ditation set is a space where;

a) elements within that space are related to each other through a particular constant angle β defined from a mirror line.
b) there exists a displacement d such that for $d = y_{i+1} - y_i$ or for $d = y_{i+1} + y_i$ this is constant. The expression $d = y_{i+1} - y_i$ derives $d = 2n$, if n is the distance to the mirror line. The smallest displacement occurs for $n = 1$ for integer values.
c) for $f(x) = (y_1, y_2, \ldots y_n, \beta)$, there exists another set $f(x) = (y_{1,1}, y_{1,2}, \ldots y_{1,n}, \beta)$ where $y_{i,j} = y_j + k$. This is called the generative property of a ditation set. Hence the same value of x can produce a similar but different set, and $k = mx + c$.
d) The property that $y_2 = -y_1$ defines reflection in the ditation space between two elements from the respective sets.

As an example, $f(x) = (x^2, 4x + 5)$ does not satisfy the condition of a multi-value function because of intersection. A simple example of a of multi-valued function is given as;

$$f(x) = (x^2 - 2x + 101, -x^2 + 2x - 101)$$

Here y_2 is a reflection of y_1 about the mirror line $y = 0$, and consequently we have $\beta = 90$. That is, $f(x) = (y_1, y_2)$. It satisfies the basic condition of no intersection between $y_1, y_2$for a given value of x and the functions are not an inverse relationship. The major difference here to the standard definition of a multivalued function is that y_1 *and* y_2 are distinct different functions under particular conditions. If $y_2 = -x^2 + 2x - 101$, and we have $y_1 = x^2 - 2x + 101$, then $y_2 = -y_1$, and they are a reflection of each other at an angle of ninety degrees. In this case $m = n$

because the equations are symmetric about the x-axis for the condition $y_2 = -y_1$.For example, an element of $f(x)$ is $f(x) = (181, -181)$. In this case $d = y_{i+1} - y_i = 181 - (-181) = 362$, or $d = y_{i+1} + y_i = 181 + (-181) = 0$, and this is constant for any x. This property is due to the reflection characteristic of the ditation set. The angle of reflection is constant, and also $\beta = \theta$.

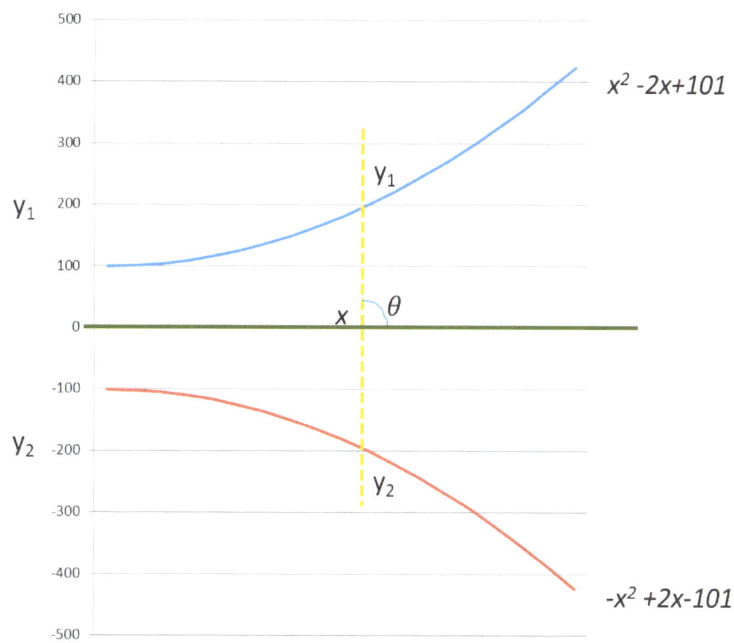

Figure 25

As stated above, the ditation set has a generative property. Therefore for the same set where $k = mx + c$, let $m = 0$ and we define;

$$f(x) = (x^2 - 2x + k, -x^2 + 2x - k)$$

This gives us a general family of the ditation set at the same angle of 90 degrees, where k can take any value. Similarly, the displacement value will still be zero. A function such as $f(x) = (x^2 + 8, x - 5)$ is not a ditation set because the two functions are not a reflection of each other, we cannot find a mirror line between them even though the condition $(y_1 \cap y_2) = \emptyset$ has been met.

Now what we have done is to provide a general framework of understanding for a particular type of function. We then apply this in the context of the prime number generation assumptions. Generally, prime number generation is considered in terms of derivation of a particular formula, but the difference here is that it is being considered in the context of a particular type of set called the algebraic sieve. Hence we can say, let there be two functions given as follows to form a multi-valued function.

Definition – Algebraic Sieve

Let $f(x) = (6x - 1, 6x + 1)$, *where* $f(x)$ *is a linear ditation set* D_t *and* $-\infty \leq Int(x) \leq \infty$.

Hence D_t is said to consist of linear ditation numbers, and we can define $y_1 = 6x - 1$, $y_2 = 6x + 1$. The function is defined as $f(x) = (y_1, y_2)$. These two functions satisfy the conditions of the definition for a ditation set, as there is no intersection and the mirror line is $y = 6x$ at some angle θ. What is key about the angle θ is that it should be the same at any value of x for the two functions.

We can choose two ways to relate to the mirror line in terms of reflection, using the point x_1 or x_2. If we take x_2, the gradient of the perpendicular is $f(x) = -\frac{1}{6}x + k$, and (a, b) is a reflection of (c, d). The perpendicular line cuts the mirror line at 90 degrees and the x-axis at the point x_2. Therefore we can relate x_2 to the two points where (a, b) is a reflection of (c, d) at 90 degrees as this forms a unique relationship. However, since $0 \leq \beta \leq 90$, we are not bound to use ninety degrees as an angle of reflection, and as such we can make a convenient choice. It is easier to use the point x_1 at $\beta = \theta$ because we will then have two unknown points y_1 and y_2, or $f(x) = (y_1, y_2)$ unlike the case of relating (a, b), (c, d) and x_2 which requires extra calculation.

Choosing a point like $x = 4$, then $y = 6(4) = 24$ on the mirror line. Since the perpendicular forms a point of intersection with $f(x) = 6x + 1$ at (c, d), then we solve to derive $(c, d) = (3.97297, 24.837857)$. Hence solving for the angle gives us $\theta = 33.0417329556178$. This angle θ is the same for $6x + 1$ and $6x - 1$ with respect to the x-axis.

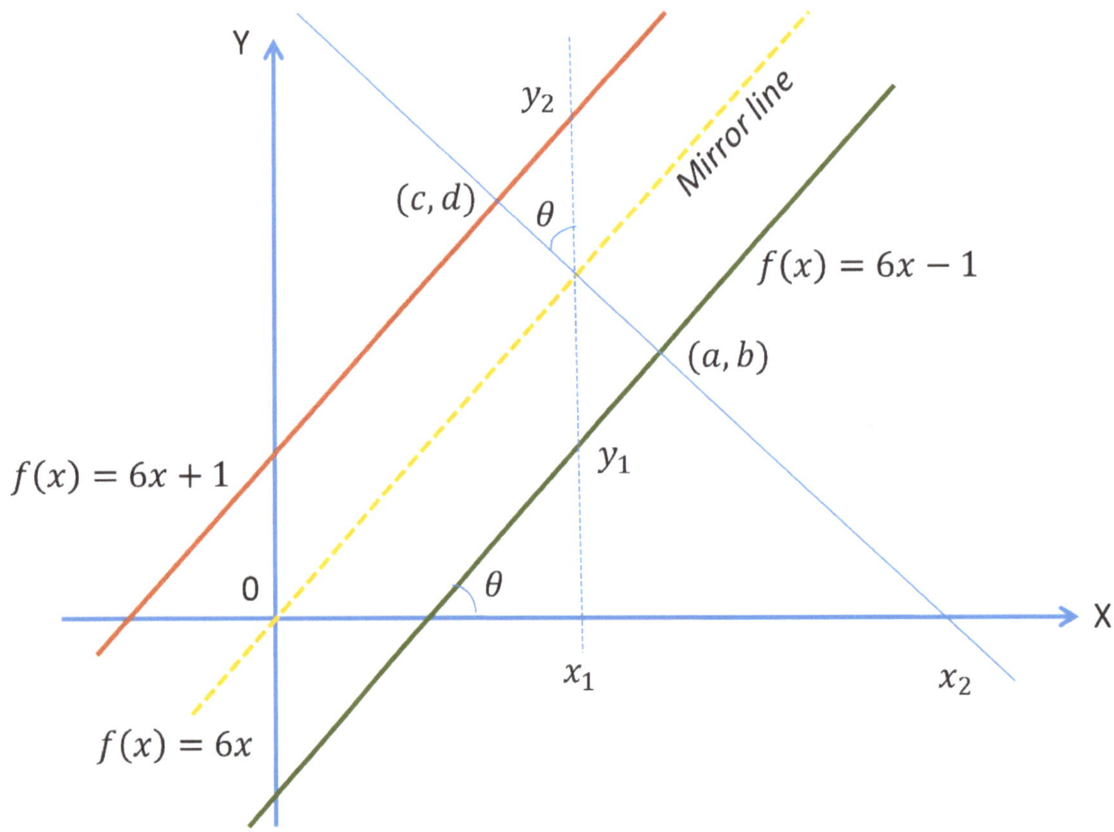

Figure 26

Hence for the set D_t, then β =33.0417329556178. In other words, this is the angle between some of the prime numbers, since D_t consists of prime numbers as well. For example, this is the angle between the two prime numbers 5 and 7 or 193 and 197. However, not all such numbers need be prime, where this is obvious from the equations and the fact that no linear relationship predicts prime numbers. We may also combine the functions into one function represented as one for convenience. That is;

$$f(x) = 6x \pm 1$$

In this example, $(6x + 1) - (6x - 1) = 2$, a constant, hence satisfying the condition for a ditation set. As has been stated, the ditation set generates similar sets at the same angle, and the difference will be a displacement of $2n$. That is, $d = y_{i+1} - y_i$ derives $d = 2n$. An example for this case would be $y = 6x + 7$ and $y = 6x + 11$. The angle is still $\beta = 33.0417329556178$ but the displacement is now $(6x + 11) - (6x + 7) = 4 = 2(2)$.

Now to demonstrate the condition $u \neq v$, and $\beta \neq \theta$ the angle of reflection between the two sets. In this case the angle of reflection between y_i and $y_{i+1} = n$ is variable. This is also a reflective property of the ditation set that is illustrated in the behavior of the values of $f(x)$ when it is negative and positive. In our case we have the function $f_1(x) = 6x - 1$ and $f_2(x) = 6x + 1$. However, for the two sets for the same value of x, we observe the reflective property such that $-f_2(x) = f_1(-x)$ is satisfied. That is (x, y_1) is a reflection of $(-x, y_2) = (-x, -y_1)$. Hence it implies $f(x) = (6x + 1, 6x - 1)$ is a ditation set, that is $y_2 = -y_1$. If one is to graph the values, the pattern structure would not be so evident that $y_2 = -y_1$.

	y_1	y_2
x	$6x - 1$	$6x + 1$
-4	-25	-23
-3	-19	-17
-2	-13	-11
-1	-7	-5
0	-1	1
1	5	7
2	11	13
3	17	19
4	23	25

Table 11

When x is negative, then y_2 is a reflection of y_1 at two points, $y_2 = u$ and $y_2 = v$. The first point defines the reflection angle β between the two sets and the second point where the entire positive values of y_1 are reflected to y_2. Therefore the reflection of (5,31) is (-5, -31).

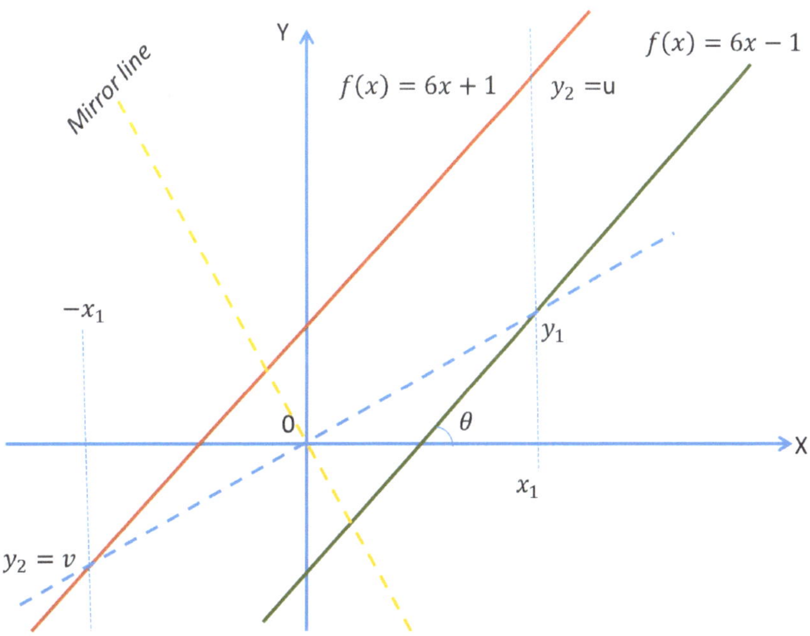

Figure 27

The graphic shows how y_1 relates to $y_2 = u$ and $y_2 = v$. The mirror line rotates about the origin for a variable θ.

2.5 Prime Number Definition

As already observed, the major thrust of the definition of the prime number is divisibility by 1 and itself, and the fact that it is a natural number. The definition was and still is limited to this context. We have shown that the definition is heavily influenced by historical circumstances where only natural numbers were known. After natural numbers came the real numbers, and these result in a linear approach to the concept of numbers. This linear approach was also influenced by the geometric context of the real number line where prime numbers tended to be studied in terms of sequence, hence the emphasis on the gap theory.

An alternative context and definition is stated, where this is a non-linear approach that emphasizes on relationship between the different types of numbers. Secondly the approach is not confined to the arithmetic context of the definition, but is based on set properties with an algebraic context rather than a geometric influence.

Definition

The Prime Number Axiom

An algebraic prime number is any number p such that;

a) $p \in D_t$.
b) it is divisible by one and minus one; and
c) it is divisible by itself, and by its magnitude.

Note the use of the qualification "algebraic" is a critical distinction from the standard notion of what is meant by a prime number. Rather than challenging the definition of the current paradigm on what is meant by a prime number, a new paradigm is defined. There are five major consequences of this definition.

a) Prime numbers are now part of a given set, the ditation set. This makes a big difference in how they are studied in terms of distribution and behavior.
b) An algebraic prime number can be negative, that is, it is now generally considered as an integer rather than being limited to being a natural number.
c) The prime number now attains an algebraic description.
d) 1 is an algebraic prime number. There is no need to debate the issue.
e) 2 and 3 are not algebraic prime numbers.

Hence for example, $-19 \in f(x)$, it is divisible by $1, -1, -19$, and 19 which is also its magnitude. This is the algebraic definition of the prime number, since it is part of the set $f(x)$. The set $f(x)$ is not defined to consist of prime numbers only, this is the major departure from tradition in conceptualizing the prime number.

The use of this type of function $f(x) = (y_1, y_2)$ in describing prime numbers has several consequences:

a. This results in a strict description of the pseudoprimes as being structurally related to primes. Pseudoprimes are a special type of numbers. They are not regarded as composite numbers because their behavior is different from composite numbers. The standard notion is that a pseudoprime is a composite.
b. It provides an opportunity to develop a sieve structure for deriving primes according to gap or classification.
c. It allows for the derivation of primality tests.
d. It develops a gap base system for prime numbers.

e. It forms a basis for structure factorization procedures.

It also causes a shift from considering a solution based on a one-to-one relationship. The two formulas allow us to relate the equation $p = 6x \pm 1$ to describe a linear relationship between the primes, and in so doing, also providing a simple sieve that conforms to a classification theory based on gaps.

The most significant outcome of applying the above definition is that it leads to the following postulate. The postulate is stated as a point of clarification because of the impact it has upon the current structure of prime number theory. However, the term algebraic prime is used to denote the different approaches.

Postulate 1

a. 1 is a prime number
b. 2 and 3 are not prime numbers

Practically, 2 and 3 are not algebraic primes, they are composites. The postulate is based on the conceptual structure of the model for prime number generation; it is not based on the conventional definition of a prime number that focuses on the natural number. What the model does is to solve the argument on whether 1 is a prime number or not and in addition throws a new argument that actually, even though 2 and 3 have no factors besides themselves and one, they do not satisfy all the characteristics for a prime number, or in particular an algebraic prime number. They satisfy only part of the definition of the prime number.

The postulate has major consequence upon the following theorem and conjecture. These are;

1. The fundamental theorem of arithmetic which states that all numbers can be written as a product of primes. Products with a 2 or 3 present a contradiction to the above definition.
2. Goldbach's Conjecture (made in a letter by Goldbach to Euler in 1742) that states every even integer greater than 2 can be written as the sum of two primes. For example 8=5+3, but 3 according to this definition is not a prime.

Since the fundamental theorem has been proved to be true, this becomes an immediate contradiction unless we recognize the fact that a truth is correct depending on the conceptual framework in which it was developed. Similarly, Goldbach's Conjecture does not hold in this framework. Consequently, it is easier to define a new structure of a prime number called the algebraic prime number, so that the above theorem and conjectures can continue to hold within a contextual criterion. The postulate brings about a standard in terms of treatment of the two paradigms; that the algebraic prime number is then assumed to be a general case, while the definition based on the natural number as a special case. The postulate is stated in order to emphasize the existence of two lines of thought in regard to the definition of the prime number, and puts the challenge to the current paradigm.

1	**7**	**13**	**19**	**25**	**31**
2	8	14	20	26	32
3	9	15	21	27	33
4	10	16	22	28	34
5	**11**	**17**	**23**	**29**	**35**
6	12	18	24	30	36

Table 12

In terms of pattern methodology and observations made herein, there is ample evidence for the validity of an alternative approach. For example, using the function $f(x)$, the table shows the possible position of occurrence of a prime number in yellow, and there are only two such strips consistently up to infinity since $-\infty \leq Int(x) \leq \infty$. In terms of pattern, the prime number does not just occur anywhere even though it is random. It occurs at specific positions within the number grid that are described by a theorem called UNDT, and one cannot argue against such evidence. From the graphic, it is also clear that 2 and 3 are not at the same pattern position of occurrence of prime numbers.

Consequently, the universal number distribution theorem (UNDT) becomes the new framework for interpreting and understanding the behavior of the integer number. This theorem also provides a classification system and distribution system of all integer numbers, positive and negative, prime and composite. It is based on an algebraic structure to define the relationship between the three classic types of numbers; the composites, the pseudoprimes and the prime numbers.

Generally, a pseudoprime is any number that shares a similar characteristic with a prime number. In the context of UNDT, a pseudoprime shares the algebraic property that it is also part of the set of numbers generated by $6x \pm 1$. This derives a conjecture because the primes and pseudoprimes exist in the same structure.

Randomness Conjecture

If primes are random, then pseudoprimes are random.

Hence the pseudoprime is also a special type of number that we can study and derive particular properties. Its prominent role is in the factorization theory. It will later be shown that the randomness of the prime numbers depends on the randomness of the pseudoprimes, which implies the relationship between the two of them is critically important. Hence this is a related randomness. Composites are any numbers that are not part of this set, but under UNDT, they are also random though their randomness is independent of the randomness of the primes or pseudoprimes.

A question of interest is which number generates the other, the prime number or the pseudoprime number? This is due to the fact that the two numbers exist in the same space, and it appears that prime numbers occupy a space that has been skipped by the pseudoprime generation. The general assumption here is that pseudoprimes generate the prime numbers, a prime number existence is actually a failure for a pseudoprime to be found. Quite a change of thought this is! This is also formulated as a postulate.

Postulate 2

In the algebraic sieve, the failure condition for a pseudoprime defines the existence of a prime number.

The postulate creates an environment of mutual existence between the two types of numbers, which implies that the study of one is not complete without the other. It results in the distribution theory of integers because the composite then are defined as a complementary set to these two.

The Classification Framework

Rather than focus on gap between primes as a key method of study, the approach has been the development of classification techniques that unlock their mystery.

The fundamental discovery is the gap base system of numbers that are prime or pseudoprime. The name pseudoprime has been retained though really it is a description of a unique set of numbers. The name has been retained because the numbers are indeed related to the prime numbers.

OUTCOMES

The section defines the gap parameter between prime numbers by developing a concept of the delta family and demonstrates the existence of the gap base classification system:

1. Gap base 2
2. Gap base 4
3. Gap base 6

It also illustrates how this can be used to explain the randomness of the prime numbers.

3 The Basis for Prime Families

Once we accept the existence of a multi-valued function as defined, then the next step is to understand how this generates random behavior. Consider the following numbers produced by the given functions for $x = 1, 2, 3, 4$. Then,

> **A** = {5, 7, 9, 11), where these are elements of the set $y = 2x + 3$.
> **B** = {2, 7, 12, 17}, where these are elements of $y = 5x - 3$.
> **C** = {1, 4, 9, 16}, where these are elements of the set $y = x^2$.

The equations above are not necessarily following the definition of a multi-valued function, since randomness is not necessarily a function of these only. Now if we consider a union of the above sets, where we place the condition of taking the first four elements only to form such a set **U**, then we have;

$$\boldsymbol{U} = \{1, 2, 4, 5, 7, 9, 11, 12, 13, 16, 17\}$$

However, the set ***U*** does not consist of predictable elements. This we note from the gap between the elements.

1	2	4	5	7	9	11	12	13	16	17
Gap	1	2	1	2	2	2	1	1	3	1

Table 13. Showing the random nature because of underlying sets.

From this, we then consider the set ***U*** to consist of random elements. Random behavior is recognized by the inability of a pattern to demonstrate a systematic repetition. This is the primary framework around which the prime numbers have been studied and continue to be studied. Now while prime numbers have been observed to be random in occurrence, they also demonstrate predictable behavior patterns. This then is an indication of some underlying pattern, and this is what has given hope to mathematicians to keep on looking for an algorithm that will describe such an occurrence.

If we study the set ***U***, we realize that the numbers so generated are actually random. Now if we call the sets ***A***, ***B***, and ***C*** as underlying sets that defined the exact nature of the elements, then set ***U*** becomes the observable set. This is the type of challenge that is presented by the prime numbers. We can derive the set consisting of prime numbers, and it forms the observable set P. One of the alternative approaches is to determine if there are underlying sets that can assist us to get back to the original description of the elements, hence the concept of the prime family.

We can also make the following observations about the sets ***A***, ***B***, and ***C***:

a. The elements of each set can be described by a dependent variable.
b. The elements of each set are predictable.
c. The gap between the elements is consistent with the equation describing the elements.

Assumption

The sequential incidence of a prime number occurs at the underlying set.

That is, there are two sets to consider when studying prime numbers; the underlying set and the observable set. Therefore the next prime number occurs within the underlying set rather than the observable set whereas we tend to look at the observable set in order to study randomness and prime number sequence.

3.1 The Delta Prime Family

The underlying sets in this theory are called prime families. We must then assume the following definition.

Definition

Let P_u be the observable set of prime numbers consisting of a family of subsets $f_1, f_2, \ldots f_n$, where $f_1 \cap f_2 \cap \ldots \cap f_n = \emptyset$

That means in the set of prime numbers we expect to find a subset that we can call a family which is unique in its own way. Consequently, this is a form of a multi-valued function.

The concept of the prime family is based upon gaps. The nth prime gap g_n is defined as the difference between the $(n+1)^{th}$ and the nth prime number, that is, $g_n = p_{n+1} - p_n$. For the purpose of distinguishing concepts and clarity, this is also referred to as the natural gap of the set of prime numbers. The set of such prime numbers forms a space P_s. A different type of space is defined as follows.

Definition

a) *The delta prime gap d_g is the difference between a prime p_n and the previous prime p_{n-1} in a given group of sequential primes*

b) *The set of prime numbers p_i where the gap $d_g = k$ is constant is called a delta prime space.*

For the delta space there is no interest in establishing the nth prime gap because in such a space, the gap is a constant that defines a prime family. Therefore, the prime gap is only considered in the context of magnitude rather than as the number of composites before the next prime. For example consider the following prime numbers from 1 to 50 and the gaps between them. These are sequential primes. For example (5-1) = 4, hence 4 is a delta gap.

Number	1	5	7	11	13	17	19	23	29	31	37	41	43	47
Gap	2	4	2	4	2	4	2	4	6	2	6	4	2	4

Table 14

The condition for finding the gap is that it must be based on the previous prime number rather than the next prime number. This is another deviation from the norm as most standard analysis considers the gap to the next prime number. If we considered the gap as being the next prime number, then we would have the following result.

Number	1	5	7	11	13	17	19	23	29	31	37	41	43	47
Gap	4	2	4	2	4	2	4	6	2	6	4	2	4	6

Table 15

The concept of the next gap is probably based on the fact that we had an interest to find the next prime number. It is referred to as the natural gap of the prime numbers. The concept of the previous gap is based on classification methodology rather than the next prime. The previous prime number is already known, and it becomes the reference to the current known prime number. This is called the delta gap.

Now we can re-arrange these according to the same gap difference, and this forms the basis of the family concept. The only occurring gaps are 2, 4 and 6 in this set of primes.

Family *Gap is two*	1	7	13	19	31	43	f_1
	2	2	2	2	2	2	
Family *Gap is four*	5	11	17	23	41	47	f_2
	4	4	4	4	4	4	
Family *Gap is six*	29	37					f_3
	6	6					

Delta Space

Figure 28

The delta space is then given by $f_1 \cup f_2 \cup f_3$and consists of three families. In this space the prime numbers are arranged according to the same gap consecutively.

Definition of Delta Element

a) A prime family (p, d_g) is a set of consecutive prime numbers with the same delta gap where $d_g = 2k$, for $1 \leq k \leq \infty$.

b) A pseudoprime family (p, d_g) is a set of consecutive pseudoprimes with the same delta gap where $d_g = 2k$, for $1 \leq k \leq \infty$.

That is, the same operation that we do for the prime numbers, we can also define and carry out for the pseudoprimes. We observe that the notion $f_1 \cap f_2 \cap \ldots \cap f_n = \emptyset$ holds as there is no intersection in terms of gaps, the sets are unique and form underlying sets. This definition is very important in the sense it introduces a new concept in defining a prime number or pseudoprime. For example, it tells us that the prime number 7 and the delta gap 2 are not separable but represent on entity.

Therefore elements of the delta space are denoted (p, d_g). From the above example, then;

1. (19,2) and (31,2) are in the same family
2. (19,2) and (29,6) are in the same delta space defined from 1 to 50 but in different families.
3. $(p, 6) = (29{,}37)$ or this can be written as $(p, 6) = \{(29{,}6), (37{,}6)\}$.

In the so-called delta prime space, d_g is even and constant because of re-arrangement of the primes, whereas g_n is a random occuring variable, and g_n concerns the next prime while d_g concerns the previous prime. For example consider the following primes with gap 2. This is an example of a delta space.

Delta Gap	Prime Number	Singular Gap	Gap Construction
2	400723	42	7
2	401311	588	98
2	401539	228	38
2	401629	90	15
2	401671	42	7
2	401773	102	17
2	401959	186	31
2	402091	132	22
2	402133	42	7
2	402139	6	1
2	402223	84	14
2	402331	108	18
2	402343	12	2

2	402361	18	3
2	402529	168	28
2	402583	54	9
2	402763	180	30
2	402769	6	1

Table 16

The gap that classifies a family is called a delta gap, while the gap that between prime numbers of the same family is called a singular gap. The singular gap is also called the "gap in the gap". The singular gap may be constructed from prime numbers, pseudoprimes or composites where this is referred to as gap construction. The yellow shows all primes numbers constructed from a gap based on prime numbers, and the rest of the singular gaps are constructed from composites.

Now that we have a new construction (p, d_g) to define the prime number, it becomes the basis of prime number classification.

3.2 Gap Classification of Prime Numbers

Consider the set of prime numbers from 1 to 10 000. We can find the delta gap for each of these prime numbers, and then sort them out in ascending order. This then derives a position of the gaps from 1 to 10 000. This is presented as a graph.

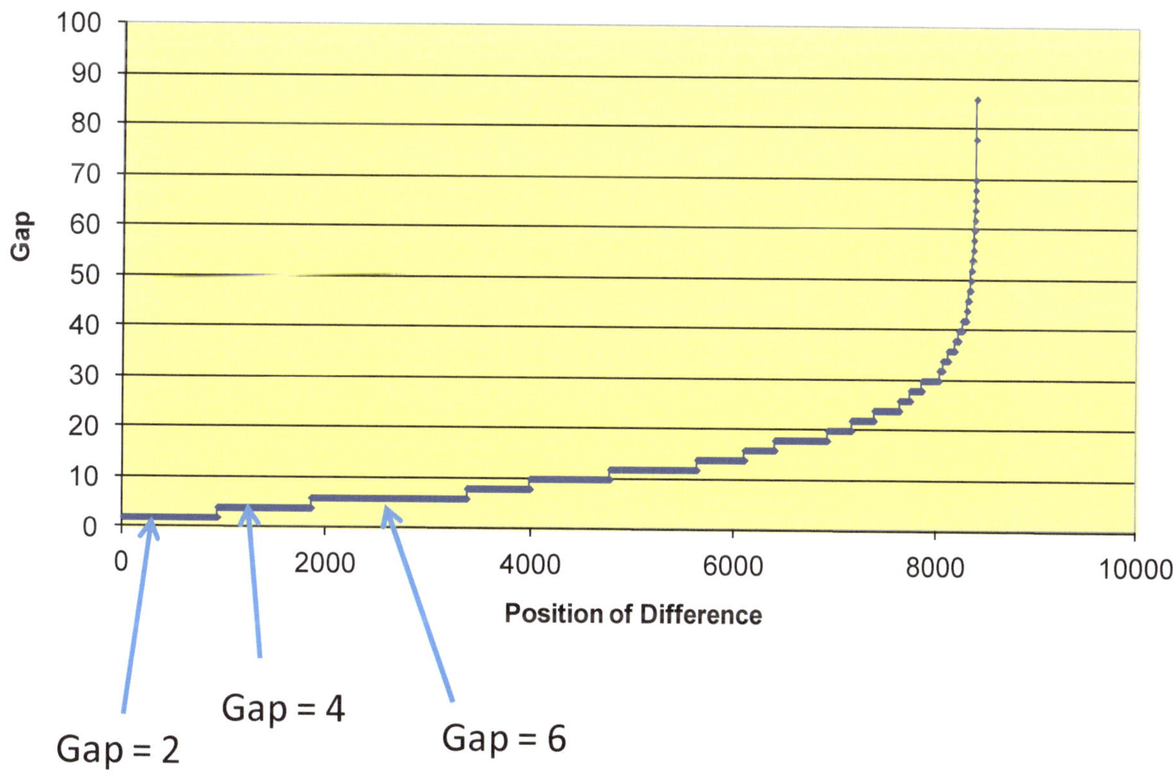

Figure 29

It will be observed from the graph that the pattern *2-4-6* repeats itself as *8-10-12* and so on. This is the first and most important observation from defining prime numbers using the concept of family. It was also my first major discovery in terms of classification. At first it was a puzzling observation as it was highly unexpected. It led to the conclusion that gaps behave as a family as well. That is, we have category family of gaps;

- 2, 8, 14, 20, . . .
- 4, 10, 16, 22, . . .
- 6, 12, 18, 24, . . .

The difference between each gap in the family is six. Since the initial gaps are *2-4-6*, this led to the concept of defining a gap base.

The next surprise is that the gaps not only act in groups, but are defined from an equation perspective as well. Now if all the primes of a given family (p, d_g) are plotted on a graph for the same number range, the surprising results that we get a relatively linear relationship. The delta gaps are plotted up to 48 to show the progressive effect.

In this relationship, we note that,

a. the set of primes $(p, 2)$ are almost parallel to $(p, 4)$, the gradient is about equal, and
b. the set of primes $(p, 6)$ are at a smaller gradient with respect to the other two families.
c. The pattern *2-4-6*, $2n - 4n - 6n$ for $n > 1$ is evident in the graph behavior.

This demonstrates that prime numbers behave as a family in terms of their distribution. That is, they do not normally follow the sequential distribution pattern that is expected of them. This pattern of behavior is defined as a spectral distribution of primes, and the graph is called spectral lines of prime numbers.

From the graph, we can clearly see the underlying sets for a given range. Yes indeed, prime numbers are in a sequence, the next prime is greater than the previous prime, but they are not in the same group or category because of that. Therefore, the aspect of being sequential is not an overriding factor in understanding their distribution behavior. The graphics demonstrate a parallel behavior in terms of the distributive property.

Fundamental Gap behaviour

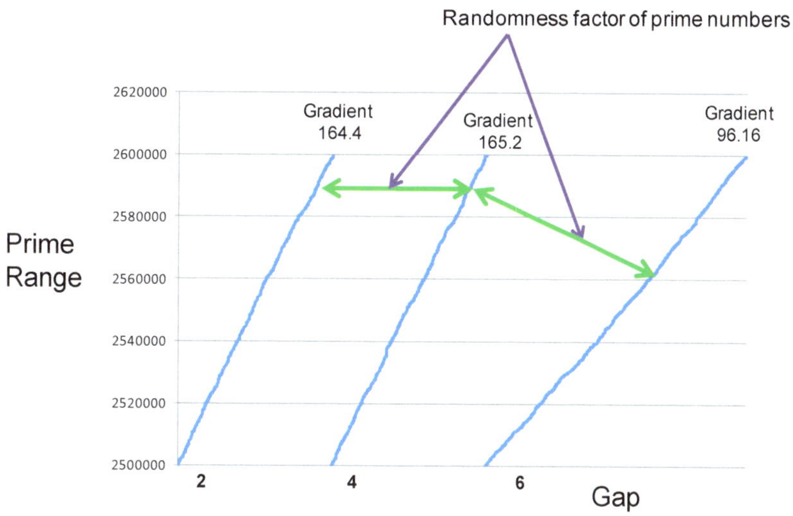

The above shows a plot of the prime numbers for a given family. A similar pattern will be obtained for **any** dimension, so it is representative. The gradient is obtained from a best linear fit on the plot. This pattern, called a **spectral pattern** of the prime numbers, is responsible for the random behaviour of the primes though they have regular properties.

Figure 30

GAP THEORY
Spectral Distribution of the Prime Numbers

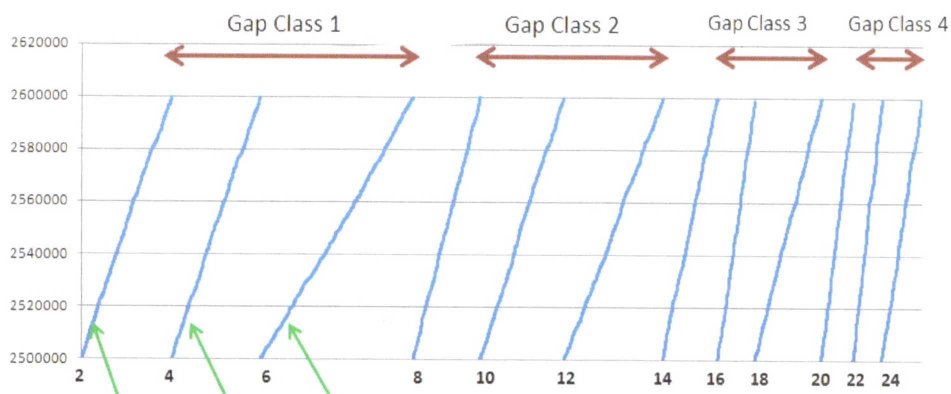

6a-1

6a+1

6a+1

or

6a-1

The spectral lines demonstrates the gap structure of prime numbers between 2.5 to 2.6 million. Gap 8 is also described by 6a+1, gap 10 by 6a-1, and gap 12 by 6a+1 or 6a-1; each gap class is described in a similar manner. **The spectral lines show that prime numbers are linearly related by gap.**

Note that “2” and “4” components are almost parallel in behavior.

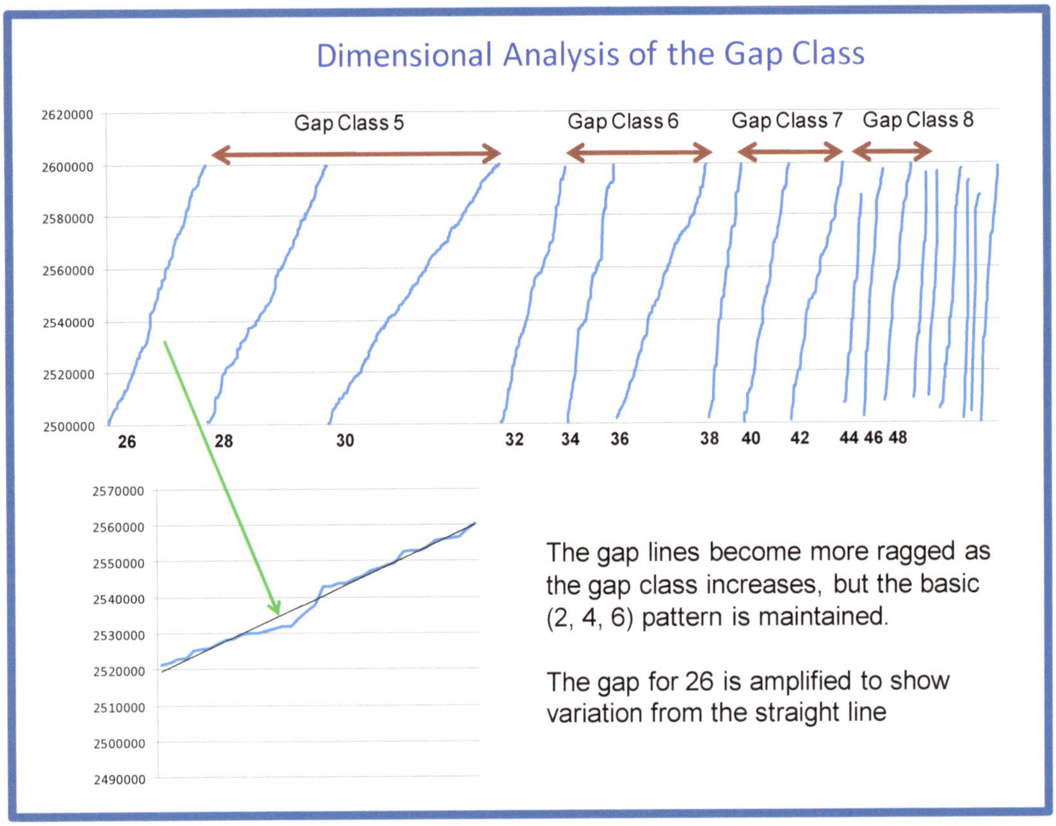

Figure 31

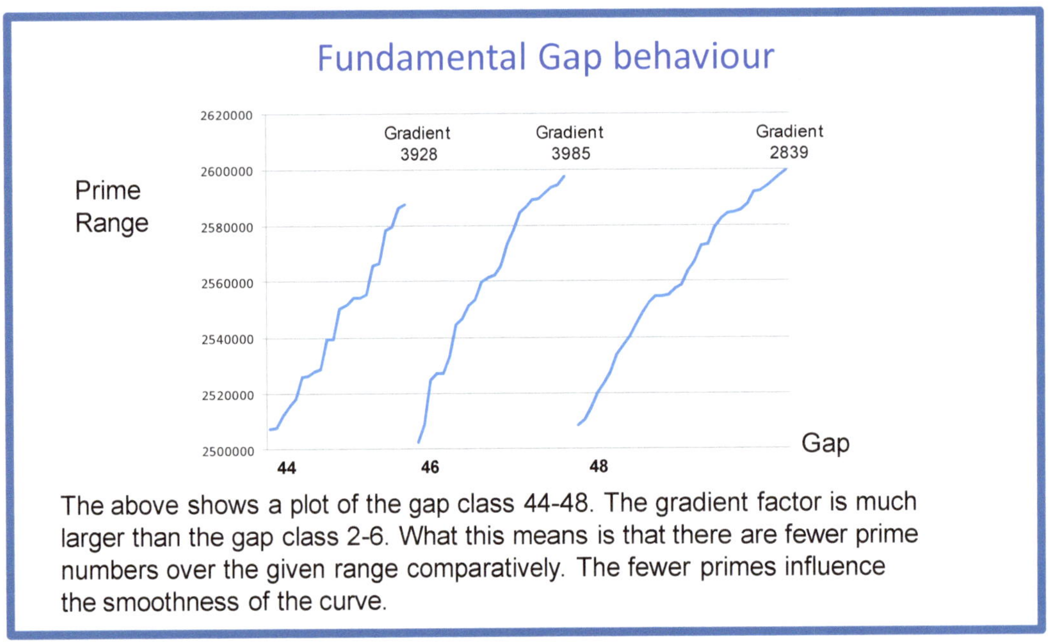

Figure 32

We can further define each *2-4-6*, *8-10-12* and so on as a gap class to enhance the classification criteria for prime numbers. In each case, the next three groups give the next class of prime gaps. Now as the class number goes up, the straight line aspect of the distribution of the primes along the line is lost; they become more rugged because of fewer primes. However, the relationship that exists in terms of the base structure of the gaps is still very strong. If we increase the elements of the prime space for a higher gap class, then the curves will also be smoother.

3.3 Source of Randomness of Primes

The major consequence of the function $f(x) = (6x - 1, 6x + 1)$ is that prime numbers and pseudoprimes are structurally related. The construct creates a positional relationship between the two types of numbers such that if a position is not occupied by a prime number, then it is occupied by a pseudoprime.

Therefore we now consider the set of numbers defined by such a function in the context of a classification system based on a gap base system.

Definition

Let gap 2, gap 4, gap 6 construct a gap-base system for classification and generation of any type of prime numbers and pseudoprimes.

That is, when we apply it to the prime numbers only, we can define a prime number to be in gap base 2, gap base 4, or gap base 6. This is called the gap base classification of prime numbers.

Type	Gap	Comment
Gap Base 2	2, 8, 14, 20, 26.....	Regular difference of 6
Gap Base 4	4, 10, 16, 22, 28, ...	Regular difference of 6
Gap Base 6	6, 12, 18, 24, 30, 36, ...	Regular difference of 6
Below shows combining gap base 6 into gap base 2 and 4		
Gap Base 2	2, 6, 8, 12, 14, 18, 20, 24, 26,	Irregular difference
Gap base 4	4, 6, 10, 12, 16, 18, 22, 24, 28,	Irregular difference

Table 17

The occurrence of random prime number behavior is partly contributed by gap base 6 that comes into gap base 2 and 4 as shown by the table. The gap base 6 is said to emphasize the random behavior of prime numbers. The prime numbers would still behave in a random fashion without gap base 6, so it is seen as a contributory factor rather than as the sole factor.

Therefore the two primary factors that contribute to the randomness of the prime numbers are:

a) the structural relationship between the prime numbers and the pseudoprimes
b) the appearance of gap 6 in either gap base 2 or 4.

However, the relationship between primes within a gap base is reasonably linear as demonstrated by the spectral lines. The figure shows that the pattern for gap base 2 and 4 are similar, though gap base 2 has a (4,2) pattern and gap base 4 has a (2,4) pattern.

GB2	Pattern	GB2	Pattern	GB4	Pattern	GB4	Pattern	GB6
2		2		4		4		6
8	6	6	4	10	6	6	2	12
14	6	8	2	16	6	10	4	18
20	6	12	4	22	6	12	2	24
26	6	14	2	28	6	16	4	30
32	6	18	4	34	6	18	2	36
38	6	20	2	40	6	22	4	42
44	6	24	4	46	6	24	2	48
50	6	26	2	52	6	28	4	54
56	6	30	4	58	6	30	2	60
		32	2			34	4	
		36	4			36	2	
		38	2			40	4	
		42	4			42	2	
		44	2			46	4	
		48	4			48	2	
		50	2			52	4	
		54	4			54	2	
		56	2			58	4	
		60	4			60	2	

Figure 33

Consequently we define two sets; GB-2 to define gap base 2, and GB-4 to define gap base 4.

Fundamental Gap Theorem

Let $-\infty \leq x \leq \infty$, *where* x *is an integer, then*

a. $f(x) = 6x + 1, f(x) \in GB2$ *and* $f(x)$ *is a prime number or pseudoprime.*
b. $f(x) = 6x - 1, f(x) \in GB4$ *and* $f(x)$ *is a prime number or pseudoprime.*
c. $f(x) = 6x \pm 1, f(x) \in GB2$ *or* $f(x) \in GB4$ *defines the set GB6.*
d. $f(x) \neq 6x \pm 1$ *defines a composite.*

This theorem defines a mathematical construct that is called an algebraic sieve, where this is formed from linear ditation numbers. The fundamental gap theorem is of distinction in the sense that it defines a mathematical construct resulting in the following.

a) It establishes a framework for defining precisely three types of integer numbers that can be described using an algebraic relationship. This provides an opportunity for a unified and integrated approach to interpreting and understanding integers in contrast to the emphasis of real number properties only.
b) It relates all the three types of numbers into the types of gaps that exist between them. This is indeed a unique feature of the theorem, because the gap theory has only been applied to prime numbers to the exclusion of pseudoprimes or composites. But now we use the gap theory to define the composite number as well.
c) It makes the prime number to be no more studied in an isolated manner. For example the pseudoprime has exactly the same prominence as the prime number, they belong to the same set!
d) A prime number can be defined using the notion (p, d_g) to define a prime family. The same notion can be applied to pseudoprimes, they can be classified in terms of their gaps to form families.
e) It does not recognize 2 and 3 as prime numbers. In other words, the theorem extends the meaning and definition of a prime number. Furthermore, since x can be negative, it implies that the prime number can be a negative entity.
f) It gives us a precise algebraic description of the composite number. It does not tell us the gap theory with regard to composite numbers, but allows us to recognize their definition in the context of a unified approach.

	GB-4 Primes only
	GB-2 Primes only
	GB-6 Primes only
	Pseudoprimes

X	GB-4	GB-2
1	5	7
2	11	13
3	17	19
4	23	25
5	29	31
6	35	37
7	41	43
8	47	49
9	53	55
10	59	61
11	65	67
12	71	73
13	77	79
14	83	85
15	89	91
16	95	97
17	101	103
18	107	109

GB-6
29
37
53
59
67
79
89

Natural gap: 2+6(n-1)

Natural gap: 4+6(n-1)

Natural gap: 6n

Figure 34

Therefore it is fairly easy to find out what gap base a prime number is in. We just use one of the equations starting with 2 or 4. Since gap base 6 uses two equations, this implies a given prime number will either be in gap base 2 or gap base 4. That is, gap base 6 is a subset of gap base 2 and gap base 4. The gaps can be defined by the general equations for each gap base.

1. *Gap base 2: The equation is gap* $g(m) = 2 + 6(m-1)$, *and for gap base 6 then* $g(m) = 6m$, *where* $1 \leq m \leq \infty$.
2. *Gap base 4: The equation is gap* $g(m) = 4 + 6(m-1)$, *and for gap base 6 then* $g(m) = 6m$, *where* $1 \leq m \leq \infty$.

m	GB2	GB4	GB6
1	2	4	6
2	8	10	12
3	14	16	18
4	20	22	24
5	26	28	30
6	32	34	36
7	38	40	42

Table 18

Consequently, we can work with two gap bases, gap base 2 and 4 in doing a general classification of prime numbers. We refer to gap base 6 if there is a particular angle being researched or addressed. The interesting mystery is that while the gap bases have different equations to define their gap distribution, gap base 6 numbers are also generated by the same equations that define gap base 2 and 4.

The table below shows that gap base 6 starts from the prime number 23, and also 31. That is 29 and 37 are the first gap base 6 numbers. We also note that if arranged in order, and we have 31 numbers for each gap base, then the gap base 6 numbers increase faster than then those in gap base 2 and 4. On the other hand, gap base 2 and 4 increase almost at the same rate. A statistical analysis over a much larger range also reveals the same result.

It is not clear why gap base 6 numbers increase at a faster rate than gap base 2 and 4 numbers, but it does suggest that the gap base 6 numbers have more pseudoprimes in between them than the other gap bases.

GB-4	Gap	GB-2	Gap	GB-6 in 4	Gap	GB-6 in 2	Gap
5	4	7	2	29	6	37	6
11	4	13	2	53	6	67	6
17	4	19	2	59	6	79	6
23	4	31	2	89	6	157	6
41	4	43	2	137	6	163	6
47	4	61	2	173	6	211	12
71	4	73	2	179	6	223	12
83	4	97	8	239	6	277	6
101	4	103	2	257	6	337	6
107	4	109	2	263	6	373	6
113	4	127	14	269	6	379	6
131	4	139	2	359	6	439	6
149	10	151	2	389	6	541	18
167	4	181	2	449	6	547	6
191	10	193	2	479	12	577	6
197	4	199	2	509	6	607	6
227	4	229	2	521	12	613	6
233	4	241	2	563	6	631	12
251	10	271	2	569	6	673	12
281	4	283	2	593	6	733	6
293	10	307	14	599	6	739	6
311	4	313	2	653	6	757	6
317	4	331	14	659	6	997	6
347	10	349	2	683	6	1009	12
353	4	367	8	809	12	1039	6
383	4	397	8	947	6	1069	6
401	4	409	8	953	6	1087	18
419	10	421	2	977	6	1123	6
431	10	433	2	983	6	1129	6
443	4	457	8	1019	6	1213	12
461	4	463	2	1103	6	1237	6
31		31		31		31	

Table 19

We can conclude by comparing this framework to the definition of odd numbers, since all primes are odd. The odd numbers are defined by $2n-1$. We note the following pattern from such a set.

n	*2n-1*
0	-1
1	1
2	3
3	5
4	7
5	9

6	11
7	13
8	15
9	17
10	19
11	21
12	23
13	25
14	27
15	29
16	31
17	33
18	35
19	37
20	39

1st Occurrence	2nd Occurrence
Gap Base 4	**Gap Base 2**
5	7
11	13
17	19
23	25
29	31
35	37

Table 20

The set contains both gap base 2 and gap base 4 primes occurring at regular and specific intervals. There is a specific position at which the probability of a prime can occur. The other odd numbers produced by $2n - 1$ belong to another special set called Class 1A. In terms of pattern methodology, we note that prime occurrence is at a specific position on the grid, and we note that 1 is actually part of this set in the context of this approach. If we separate the result into two groups, the first occurrence and second occurrence, then we get the gap base 2 and gap base 4 format.

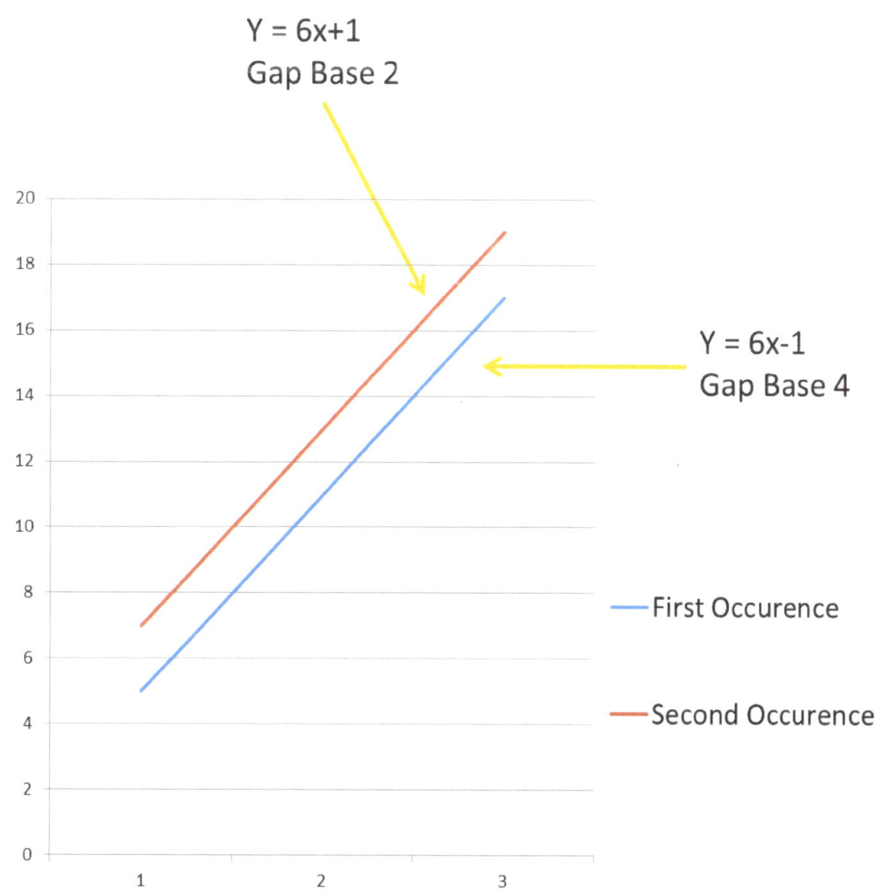

Figure 35

Again we get our multi-valued relationship, $6x \pm 1$. The example also shows the difference one gets in terms of emphasis on pattern methodology as an investigative tool in mathematics in contrast to the analytical approach that focused mainly on the strict definition of a prime number. Consequently odd numbers can be classified as;

a) Prime, and $f(x) = 6x \pm 1$
b) Pseudoprime, and $f(x) = 6x \pm 1$
c) Composite, where these are in Class 1A and $f(x) \neq 6x \pm 1$. Class 1A consists of odd numbers that are not prime or pseudoprime.

Consequently, there are three types of odd numbers, rather than two types as prime and non-prime.

3.4 Delta Signatures

In the gap base system, the pseudoprimes are given a zero gap with respect to the prime numbers though one can find the gap between pseudoprimes the same way as in the case of prime numbers. The vertical axis is defined as a pseudoprime axis because all the pseudoprimes are represented on it. The horizontal axis represents the prime number gap, the delta gap. Consequently this is called a delta signature for a given range $[a, b]$ defines a relationship between the prime numbers and pseudoprimes.

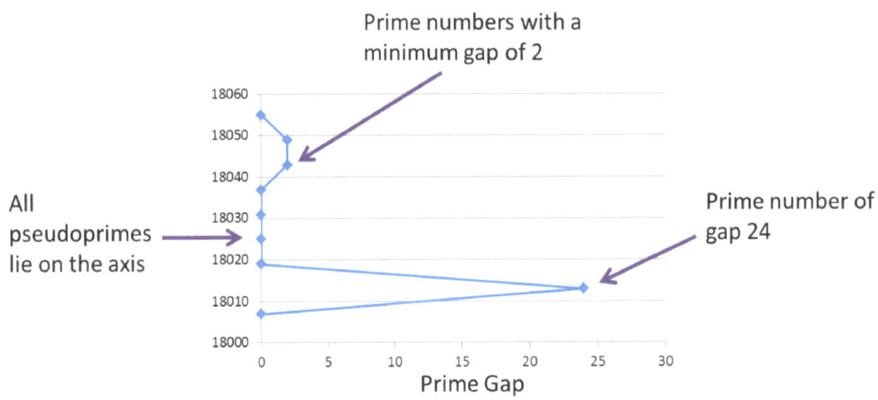

Figure 36

The delta signature is therefore defined to have the following features.

a) A range [a,b] given in terms of x such that $(b - a) = 41$.
b) A vertical pseudoprime axis with range [a,b], where a single point on the axis denotes a number that is pseudoprime.
c) A horizontal axis that defines a prime gap.
d) A signature space defined by the two axes.
e) A prime number in the signature space denoted as a peak of a curve.
f) A signature can consist of gap base 2 or gap base 4 numbers only, or a combination of both.
g) Signature space can be paired.

Definition

The delta signature is determined by the number and sequence of points on the pseudoprime axis and the number of peaks of the prime numbers.

The value of $(b-a) = 41$ is chosen as conveniently and sufficiently small to contain a significant amount of information. For either gap base 2 or 4, this will imply that there are $(b-a+1)$ numbers that used to define the signature, where these numbers are either prime or pseudoprime. That is, 42 numbers are used to define a signature. This is a standard given for a signature, but this can vary depending on application. Also, the range [a,b] can be defined as a variable to introduce a certain level of complexity.

The minimum gap in a gap base 2 is 2 and it is 4 for the gap base 4. A graph is given for a range [2,40].

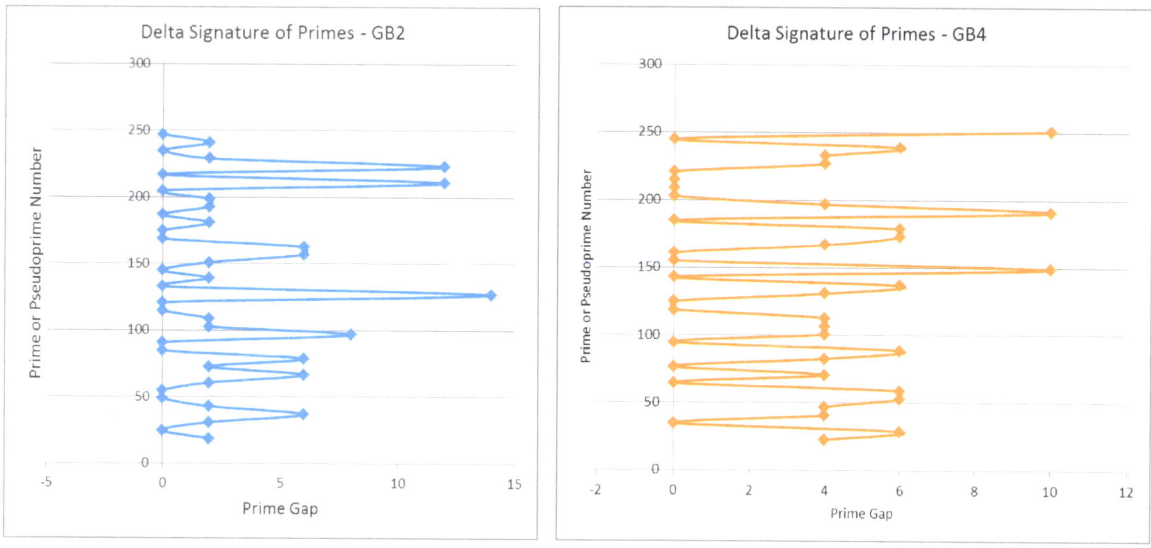

Figure 37

If the signature is defined by a single gap base, the GB2 gap will only consist of gaps 2, 8, 14, and so on while the GB4 will consist of 4, 10, and 16. However the gaps 6, 12, 18 and so on will occur for both gap bases. The following is hypothesized. The prime signatures are said to be paired if they include both gap bases for the given range [a,b].

Delta Signature Hypothesis

A delta signature in a given range $[a, b]$ is unique.

The basis of the hypothesis is the fact that prime numbers are random in their distribution. This leads to the following assumption, that the delta signature can actually be used to present a number.

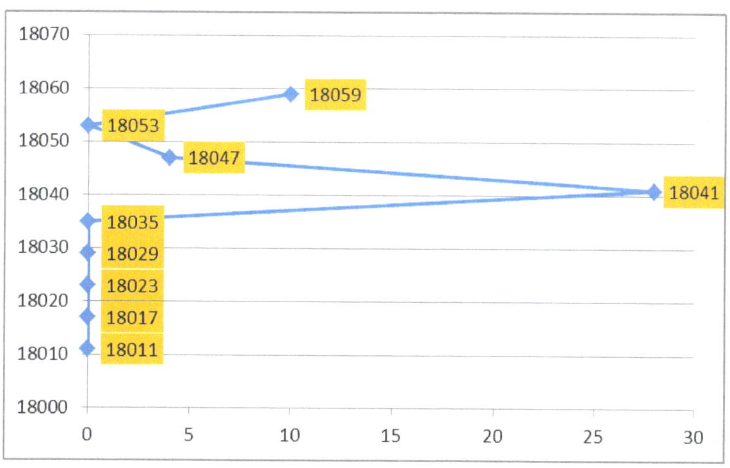

Figure 38

For example, consider the prime numbers 18041, 18047 and 18059. The rest are pseudoprimes as they lie on the horizontal axis. The product of the prime numbers is 5,879,756,255,693 and the sum of the pseudoprimes is 108150. Hence we can define an entity based on sum and product $W_d = (108150, 5879756255693)$, where this has the implication that the delta signature may be used to represent what may be referred to as a delta signature number. Whilst the above example has consecutive numbers, this does not have to be the case always.

Definition

Let W_d be defined by five or less prime numbers, and five or less pseudoprimes in sequential order then W_d is a signature number in range [a,b] of the delta signature.

The difference between a delta signature and the delta signature number is the fact that the delta signature is always consecutive. On the other hand, the delta number consists of sequential rather than consecutive numbers. Secondly, the information of a signature number comes from the same range of the delta signature.

For example, consider the following is a delta signature consisting of both gap base 2 and 4, where [a,b]=[3002,30042].

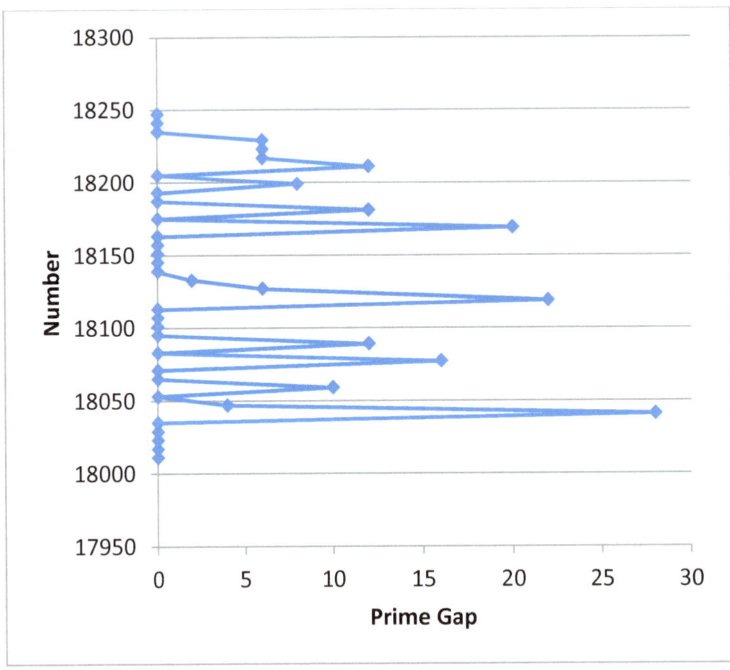

Figure 39

As an example, we can construct the delta signature number from two prime numbers and three pseudoprimes. The prime numbers are 18047 and 18169, and the pseudoprimes are 18017, 18139, and 18235. Therefore;

$$W_d = (54391, 327895943)$$

This is not the only signature number that we can construct. Other numbers as well can be constructed, for example;

$$W_d = (90625, 5956865237101)$$

One can produce all the possible signature numbers for a given space [a,b], and all of them will be unique.

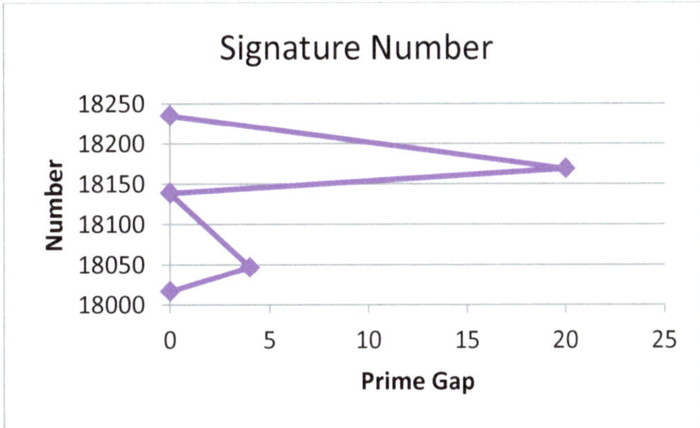

Figure 40

This leads to the following.

Signature Derivative Theorem

For a given space [a,b], then all the signature numbers W_d are unique.

Proof

Let W_d be defined by prime numbers $p_1, p_2, \cdots p_5$ and pseudoprimes $q_1, q_2, \cdots q_5$, then any product in the space [a,b] of the primes will result in a pseudoprime $q_6, q_7, \cdots q_n$, and any sum of $q_1, q_2, \cdots q_5$ will derive pseudoprimes $q_{s1}, q_{s2}, \cdots q_{sn}$ where for product or sum $q_i \neq q_j$. If $q_i \neq q_j$ then the result is unique, hence any W_d is unique. ■

It follows then that if any W_d is unique, then for a given space [a,b], the delta signature is unique. Hence this proves the delta signature hypothesis. Also, since W_d is defined within the delta signature, this implies it can be used to determine the delta signature primes and pseudoprimes when two variables are known; the range [a,b] and W_d. Therefore W_d acts like an internal signature of the delta signature.

The signature concept can be further made complex by assuming that the range is a variable where for a number W, then its prime factors are a, b. Hence to find W_d the space $[a, b]$ one has to first factorize W. As a result, knowing W and W_d one can also determine the set that defines the delta signature. Hence this derives the construct $[W, W_d]$.

Prime Patterns

OUTCOMES

The emphasis here is application. That is, to understand the application of the gap base system in the distribution of integers such as;

1. Twin primes
2. Symmetric primes
3. Asymmetric primes as defined by Fletcher, Lindgren and Pomerance (1966)

4 Twin Primes

The twin prime existence gives credence to the UNDT classification structure. All twin primes begin at gap base four, which implies the next prime is then in gap base 2. But the important observation is that the occurrence is between the two gap bases. The difference between any two values in a twin pair with the same gap base is a multiple of six.

Twin Pairs	
GB4	GB2
5	7
11	13
17	19
29	31
41	43
59	61
71	73
101	103
107	109
137	139
149	151

$6x+1$	$6x+5$
GB2	GB4
7	11
13	17
19	23
25	29
31	35
37	41
43	47
49	53
55	59
61	65
67	71
73	77
79	83
85	89
91	95
97	101
103	107
109	113
115	119
121	125
127	131
133	137
139	143
145	149
151	155

$6x+1$	$6x-1$
GB2	GB4
5	7
11	13
17	19
23	25
29	31
35	37
41	43
47	49
53	55
59	61
65	67
71	73
77	79
83	85
89	91
95	97
101	103
107	109
113	115
119	121
125	127
131	133
137	139
143	145
149	151

Table 21

Similarly, the smallest difference between any two twin pairs is 4 because of the gap base structure. This type of statement would be difficult to make in a non-UNDT environment as there is no basis for making it except through observation that it is so. The standard representation for the UNDT framework is taken as $6x \pm 1$ but as noted before, other representations are equally valid. For example we can have the gap between the bases as 4 instead of 2 by using $6x + 1$ and $6x + 5$. The structure from this presentation clearly shows the minimum difference between two twin prime pairs as 4, as shown by 103 and 107 for the same value of *x*.

91	95
97	101
103	107
109	113
115	119

Table 22

The general theory is that there are an infinite number of twin primes.

5 Symmetric And Asymmetric Primes

One can use the gap base system to study the type of distribution that we have in respect of symmetric and asymmetric prime numbers. A symmetric prime number has the form of a reflection around the digit at the center. Hence it increases in steps of 3,5,7,9... in size. The one digit prime numbers are not included as no axis of symmetry may be defined around a given digit. Since 2 is not part of the 3-5-7-... formation, it implies 11 is not regarded as a symmetric prime number. The symmetric prime numbers are also referred to as palindrome. The following are found from the first 100,000 prime numbers.

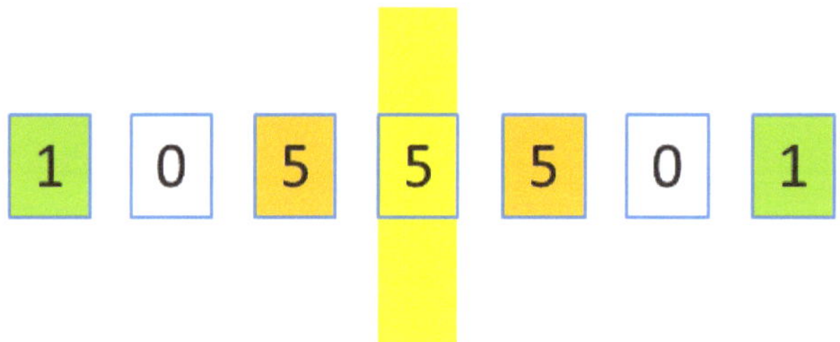

Figure 41

An example is shown with "5" as the center and a reflection of 105 to 501 about "5". The numbers can be categorized as gap base 2 and 4.

Gap base 2 symmetric prime numbers

151 181 313 373 727 757 787 919 10501 11311 12421 12721 13831 14341
15451 16561 17971 18181 18481 19891 30103 30403 30703 31513 32323 34543 34843
35053 35353 37273 37573 38083 70207 70507 71317 71917 72727 73237 74047 77377
77977 78487 78787 90709 93139 93739 94849 95959 96469 96769 97579 97879 98389
98689 1008001 1035301 1043401 1062601 1065601 1092901
1114111 1117111 1160611 1163611 1177711 1190911 1193911
1196911 1201021 1212121 1215121 1218121 1242421 1245421
1250521 1253521 1278721 1280821 1286821

Gap base 4 symmetric prime numbers

101 131 191 353 383 797 929 10301 10601 11411 12821 13331 13931
14741 15551 16061 16361 16661 17471 19391 19991 30203 30803 31013 32423 33533
35153 35753 36263 36563 38183 38783 39293 70607 72227 73037 73637 74747 75557
76367 76667 77477 78887 79397 79697 79997 91019 93239 94049 94349 94649 94949
96269 97379 1003001 1022201 1028201 1055501 1074701 1082801
1085801 1093901 1120211 1123211 1126211 1129211 1134311
1145411 1150511 1153511 1175711 1178711 1180811 1183811
1186811 1208021 1221221 1235321 1243421 1257521 1262621
1268621 1273721 1276721 1281821 1287821

For this given range, there are more gap base 4 numbers that are symmetric compared to gap base 2. This is 79 against 87. The numbers 5 and 7 have not been included as symmetric. The average gap to the next symmetric prime for gap base 4 is 14974.53 and for gap base 2 its 16495.76, which means this is about the same.

GB2	GB4
151	11
181	101
313	131
373	191
727	353
757	383
787	797
919	929

GB2	GB4	GB2	GB4	GB2	GB4	GB2	GB4
10501	10301	30103	30203	70207	70607	90709	91019
11311	10601	30403	30803	70507	72227	93139	93239
12421	11411	30703	31013	71317	73037	93739	94049
12721	12821	31513	32423	71917	73637	94849	94349
13831	13331	32323	33533	72727	74747	95959	94649
14341	13931	34543	35153	73237	75557	96469	94949
15451	14741	34843	35753	74047	76367	96769	96269
16561	15551	35053	36263	77377	76667	97579	97379
17971	16061	35353	36563	77977	77477	97879	
18181	16361	37273	38183	78487	78887	98389	
18481	16661	37573	38783	78787	79397	98689	
19891	17471	38083	39293		79697		
	19391				79997		
	19991						

Table 23

On close observation we note that the pattern on the symmetric primes follows 1-3-7-9 pattern, skipping 5 as this is a pseudoprime ending.

For the asymmetric prime numbers, the result is very different. A definite pattern is observed, but interpretation and meaning differs because of the application of pattern methodology. These are numbers that are defined Fletcher et al [7] as follows.

Definition

Let $n > 1$ and suppose that both $2n + 1$ and $4n + 3$ are primes. If $4n + 5$ is not prime,

a. $4n + 3$ is asymmetric.
b. $4n + 1$ is asymmetric.

This actually defines a sieve method because of the decision criteria where both $2n + 1$ and $4n + 3$ are primes for a given set of prime numbers. Evidence of this is the fact that can apply pattern methodology techniques and find relevant equations to describe the numbers.

The first step of the analysis is to determine the gap base of the numbers.

Gap base 4 asymmetric prime numbers

23 47 83 167 173 263 359 383 389 467 479 503 509 557 563
587 653 719 797 839 863 887 907 971 983 1103 1187 1259
1283 1307 1367 1439 1499 1511 1523 1571 1637 1733 1823 1907 1913
2039 2063 2099 2207 2411 2447 2459 2543 2579 2663 2819 2879 2903
2927 2963 3023 3089 3203 3491 3593 3623 3779 3803 3863 3923 3947
3989 4007 4013 4073 4079 4139 4283 4349 4373 4391 4493 4583 4679
4703 4919 5003 5039 5087 5189 5381 5387 5399 5471 5483 5507 5717
5807 5843 5927 5939 6047 6173 6263 6317 6323 6389 6599 6653 6719

[7] Fletcher, Lindgren, Pomerance. *Symmetric and Asymmetric Primes*. Journal of Number Theory No.58, 89-99. 1996 Academic Press, Inc.

6857 6863 6899 6983 7013 7079 7187 7247 7283 7499 7517 7523 7607
7643 7703 7727 7817 7823 7883

Gap base 2 asymmetric prime numbers

1579 2203 2767 5323 5623 6607 6619 7027 7159 7699

Hence we establish in terms of pattern methodology the following in reference to the gap base system.

				GB2	GB4	n for $4n+3$	n for GB4
n	$4n+3$	$2n+1$	$4n+5$	$6n+1$	$6n+5$	$3n-1$	$2n-1$
1	7	3	9	7	11	2	1
2	11	5	13	13	17	5	3
3	15	7	17	19	23	8	5
4	19	9	21	25	29	11	7
5	23	11	25	31	35	14	9
6	27	13	29	37	41	17	11
7	31	15	33	43	47	20	13
8	35	17	37	49	53	23	15
9	39	19	41	55	59	26	17
10	43	21	45	61	65	29	19
11	47	23	49	67	71	32	21
12	51	25	53	73	77	35	23
13	55	27	57	79	83	38	25
14	59	29	61	85	89	41	27

Table 24

The green coloring shows numbers that will satisfy the condition $4n+3$ as well as the condition $6n+5$ that derives gap base numbers. Therefore an exact relationship can be established where we define n for gap base 4 and n for $4n+3$.

Asymmetric Criteria Rule

Every occurrence of n being odd defines a chance of the gap base 4 number forming an asymmetric prime through the following equations;

i. $3n-1$, *where this is n for $4n+3$*
ii. $2n-1$, *where this is n for gap base 4*

Using consistency and the definition, it is evident that the asymmetric primes as per criteria are from the set given by gap base 4. Therefore the gap base 2 numbers are rejected and assumed not to be consistent with the criteria. There are other numbers as well that fall by the wayside as shown by the decision criteria which can be simplified as [Prime, Prime, Not Prime] from the definition. The "not prime" is the same as implying a pseudoprime occurrence. By applying this criteria, it is clear that 11 is not the first asymmetric prime number because 13 is prime.

	4n+3	2n+1	4n+5	
n	Prime	Prime	Prime	Does not meet criteria
2	11	5	13	

	4n+3	2n+1	4n+5	
n	Prime	Prime	Pseudoprime	Meets criteria
296	1187	593	1189	

	4n+3	2n+1	4n+5	
n	Prime	Not Prime	Not Prime	Does not meet criteria
275	1103	551	1105	

	4n+3	2n+1	4n+5	
n	Prime	Prime	Prime	Does not meet criteria
86	347	173	349	

	4n+1	2n+1	4n+5	
n	Prime	Prime	Pseudoprime	Does not meet criteria
140	563	281	565	

	4n+3	2n+1	4n+5	
n	Prime	Not Prime	Not Prime	Does not meet criteria
226	907	453	909	

Table 25

As indicated, several other numbers do not meet the decision criteria or sieve requirements such as 173 and 1103. The number 563 appears to meet the criteria if we relax certain conditions, like if we assume that the only condition must be [Prime, Prime, Not Prime]. However, 907 does not meet the criteria altogether because it has the format [Prime, Not Prime, Not Prime]. An additional condition that can be added as a consequence of the definition is that as follows. The rule allows for conceptual consistency in the definition of the asymmetric prime number, so that we can weed out further numbers that don't match but seem to belong.

Asymmetric Gap Rule

The gap between any two asymmetric numbers is a multiple of 12.

Consider the example below.

	4n+3	Gap not 12	Gap Base
GB6	467	78	GB4
GB6	563	6	GB4
	907	20	GB2
	1579	8	GB2
	2203	104	GB2
	2767	104	GB2
	5323	236	GB2
	5623	116	GB2

4n+3	Gap not 12	Gap Base
6607	8	GB2
6619	20	GB2
7027	44	GB2
7159	80	GB2
7699	56	GB2

Table 26

That being the case, the numbers above do not fit into the asymmetric rule because they are not multiples of twelve. The gap is calculated with respect to the previous prime. Secondly, the rest of the numbers are gap base 2 except for 467 and 563. However the asymmetric numbers are gap base 4, hence they present a contradiction.

For example 563 fits into the criteria [Prime, Prime, Not Prime] for asymmetric numbers but is disqualified by its inconsistency to the asymmetric gap rule. The above group of numbers is defined by 4n+3 but does not have a gap of 12 with the previous prime number. The following group of numbers is of 4n+5 but also does not have a gap of 12 to the previous prime number. Consequently they do not belong to the asymmetric group.

Gap base 6 in GB4		Gap base 6 in GB4		Gap base 6 in GB4	
4n+5	Gap not 12	**4n+5**	Gap not 12	**4n+5**	Gap not 12
173	6	4013	6	7013	30
389	6	4073	66	7517	18
509	6	4349	66	7817	90
557	54	4373	90		
653	66	4493	102		
797	78	5189	102		
1637	66	5381	294		
1733	162	5717	210		
1913	6	6173	126		
3089	66	6317	54		
3593	102	6389	66		
3989	42	6653	54		
		6857	138		

Table 27

The following table gives all the values that are generated by 4n+3 but also the gap is 12 or a multiple. What can be observed is that the numbers that are defined by 4n+3 and 4n+5 all belong to gap base 6. Remember that there are three gap bases; 2 4, and 6. Therefore the numbers that are described as asymmetric are gap base 6 but they have a common gap between them that is a multiple of 12. It is like saying we make them to be gap base 12 numbers. Therefore in terms of gap base 6, the numbers 173, 467, 563 are in the same set as 23.

Gap base 6 in GB4		Gap base 6 in GB4		Gap base 6 in GB4		Gap base 6 in GB4	
4n+3	*Gap 12*	*4n+3*	*Gap 12*	*4n+3*	*Gap 12*	*4n+3*	*Gap 12*
23	12	1511	12	3779	156	5927	84
47	24	1523	12	3803	24	5939	12
83	36	1571	48	3863	60	6047	108
167	84	1823	252	3923	60	6263	216
263	96	1907	84	3947	24	6323	60
359	96	2039	132	4007	60	6599	276
383	24	2063	24	4079	72	6719	120
479	96	2099	36	4139	60	6863	144
503	24	2207	108	4283	144	6899	36
587	84	2411	204	4391	108	6983	84

719	132	2447	36	4583	192	7079	96
839	120	2459	12	4679	96	7187	108
863	24	2543	84	4703	24	7247	60
887	24	2579	36	4919	216	7283	36
971	84	2663	84	5003	84	7499	216
983	12	2819	156	5039	36	7523	24
1103	120	2879	60	5087	48	7607	84
1187	84	2903	24	5387	300	7643	36
1259	72	2927	24	5399	12	7703	60
1283	24	2963	36	5471	72	7727	24
1307	24	3023	60	5483	12	7823	96
1367	60	3203	180	5507	24	7883	60
1439	72	3491	288	5807	300		
1499	60	3623	132	5843	36		

Table 28

In this case we have 116 numbers that are gap base 4 which qualify and 2 in gap base 4 that have been disqualified because they don't fit the criteria. There are also 13 that are in gap base 2 that also don't qualify. The gap between the gap base 4 primes also at traces a curve.

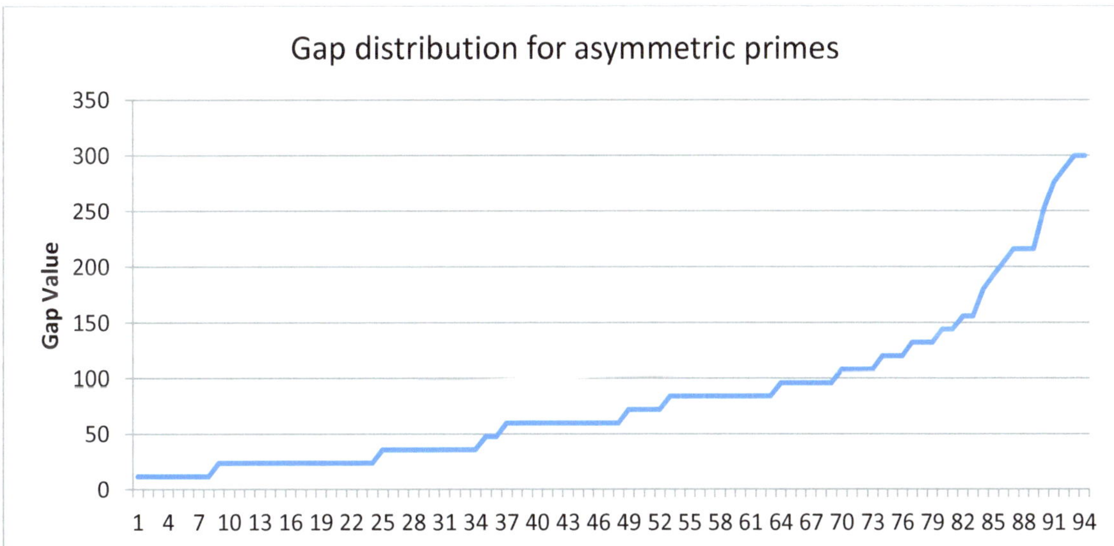

Figure 42

The horizontal straight part of the curve reflects the values 12, 24, 36, 60, 72, 84, 96 respectively.

The above analysis also implies that there are primes that are neither symmetric nor asymmetric if they do not fit the criteria defined, and this is quite a substantial number. It is also not clear why the prime numbers are defined as asymmetric as in contrast to symmetric primes when there are other prime numbers that may also non-symmetric.

As indicated in the introductory observations, the focus tends to be on one number than a set of numbers that defines a solution. For example, in the so called asymmetric primes, the construction is that these define a sieve relationship, and that the answer is a criteria [Prime, Prime, Pseudoprime]. Hence for example, the correct answer is not 23, but (23, 11, 25). This is in the same sense that we define the format $(6x - 1, 6x + 1)$ where we can have

[Prime, Prime] to define the twin prime for example. While this is a dual relationship, the other can be seen as a triplet relationship rather than making it unitary.

Consequently, the implication is that we can then have different sieve conditions as follows.

a) [4n+3,2n+1, 4n+1], and Sieve A_1 =[Prime, Prime, Pseudoprime]
b) [4n+3,2n+1, 4n+1], and Sieve A_2 =[Pseudoprime, Prime, Prime]
c) [4n+3,2n+1, 4n+1], and Sieve A_3 =[Prime, Prime, Prime]
d) [4n+3,2n+1, 4n+1], and Sieve A_4 =[Prime, Pseudoprime, Prime]

This then gives a generality in terms of different solutions. The table gives some examples. It is assumed that examples for the first sieve have already been covered. The Sieve A_4 is a twin prime condition with a pseudoprime in between. The occurrence of this one is very much less frequent than the other sieve conditions. It could be interesting how many satisfy this condition for the first 100 000 primes.

Sieve	Example
Sieve A_2	(27, 13, 19), (35, 17, 37), (87, 43, 89), (95, 47, 97), (135, 67, 137)
Sieve A_3	(11, 5, 13), (59, 29, 61), (107, 53, 109), (179, 89, 181)
Sieve A_4	(71,35, 73), (227, 113, 229)

Table 29

This can even be conceptualized as a cryptographic key, where the twin prime (227, 229) is unlocked by 113. Interest may then focus on finding all such twin primes that have a specific key that contextualizes them. The same line of thought applies to Sieve A_1 and A_2. For example in A_1 for (83, 41, 85) then 85 is defined as the key for (83,41).

The Sieve A_3 has the first and last numbers as twin primes, and in between is another prime number. However, for A_3 a different approach may be used since the key was defined because of a pseudoprime occurrence. Instead here we define a particular sequence combination since the format is [Prime, Prime, Prime]. This may be defined as a key in its entirety since it derives a unique sequence. Actually it has the same consequence as factorization of a particular number W, for example $W = 107 \cdot 53 \cdot 109$. This does not mean we cannot define the same approach as we did for A_4 since this criteria has a twin prime definition. That is, for (179, 89, 181) then 89 can be a key for the twin prime (179, 181).

It is interesting to note that while both sieve A_3 and A_4 are based on a twin prime structure, the following is equally true, that Sieve $A_3 \cap$ Sieve $A_4 = \emptyset$.

This implies a complex cryptographic application may be developed where for example we have a combination such as suggested below.

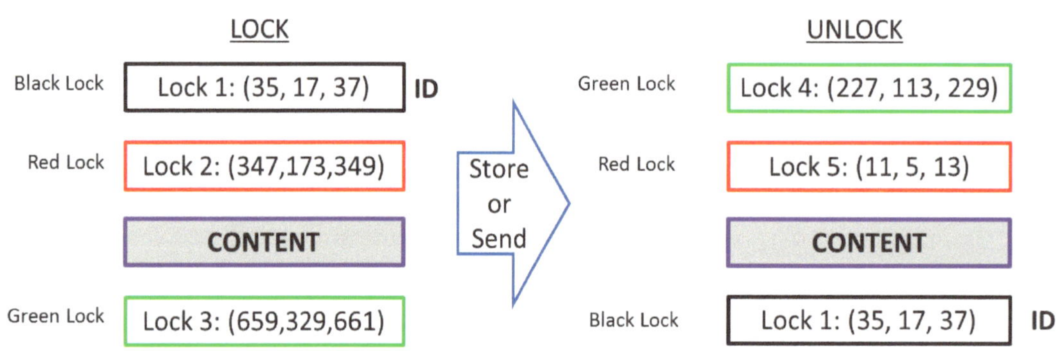

Figure 43

Therefore Lock1 is the ID that the person enters to lock the content, and is particular to the individual. The red lock is the security lock of the operation, and the green lock is additional security to ensure integrity of the operation. The content can then be stored or sent. The reverse process involves starting with the green lock that check for integrity followed by the red security lock. The ID then unlocks the content. This is similar to the username and password format, but different codes are applied except for the identity.

This is a suggestion as the author is not a cryptographer! The main advantage here is that you are able to combine both the prime numbers and the pseudoprimes in the cryptographic process, unlike the conventional use of prime numbers only. Note also that the description [Not Prime, Not Prime, Not Prime] is not as accurate as [Pseudoprime, Not Prime, Not Prime] as not all odd numbers are pseudoprime. An example is [55, 27, 57] where 55 is a pseudoprime, it satisfies the condition $6x \pm 1$. Similarly, different criteria can also be built such as $3n - 1$, $2n + 1$ and $2n - 1$ where we specify which ones must be prime as shown below.

n	*3n-1*	*2n+1*	*2n-1*
1	2	3	1
2	5	5	3
3	8	7	5
4	11	9	7
5	14	11	9
6	17	13	11
7	20	15	13
8	23	17	15
9	26	19	17
10	29	21	19
11	32	23	21
12	35	25	23
13	38	27	25
14	41	29	27

Table 30

For example we can define the criteria as [Class 2B, Prime, Prime]. The Class 2B is a UNDT classification of even numbers. This actually gives us twin prime relationship along with class 2B, as a result giving us a unique relationship between the integers.

	Class 2B	Prime	Prime
n	*3n-1*	*2n+1*	*2n-1*
3	8	7	5
9	26	19	17
15	44	31	29
21	62	43	41

Table 31

However this occurs at probable values of $6n - 3$. Now $6x - 3$ is a UNDT classification of a Class 1A number, which is an odd number. So this derives a very interesting relationship between the sets, adding further intrigue to the already unique relationship! Infact *n* can be used to specify a key to the three numbers that one is looking for, for example key 9 and 21. Practically, a person can type in any number, and the computer program can find the nearest Class 1A number, and give it as a key to the user, but hide the three numbers.

Mersenne Primes Under Classification

The section also deals with various patterns of prime numbers and how they relate to each other, where this is described in terms of sets.

Outcomes.

Different type of set structures according to gap base 2 and 4 are derived. A generality on Mersenne numbers is developed into a theoretical context.

Mersenne Theory	Comment
Mersenne Numbers – conventional definition	$M_n = 2^n - 1$
Mersenne Primes – conventional definition	$M_p = 2^p - 1$
Mersenne Primes **Gap Base 2**	$M_p = 2^p - 1$
Mersenne Numbers **Gap Base 4**	$M_n = 2^n + 1$
Mersenne Primes **Gap Base 4**	$M_p = 2^p + 1$
General Mersenne Primes generation	$M_p = 2^p \pm h$, where $h \in D_t$
There exists a sieve structure, the M(b) set $M_b = 6^n \pm h$	*For n an integer,* $M_b = 6^n \pm h$, *where* $h \in D_t$, *and* $0 < h \leq \infty$.
Primes for Gap Base 2 M(b) set	$f(p) = 6^p + h$
Primes for Gap Base 4 M(b) set	$f(p) = 6^p - h$
Exclusion Theorem *Let p be an element of* M_b, *then* $p \notin M_p$	A prime number cannot belong to the Mersenne numbers and at the same time to the set M_b.

Table 32

6 Gap Base Classification on Mersenne Primes

An analysis is undertaken of the Mersenne primes to show the usefulness of the newly developed gap base system in the way it assists in interpretation and contextualization of behavior between primes, pseudoprimes and composites.

A Mersenne number is a number of the form $2^n - 1$. When n is prime and such a number is prime, then it is referred to as a Mersenne prime and the format converts to $M_p = 2^p - 1$, to indicate that p is prime. This is actually stated as a theorem thus.

> **Theorem**: *If for some positive integer n, and $2^n - 1$ is prime, then n is prime.*

If we consider the gap classification framework, then the theorem is extended further.

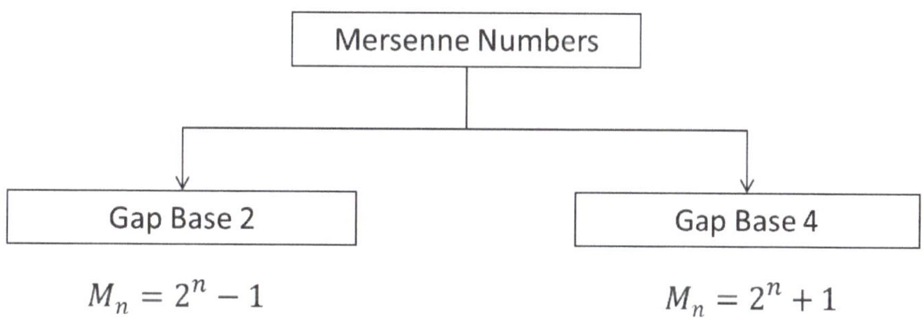

Figure 44

The standard definition of Mersenne number therefore is $2^n - 1$ and this does not include $2^n + 1$. That is, it is only limited to gap base 2 because of the fact that $M_p = 2^p - 1$ implies p is prime and this is not true in $2^n + 1$. However in the context of UNDT, there is justification for including $2^n + 1$ as it represents a particular class.

6.1 Gap Base 2 Mersenne Numbers

For example if $p = \{2, 3, 5, 7, 13, 17, 19, 31, 61, 89, 107, 127\}$ by definition these values give us Mersenne primes. Our interest lies in looking at such primes in the context of the gap classification. For $M_n = 2^n - 1$, a Mersenne number;

a) M_n is an element of gap base 2, where this holds for any n that is odd
b) M_n as an element of gap base 2 can either be a prime or a pseudoprime according to the function $f(x) = 6x + 1$. In this case n is both odd and a prime, and this is denoted M_p.
c) M_n is an element of Class 1A, where this holds for any value of n that is even. The Class 1A consists of odd numbers that are not prime.
d) The ending digit pattern for the $M_n$numbers is observed as $1 - 3 - 7 - 5$. Therefore no number ends with a 9.
e) For an M_p occurrence the pattern for the last digit is either a $1\ or\ 7$. Take for example the numbers 31, 127, 131071 and 524287. Therefore $2^p - 1$ must have an ending either in $1\ or\ 7$. Infact the pattern for the M_p is consistently (1,1) and (7,7) consecutively on the condition that we use elements of the set D_t. An example of this is the elements 1 and 31, and 127 and 2047. Observe that if 2 and 3 are included in this format, the pattern would not hold, but this supports the notion of 1 as being the first Mersenne number. Note that 25

has been included to show the consistency of the pattern as it is a pseudoprime. This is a major difference in approach in the treatment of Mersenne primes, that pseudoprimes are also regarded as generators.

Power	M_n **Number**	
1	1	M_p
5	31	M_p
7	127	M_p
11	2047	
13	1891	M_p
17	131071	M_p
19	524287	M_p
23	8388607	
25	3354431	
29	536870911	
31	2147483647	M_p

Table 33

f) In gap classification, the framework of the definition of a prime number is algebraic, rather than arithmetic (in terms of divisibility). Therefore as noted before, 1 is the first Mersenne prime, 2 and 3 do not fit the pattern of primes or pseudoprimes (that is in being elements of D_t). However from the conventional perspective, these two numbers represent Mersenne primes.

Power	M_n
2	3
3	7
5	31
7	127
11	2047
13	1891
17	131071

Table 34

g) We can determine at what values of n the pattern 1 occurs, that is, $n = 9, 13, 17$ where this derives the equation $n = f(x)$, that is $f(x) = 4x + 1$, for $0 \leq Int(x) \leq \infty$. We get a regular pattern for the gap base of $f(x)$. As already observed, the Mersenne number is always a gap base 2 number.

x	$f(x)$	**Number in GB2**	**Gap Base of** $f(x)$
0	1	1	GB2
1	5	31	GB4
2	9	511	Class 1A
3	13	8191	GB2
4	17	131071	GB4
5	21	2097151	Class 1A
6	25	33554431	GB2
7	29	536870911	GB4
8	33	8589934591	Class 1A

Table 35

We now discover that the powers $f(x)$ involved are not only GB2 and GB4, but also Class 1A. In other words, because of the restriction that n must also be prime, this class is ommitted, but this does not imply that there is no prime number generated from Class 1A that cannot be prime. The assumption is based on the fact that any number in gap base 2 can either be prime or pseudoprime.

h) Similarly, the pattern for 7 occurs for $n = 3, 7, 11$. Then the powers are given by $n = f(x)$ where this gives $f(x) = 4x + 3$, for $0 \leq Int(x) \leq \infty$.

x	$f(x)$	Number in GB2	Gap Base of $f(x)$
0	3	7	Class 1A
1	7	127	GB 2
2	11	2047	GB4
3	15	32767	Class 1A
4	19	524287	GB2
5	23	8388607	GB4
6	27	134217727	Class 1A

Table 36

Note that 7 bounces back as a Mersenne prime because 3 is in Class 1A and it is part of the pattern. This explains why the power 3 has the pattern of 7 even though it is not a prime number according to the UNDT framework.

i) For example, one of the largest known prime is $2^{43112609} - 1$, which is a Mersenne prime. We can test which equations applies.

Number	43112609	Result
$f(x) = 4x + 1$	10778152.000	Integer
$f(x) = 4x + 3$	10778151.500	Decimal

Table 37

The integer result in solving for x shows that the Mersenne prime ends with the digit 1.

6.2 Gap Base 4 Mersenne Numbers

In the context of the classification theory, we can consider a version of the numbers in gap base 4 using the expression $M_n = 2^n + 1$. This is not a formal definition of a Mersenne number, but it is a recognition of the gap base classification system. In a similar manner, we have $M_n = 2^p + 1$ when the resulting number is a Mersenne prime.

In this case, the following observations can be made from the table below.

a) M_n is an element of gap base 4, where this holds for any n that is even.
b) M_n as an element of gap base 4 can either be a prime or a pseudoprime according to the function $f(x) = 6x - 1$.
c) M_n is an element of Class 1A, where this holds for any value of n that is odd.
d) The ending digit pattern for the M_n numbers is defined as $3 - 5 - 9 - 7$. Therefore no number ends with a 1.
e) For an M_p occurrence the pattern for the last digit is either a $5 \; or \; 7$. Take for example the numbers 5, 17, 257 and 65537. Therefore $2^p + 1$ must have an ending either in $5 \; or \; 7$.
f) Now numbers ending with a 5 are divisible by 5, hence they cannot be prime. Consequently, the only numbers that can be prime are those that end with a 7.

g) We can determine at what values of n the pattern 7 occurs, that is, $n = 4, 8, 12$ where this derives the equation $n = f(x)$, that is $f(x) = 4x$, for $0 \leq x \leq \infty$. The classes 1B, 2B, and 2A defines the set of even numbers.

n	GB2 2^n-1	GB4 2^n+1	GB2 (Minus 1)	GB4 (Plus 1)
1	1	3	0	0.667
2	3	5	0.333	1
3	7	9	1	1.667
4	15	17	2.333	3
5	31	33	5	5.667
6	63	65	10.333	11
7	127	129	21	21.667
8	255	257	42.333	43
9	511	513	85	85.667
10	1,023	1,025	170.333	171
11	2,047	2,049	341	341.667
12	4,095	4,097	682.333	683
13	8,191	8,193	1,365	1,365.667
14	16,383	16,385	2,730.333	2,731
15	32,767	32,769	5,461	5,461.667
16	65,535	65,537	10,922.333	10,923
17	131,071	131,073	21,845	21,845.667
18	262,143	262,145	43,690.333	43,691
19	524,287	524,289	87,381	87,381.667
20	1,048,575	1,048,577	174,762.333	174,763
21	2,097,151	2,097,153	349,525	349,525.667
22	4,194,303	4,194,305	699,050.333	699,051
23	8,388,607	8,388,609	1,398,101	1,398,101.667

Table 38

Fundamentally, the definition of a Mersenne prime is that n must be prime, but we observe that in gap base 4, n is actually even but can give us a prime number. In terms of pattern methodology, there is no contradiction as to the fact that both gap base 2 and 4 yield a prime number for some given value of n. This is what matters.

x	$f(x)$	Number	Class for $f(x)$
1	4	17	Class 1B
2	8	257	Class 2B
3	12	4097	Class 2A
4	16	65537	Class 1B
5	20	1048577	Class 2B
6	24	16777217	Class 2A

Table 39

However, the norm recognizes n as long as it is a prime because it makes $M_p = 2^p - 1$ an elegant expression, whereas in terms of practicality of the classification system this is also a subset of the expression $M_n = 2^n - 1$, suggesting a broader view of analysis. If we use the formula $6x \pm 1$ to test for gap base 2 and 4, we see the pattern of behavior through the integer values. All of this demonstrates pattern consistency and recognition of underlying behaviors that are undisputable.

6.3 The M_g Set

In the same breath therefore, we define a general approach to such primes through the following definition. Note that the definition exhibits properties of a ditation set because such a set has generative power for deriving similar sets through some constant k. That is, as a sieve this can be generated by $f(x) = 6x \pm k$, where $k = 6m \pm 1$, and $0 \le Int(m) \le \infty$ provided $k > 0$. We set $0 \le Int(x) \le \infty$, and k is constant for a given $0 \le x \le \infty$. Therefore there are an infinite number of equations to produce the sieve, and giving us the same result but differing in the intercept value. For example when $m = 0$, and $m = 1$, then $k = 1$, $k = 5$ or $k = 7$ giving us six equations. Each value of k gives us a pair of equations.

i. $f(x) = 6x - 1$ gap base 4 and $f(x) = 6x + 1$ gap base 2
ii. $f(x) = 6x - 5$ gap base 2 and $f(x) = 6x + 5$ gap base 4
iii. $f(x) = 6x - 7$ gap base 4 and $f(x) = 6x + 7$ gap base 2

	Using **6x±1**		Using **6x±5**		Using **6x±7**		Using **6x±11**	
x	GB-4	GB-2	GB-2	GB-4	GB-4	GB-2	GB-2	GB-4
0	-1	1	-5	5	-7	7	-11	11
1	5	7	1	11	-1	13	-5	17
2	11	13	7	17	5	19	1	23
3	17	19	13	23	11	25	7	29
4	23	25	19	29	17	31	13	35
5	29	31	25	35	23	37	19	41

Table 40

This implies we can use the notion of $\pm h$ where we set $k = h$ by observing the values of $x = 0$.

Theorem

For n an integer, there exists a set $M_g = 2^n \pm h$, *where* $h \in D_t$, $0 < h \le \infty$.

Now D_t is the set of linear ditation numbers, and in this case, only the positive ones. That is, if $h \in D_t$, then $h = 6m \pm 1$. Secondly, then the Mersenne set of numbers are a subset of M_g.

n	GB4 2^n-5	GB2 2^n+5	GB2 (Plus 5)	GB4 (Minus 5)
1	(3)	7	-0.666666667	1.333
2	(1)	9	(0.333)	2
3	3	13	0	2.333
4	11	21	1.667	4
5	27	37	4	6.333
6	59	69	9.667	12
7	123	133	20	22.333
8	251	261	41.667	44
9	507	517	84	86.333
10	1,019	1,029	169.667	172
11	2,043	2,053	340	342.333
12	4,091	4,101	681.667	684
13	8,187	8,197	1,364	1,366.333
14	16,379	16,389	2,729.667	2,732
15	32,763	32,773	5,460	5,462.333
16	65,531	65,541	10,921.667	10,924
17	131,067	131,077	21,844	21,846.333
18	262,139	262,149	43,689.667	43,692
19	524,283	524,293	87,380	87,382.333
20	1,048,571	1,048,581	174,761.667	174,764
21	2,097,147	2,097,157	349,524	349,526.333
22	4,194,299	4,194,309	699,049.667	699,052
23	8,388,603	8,388,613	1,398,100	1,398,102.333

Table 41

An example is given for $h = 5$ and $h = 7$. Note the switch between gap base 2 and 4 that occurs depending on the value of h. That is, for $h = 5$, then $M_g = 2^n - 5$ generates elements of gap base 4, but for $M_g = 2^n - 7$, this generates elements of gap base 2 instead. Hence if h is an element of gap base 4, we derive elements of gap base 4, and a similar argument applies if h is gap base 2. For the additive expression, $h = 5$ generates elements of gap base 2 which is an alternative gap base to h. Similarly for $h = 7$ which is in gap base 2, then we have elements from the alternative base which is gap base 4.

	GB2	GB4		
n	2^n-7	2^n+7	GB2 (Minus 7)	GB4 (Plus 7)
1	(5)	9	-1	1.667
2	(3)	11	(0.667)	2
3	1	15	0	2.667
4	9	23	1.333	4
5	25	39	4	6.667
6	57	71	9.333	12
7	121	135	20	22.667
8	249	263	41.333	44
9	505	519	84	86.667
10	1,017	1,031	169.333	172
11	2,041	2,055	340	342.667
12	4,089	4,103	681.333	684
13	8,185	8,199	1,364	1,366.667
14	16,377	16,391	2,729.333	2,732
15	32,761	32,775	5,460	5,462.667
16	65,529	65,543	10,921.333	10,924
17	131,065	131,079	21,844	21,846.667
18	262,137	262,151	43,689.333	43,692
19	524,281	524,295	87,380	87,382.667
20	1,048,569	1,048,583	174,761.333	174,764
21	2,097,145	2,097,159	349,524	349,526.667
22	4,194,297	4,194,311	699,049.333	699,052
23	8,388,601	8,388,615	1,398,100	1,398,102.667

Table 42

A similar approach of analysis can be applied to the example. We can for example continue adding more values of h to find out what values of M_g exist, and which ones are prime. For instance, for $h = \{1,5,7\}$, $n = 1$ to 23, then;

$$M_g = \{1,5,7,11,13,17,23,31,37,59,71,127,251,257,1031,2053,4091,8191,65537,65543,131071,\\ 262139,524287,1048571,1048583\}$$

The implication is that by limiting ourselves to the expression $M_p = 2^p - 1$, we are missing out on some of the Mersenne primes from a general context. If it is assumed that such primes are in gap base 2 and 4, we can define the set of Mersenne primes generally as;

$$M_p = 2^p \pm h, \text{where } h \in D_t,$$

and p is either a prime or an even number.

This set is obviously a subset of the M_g set. It is yet to be shown that p can be an element of Class 1A in order to derive a Mersenne prime.

6.4 The M_b Prime Set

We can consider other similar prime formats to the Mersenne primes as demonstrated by the following definition. The same characteristic of a ditation set is observed for this set, that is $y_{i,j} = y_j + k$.

Theorem

For n an integer, there exists a set $M_b = 6^n \pm h$, where $h \in D_t$, $0 < h \leq \infty$.

The set M_b derives a prime number set within a general set. That is, within M_b is a set in the same sense as the Mersenne number set which contains the set of Mersenne primes. In this case, this is called the M_b prime set. In a similar fashion;

a) $f(n) = 6^n + h$ defines gap base 2, where the primes are defined as $f(p) = 6^p + h$
b) $f(n) = 6^n - h$ defines gap base 4, where the primes are defined as $f(p) = 6^p - h$

Note that this behavior is different from the set $M_g = 2^n \pm h$ in terms of how h influences the gap base of the numbers. The value of h derives a consistent gap base rather than an alternative gap base. The particular characteristic for M_b is that for a given value of h, the end digit is the same in that group for any positive primes. Secondly, unlike the M_n set, the M_b set produces only primes or pseudoprimes, there are no composites. Therefore in a way compared to M_n, the set acts like a form of a grouping sieve.

	GB4	GB2	GB4	GB2	GB4	GB2
n	h = 1	h = 1	h = 5	h = 5	h = 7	h = 7
1	5	7	1	11	(1)	13
2	35	37	31	41	29	43
3	215	217	211	221	209	223
4	1,295	1,297	1,291	1,301	1,289	1,303
5	7,775	7,777	7,771	7,781	7,769	7,783
6	46,655	46,657	46,651	46,661	46,649	46,663
7	279,935	279,937	279,931	279,941	279,929	279,943
8	1,679,615	1,679,617	1,679,611	1,679,621	1,679,609	1,679,623
9	10,077,695	10,077,697	10,077,691	10,077,701	10,077,689	10,077,703
10	60,466,175	60,466,177	60,466,171	60,466,181	60,466,169	60,466,183
11	362,797,055	362,797,057	362,797,051	362,797,061	362,797,049	362,797,063
12	2,176,782,335	2,176,782,337	2,176,782,331	2,176,782,341	2,176,782,329	2,176,782,343
13	13,060,694,015	13,060,694,017	13,060,694,011	13,060,694,021	13,060,694,009	13,060,694,023
14	78,364,164,095	78,364,164,097	78,364,164,091	78,364,164,101	78,364,164,089	78,364,164,103
15	470,184,984,575	470,184,984,577	470,184,984,571	470,184,984,581	470,184,984,569	470,184,984,583
16	2,821,109,907,455	2,821,109,907,457	2,821,109,907,451	2,821,109,907,461	2,821,109,907,449	2,821,109,907,463
17	16,926,659,444,735	16,926,659,444,737	16,926,659,444,731	16,926,659,444,741	16,926,659,444,729	16,926,659,444,743
18	101,559,956,668,415	101,559,956,668,417	101,559,956,668,411	101,559,956,668,421	101,559,956,668,409	101,559,956,668,423
19	609,359,740,010,495	609,359,740,010,497	609,359,740,010,491	609,359,740,010,501	609,359,740,010,489	609,359,740,010,503

Table 43

The colored part shows the occurrences that are prime. Note that for GB4 and $h = 1$, then only one prime number exists, the rest are pseudoprimes.

6.5 Exclusion Theorem

An exclusion theorem may be defined that relates the set M_b and M_g in the following manner.

Exclusion Theorem

Let p be an element of M_b, then $p \notin M_g$, where p is prime or pseudoprime.

Therefore for example, we do not expect the set M_b to have a Mersenne prime.

6.8 End Digit Classification

Classifying prime numbers alone as a group does not easily yield a discernible pattern unless the set of pseudoprimes is also included. One may assume that there is no classification that can take place in terms of the last digit from a cursory observation. However, a particular pattern is noticed when considering the gap base structure and the last digit of a number in the algebraic sieve. The blue shows pseudoprimes, so that it is clear that as we consider the last digit, they are also part of the pattern confirming the sieve methodology.

X	Gap Base 2	Gap Base 4
240	1447	1451
241	1453	1457
242	1459	1463
243	1465	1469
244	1471	1475
245	1477	1481
246	1483	1487
247	1489	1493
248	1495	1499
249	1501	1505
250	1507	1511
251	1513	1517
252	1519	1523
253	1525	1529
254	1531	1535
255	1537	1541
256	1543	1547
257	1549	1553
258	1555	1559
259	1561	1565

Table 44

For gap base 2, starting at $x = 244$, the digit pattern is;

$$1 - 7 - 3 - 9 - 5$$

For gap base 4, the pattern is;

$$5 - 1 - 7 - 3 - 9$$

This assumes we start at the same value of x to consider the pattern structure as this is a multi-valued function. Hence the two patterns are connected in a cyclic manner. In both cases, the pattern repeats in a consistent manner.

The following may be noted in regard to sieves.

a) We regard a sieve as a mathematical construct with specific properties that allows us to identify prime numbers, or makes it easier to get them.
b) A particular property of a sieve is the position at which a prime or related number may occur.
c) Sieves also produce a consistent pattern in their object of study.
d) It is not necessary that a sieve gets all the prime numbers. That is, a sieve defines certain conditions that make it derive certain prime numbers.

e) A sieve is also defines and identifies pseudoprimes as well as composites. Hence it is not confined to prime numbers only because this standard approach proves to be limiting in looking for relationships and describing them.

The hypothesis held here is that it is not possible to have one particular formula to define and predict a prime number, but it is possible to define and describe the prime numbers in the context of a sieve. The best known sieve is that of Eratosthenes, and the idea here has been to introduce the concept of the algebraic sieve. Studying primes and the related numbers in a sieve makes it easier to classify them and understand their patterns. We have to change our approach and thinking from a formula approach to a sieve approach.

It is further hypothesized that many types of sieve structures exist, but are not regarded as sieves. Therefore in light of the above elucidation on what a number sieve is and what we expected from it, an example is given. Consider the set of even numbers, then the sum of even numbers starting from 0 is given by $n(n+1)$.

Example

Let us now determine to design a sieve such that it takes the sum of even number at n, and adds it to some constant, in our case 7. The assumption is that this will define a position where a prime or pseudoprime may occur. In column 1, we color only the primes and ignore the pseudoprimes to show that no discernible pattern is observed, but in column 3, we color the pseudoprimes orange to show the distinct pattern we get. Therefore the regularity in the sieve is provided by the constant, and the value of the sum of even numbers at n.

	Col-1	Col-2	Col-3
n	*n(n+1) + 7*	*Difference*	*n(n+1) + 7*
0	7	0	7
1	9	2	9
2	13	4	13
3	19	6	19
4	27	8	27
5	37	10	37
6	49	12	49
7	63	14	63
8	79	16	79
9	97	18	97
10	117	20	117
11	139	22	139
12	163	24	163
13	189	26	189
14	217	28	217
15	247	30	247
16	279	32	279
17	313	34	313
18	349	36	349
19	387	38	387
20	427	40	427
21	469	42	469
22	513	44	513
23	559	46	559
24	607	48	607
25	657	50	657

	Col-4	Col-5	Col-6
n	*n(n+1) + 7*	*Difference*	*n(n+1) + 7*
26	709	52	709
27	763	54	763
28	819	56	819
29	877	58	877
30	937	60	937
31	999	62	999
32	1063	64	1063
33	1129	66	1129
34	1197	68	1197
35	1267	70	1267
36	1339	72	1339
37	1413	74	1413
38	1489	76	1489
39	1567	78	1567
40	1647	80	1647
41	1729	82	1729
42	1813	84	1813
43	1899	86	1899
44	1987	88	1987
45	2077	90	2077
46	2169	92	2169
47	2263	94	2263
48	2359	96	2359
49	2457	98	2457
50	2557	100	2557
51	2659	102	2659

Table 45

As stated before, the sieve does not necessarily derive all the primes that we need, and we also recognize that a sieve always has extra numbers that surrounds the numbers that we need or are studying.

a) Column 2 & 5: These two columns show the sieve characteristic. The difference between numbers in column 1 and 4 is an even number, for example 469-427 = 42.
b) Column 1 & 4: This shows our numbers of interest, that is, the primes. There is not much of pattern that we observe when we "throw" away the other numbers.
c) Column 3 & 6: This now shows that we do not throw away the other numbers because we recognize them as part of the occurrence of the prime number. These are the pseudoprimes, and we now see a distinct pattern, since supporting data is seen as part of the complete description of the sieve process.
d) The numbers that are not colored are of class 1A. They too have a specific regularity, given by the expression $18m + 9$ where $m = 3n - 2$ and $m > 0$.
e) The sieve only considers primes and pseudoprimes from gap base 2. All other primes in gap base 4 are completely sieved out! Sieves can focus on a particular class as part of the sieving process.

Now if we had completely focused on the fact these numbers generated are of the form $x^2 + x + 7$, and then looked for primes that result from this, then a lot of information that would give us clues in terms of distribution and behavior would have been lost. Pattern methodology therefore strongly considers surrounding data as a form of analysis. The throw away approach is also typical of the sieve of Eratosthenes. Because of the approach of cancellation, this sieve has encouraged little focus on surrounding numbers in a given sieve, whereas these may provide useful insights as has already been demonstrated.

However, if we treat $f(x) = x^2 + x + 7$ as a function, every integer value of the function is considered significant and relevant. This shows the power of interpretation, how it affects what we see intellectually. Furthermore;

a) $f(x) = 2x$ gives us the set of even numbers, and it is actually a sieve that gives us all numbers ending with the digit 0, 2, 4, 8.
b) $f(x) = 2x + 1$ gives us the set of odd numbers, and it is actually a sieve that gives us all numbers ending with the digit 1, 3, 5, 7, 9.

Hence sieves also have the power of classification besides giving particular properties. For example the above two sieves give us the class of even numbers and the class of odd umbers.

6.9 The 30-Set

In the light of the above arguments, we therefore also consider the following which are defined as algebraic power sieves, or polynomial sieves. The first is the expression for the M_b set. This generates elements in gap base 2 and 4, hence we can equate to solve for x. That is;

Gap Base 2	Gap Base 4
$2^n - h = 6x + 1$ Hence $x = (2^n - h - 1)/6$	$2^n + h = 6x - 1$ Hence $x = (2^n + h + 1)/6$

Table 46

This set changes gap base depending for a given h, hence we also have;

Gap Base 4	Gap Base 2
$2^n - h = 6x - 1$	$2^n + h = 6x + 1$

Hence $x = (2^n + 1 - h)/6$	Hence $x = (2^n + h - 1)/6$

Table 47

Therefore this acts as a complex sieve to give us elements of the set M_g. As stated before, the elements of the Mersenne numbers M_n is merely a subset of this sieve. When the value returned of x is a decimal, that defines a non-occurrence as demonstrated by the previous arguments. Numbers in this set were observed to follow a certain end digit pattern. In the model demonstrated below, the cycle indicates the ending digit in its sequence for a given value of h in gap base 2 and 4, and what digit is missing in that cycle. For example if we have $12^n \pm 61$, we get the following values for GB2 and GB4 respectively. Negative values such as -49 are ignored in considering the cycle.

h	n	GB4	GB2
61	1	73	-49
61	2	205	83
61	3	1789	1667
61	4	20797	20675
61	5	248893	248771
61	6	2986045	2985923
61	7	35831869	35831747

	GB2	GB4		
Cycle Digits	3,7,5,1	3,5,9,7	61	GB2
Missing Digits	9	1		

Table 48

Then starting at $n = 1$, we look for the end digits that result, and from that determine which digit is missing since the total must always be five digits from $1 - 7 - 3 - 9 - 5$ in gap base 2 and $5 - 1 - 7 - 3 - 9$ in gap base 4. Now 61 is from GB2, but in terms of pattern methodology, we group h according to gap base to observe the result.

Consider the following pattern derived from $2^n \pm h$, then the missing digit pattern is as follows.

	GB2	GB4	
cycle	1,3,7,5	3,5,9,7	h = 1
missing	9	1	
	GB4	GB2	
cycle	3,1,7,9	7,9,3,1	h = 5
missing	5	5	
	GB2	GB4	
cycle	1,9,5,7	9,1,5,3	h = 7
missing	3	7	
	GB4	GB2	
cycle	5,1,3,7	3,5,9,7	h = 11
missing	9	1	
	GB2	GB4	
cycle	3,9,1,5	5,7,1,9	h = 13
missing	7	3	
	GB4	GB2	
cycle	5,7,1,9	9,1,5,3	h = 17
missing	3	7	
	GB2	GB4	
cycle	3,5,9,7	1,3,7,5	h = 19
missing	1	9	

	GB4	GB2	
cycle	9,1,5,3	5,7,1,9	h = 23
missing	7	3	
	GB2	GB4	
cycle	7,9,3,1	7,9,3,1	h = 25
missing	5	5	
	GB4	GB2	
cycle	3,5,9,7	1,3,7,5	h = 29
missing	1	9	
	GB2	GB4	
cycle	1,3, 7, 5	3, 5, 9, 7	h = 31
missing	9	1	
	GB4	GB2	
cycle	9,3,1,7	7,9,3,1	h = 35
missing	5	5	
	GB2	GB4	
cycle	5,9,7,3	1,3,7,5	h = 49
missing	1	9	
	GB2	GB4	
cycle	1,9,5,7	9,1,5,3	h = 77
missing	3	7	

Table 49

In this analysis approach, we could have ignored the missing digits as inconsequential. However, if we include them as part of the observation, then we see an interesting pattern emerging within the missing digits themselves as they seem to commute. The following further observations can be made with regard to this set.

a) It can be noted that for the M_g set, the gap base of the number is not consistent with the sign of gap base 2 or 4 in the equation, but rather it is consistent with the value of h.

b) If h is gap base 2, then $2^n - h$ is gap base 2, and $2^n + h$ will be gap base 4.

c) If h is gap base 4, then $2^n - h$ is gap base 4, and $2^n + h$ will be gap base 2.

d) All cycles consist of four digits that define how the number must end. Therefore whilst there are five numbers that define a gap base end digit, each cycle only uses four, and the other one is said to be missing. It is not clear why the pattern behaves in that manner, that is, why is there a missing digit on a consistent basis.

e) If for a given cycle a digit is missing in gap base 2, then it appears in gap base 4. This seems to be a behavior with a correlation characteristic.

f) Every pseudoprime that is a multiple of five has a 5 missing in both gap bases of the cycle end digits. Other pseudoprimes behave like the prime numbers in their end digit.

g) In terms of missing end digits, the system consists of only the following paired behavior;

- (1,9) and (9,1)
- (3,7) and (7,3)
- (5,5)

If h is kept in sequence, then it appears that these pairs occur at random according to the prime or pseudoprime number occurrence. Therefore the cycles behave randomly in distribution.

h) Therefore there is no way of systematically predicting how the digit will end for a given value of h in $2^n \pm h$.

Let us now assume a different approach, where we arrange according to the following: GB2-GB4, and the pairs (1,9) (9,1), (3,7) and so on.

	GB2	GB4	
cycle	1,3,7,5	3,5,9,7	h = 1
missing	9	1	
cycle	1,3, 7, 5	3, 5, 9, 7	h = 31
missing	9	1	
cycle	3,7,5,1	3,5,9,7	h = 61
missing	9	1	
cycle	7,5,1,3	3,5,9,7	h = 91
missing	9	1	

	GB2	GB4	
cycle	3,9,1,5	5,7,1,9	h = 13
missing	7	3	
cycle	1,5,3,9	5,7,1,9	h = 43
missing	7	3	
cycle	5,3,9,1	5,7,1,9	h = 73
missing	7	3	
cycle	5,3,9,1	5,7,1,9	h = 103
missing	7	3	

	GB2	GB4	
cycle	1,9,5,7	9,1,5,3	h = 7
missing	3	7	
cycle	7, 1, 9, 5	9,1,5,3	h = 37
missing	3	7	
cycle	1,9,5,7	9,1,5,3	h = 67
missing	3	7	
cycle	1,9,5,7	9,1,5,3	h = 97
missing	3	7	

	GB2	GB4	
cycle	3,5,9,7	1,3,7,5	h = 19
missing	1	9	
cycle	5,9,7,3	1,3,7,5	h = 49
missing	1	9	
cycle	9,7,3,5	1,3,7,5	h = 79
missing	1	9	

	GB2	GB4	
cycle	7,9,3,1	7,9,3,1	h = 25
missing	5	5	
cycle	9,3,1,7	7,9,3,1	h = 55
missing	5	5	
cycle	3,1,7,9	7,9,3,1	h = 85
missing	5	5	

Table 50

Now a definite pattern appears. It is noted that all GB4 cycle values are consistently the same for a given value of h. For example from $h = 1$, then (3, 5, 9, 7) is a consistent cycle group.

A similar assignment can be done on the GB4-GB2 group. This time we note that all GB2 cycle groups are also consistently the same for a given h. What is common between the two results is the fact that the consistent cycle group is generated by the equation $2^n + h$ and the random cycle group is generated by $2^n - h$. The missing digits form a key approach in identifying the pattern in the sieve system.

	GB4	GB2	
cycle	3,1,7,9	7,9,3,1	h = 5
missing	5	5	
cycle	9,3,1,7	7,9,3,1	h = 35
missing	5	5	
cycle	3,1,7,9	7,9,3,1	h = 65
missing	5	5	
cycle	3,1,7,9	7,9,3,1	h = 95
missing	5	5	

	GB4	GB2	
cycle	5,1,3,7	3,5,9,7	h = 11
missing	9	1	
cycle	3,7,5,1	3,5,9,7	h = 41
missing	9	1	
cycle	7,5,1,3	3,5,9,7	h = 71
missing	9	1	

	GB4	GB2	
cycle	5,7,1,9	9,1,5,3	h = 17
missing	3	7	
cycle	7, 1,9,5	9,1,5,3	h = 47
missing	3	7	
cycle	1,9,5,7	9,1,5,3	h = 77
missing	3	7	

	GB4	GB2	
cycle	9,1,5,3	5,7,1,9	h = 23
missing	7	3	
cycle	1,5,3,9	5,7,1,9	h = 53
missing	7	3	
cycle	5,3,9,1	5,7,1,9	h = 83
missing	7	3	

	GB4	GB2	
cycle	3,5,9,7	1,3,7,5	h = 29
missing	1	9	
cycle	5,9,7,3	1,3,7,5	h = 59
missing	1	9	
cycle	9,7,3,5	1,3,7,5	h = 89
missing	1	9	

Table 51

What is even more interesting is the common property between the two groups observed from the sieve. The sieve results in what has been defined as the "30" set of M_g. This set starts with an element of GB2 or an element of GB4, and the difference to the next prime or pseudoprime is 30. For example, if we take GB2, and set $h = 1$, then the rest of the elements for h less than 100 is $h = \{1, 31, 61, 91\}$. If we take $h = 17$ in GB4, and we establish the set for h less than 100 we get $h = \{17, 47, 77\}$. The notation for such a set structure is H_{30}. The set is identified from the common property that the missing digit is the same for a given H_{30}.

Hence the general format for finding a "30" set is given by a definition.

Definition

Let $H_{30}(x) = \{h_0, h_1, h_2, h_3, \cdots h_n\}$, where h_0 is the initial prime or pseudoprime such that $h_n = 30n + h_0$ and $0 \leq n \leq \infty$ where $x = 2$ or $x = 4$ denotes gap base 2 or 4 respectively.

It is referred to as a normal H_{30} set if it contains both primes and pseudoprimes for h_n. One can consider a selective H_{30} set where h_n consists of prime numbers only, or of pseudoprimes only. The pattern structure remains the same, but the difference between the h values will be in multiples of 30. For example, consider the set;

$$H_{30}(4) = \{17, 47, 77, 107, 137, 167, 197, 227, 257, 287\}$$

As a selective prime set consisting of primes only, this would become;

$$H_{30}(4) = \{17, 47, 107, 137, 167, 197, 227, 157\}$$

Clearly we have $(107 - 47) = 60$, which is a multiple of 30. As a selective set for pseudoprimes only it would have two elements, that is;

$$H_{30}(4) = \{77, 287\}$$

The difference is $(287 - 77) = 210 = 7 \times 30$, a multiple as indicated.

A further point of interest is noted in the missing digit pattern; they also form a pattern of their own that is repetitive. That is, if h is gap base 2, then we are dealing with a gap base 2 group, and we look at the missing digits.

Gap Base	Pattern
GB2	1-7-3-9-5
GB4	5-1-7-3-9

Table 52

This pattern is exactly the same as the gap base pattern for GB2 and GB4 in the standard algebraic sieve. What a surprising result! As if this is not enough, the following is observed.

The missing digit is always the same as the right digit of a given h.

In some way, the observations also provide a technique of finding the last digit of a given element of the M_g set. Probably the most important result that the sieves have established is this.

The elements of GB2 or GB4 have specific subsets defined by H_{30}.

That is, one can say for example the primes 7, 37, 67 and 97 belong to the same set, otherwise there was no reason to say so before. This implies that the set M_g actually covers the whole of GB2 and GB4 but in a re-arranged manner.

It is also noted that number arrangement in terms of digit-end has been mostly recognized as odd or even, but never within a set for example such as considering odd numbers only. However, the odd-even arrangement does not have any great complexity, and finding equations to describe them is fairly easy.

Observation

The existence of the H_{30} is an example of a very complex yet accurate and neat arrangement of numbers. It is as if it represents some code of some sort. One may even wonder if it cannot connect to genetic arrangement of the *DNA*.

However, it is a very good illustration of how simplicity can derive or express great complexity.

The above have given a detailed illustration of what is meant by the sieve properties, especially if we do not use the "throw away" approach.

A similar approach can be done with regard to the set M_b though this set has an entirely different behavior. However, the set is less complex in behavior as we can notice from the derived numbers.

n	GB4 h = 1	GB2 h = 1	GB4 h = 5	GB2 h = 5	GB4 h = 7	GB2 h = 7
1	5	7	1	11	(1)	13
2	35	37	31	41	29	43
3	215	217	211	221	209	223
4	1,295	1,297	1,291	1,301	1,289	1,303
5	7,775	7,777	7,771	7,781	7,769	7,783
6	46,655	46,657	46,651	46,661	46,649	46,663
7	279,935	279,937	279,931	279,941	279,929	279,943
8	1,679,615	1,679,617	1,679,611	1,679,621	1,679,609	1,679,623
9	10,077,695	10,077,697	10,077,691	10,077,701	10,077,689	10,077,703
10	60,466,175	60,466,177	60,466,171	60,466,181	60,466,169	60,466,183
11	362,797,055	362,797,057	362,797,051	362,797,061	362,797,049	362,797,063
12	2,176,782,335	2,176,782,337	2,176,782,331	2,176,782,341	2,176,782,329	2,176,782,343
13	13,060,694,015	13,060,694,017	13,060,694,011	13,060,694,021	13,060,694,009	13,060,694,023
14	78,364,164,095	78,364,164,097	78,364,164,091	78,364,164,101	78,364,164,089	78,364,164,103
15	470,184,984,575	470,184,984,577	470,184,984,571	470,184,984,581	470,184,984,569	470,184,984,583
16	2,821,109,907,455	2,821,109,907,457	2,821,109,907,451	2,821,109,907,461	2,821,109,907,449	2,821,109,907,463
17	16,926,659,444,735	16,926,659,444,737	16,926,659,444,731	16,926,659,444,741	16,926,659,444,729	16,926,659,444,743
18	101,559,956,668,415	101,559,956,668,417	101,559,956,668,411	101,559,956,668,421	101,559,956,668,409	101,559,956,668,423
19	609,359,740,010,495	609,359,740,010,497	609,359,740,010,491	609,359,740,010,501	609,359,740,010,489	609,359,740,010,503

Table 53

When $h = 1$, numbers end with a (5,7) pattern and when $h = 5$, they end with a (1,1) pattern. Infact, the only patterns occurring are;

a) (1,1)
b) (3,9) and (9,3)
c) (5,7) and (7,5)

This leads to the following definition.

Definition

Let $h > 0$, then for a given value of $h = 6x \pm 1$, there exists a pattern (a, b) to construct primes and pseudoprimes of an h-subset p ending in 'a' and an h-subset q ending in 'b'.

Hence for every value of h there are two subsets with specific digit ending. However, there are no (1,9), or (5,9) patterns for example. The inherent condition of the sieve is as observed;

a) Only one combination is chosen per group, for example only (1,1) or (3,9).
b) The chosen group commutes to form another group, hence (7,5) commutes to form (5,7).
c) The appearance of a particular group follows the random occurrence of the prime numbers. The first number to occur is from gap base 4, hence if the ending digit is 3, as a rule the corresponding ending digit for the gap base 2 number will be 9. For that particular value of h, all the numbers will then end in the digit pattern (3,9).
   This is a random yet regular behavior.

h	GB4 End Digit	GB2 End Digit	GB2	GB4	GB2	GB4	GB2	GB4	GB2	GB4	GB2	GB4
1	5	7	1	5	7	5	3	5	9	5	5	5
5	1	1	1	1	7	1	3	1	9	1	5	1
7	9	3	1	7	7	7	3	7	9	7	5	7
11	5	7	1	3	7	3	3	3	9	3	5	3
13	3	9	1	9	7	9	3	9	9	9	5	9
17	9	3										

19	7	5
23	3	9
25	1	1
29	7	5
31	5	7
35	1	1

Table 54

It is fairly easy to predict the ending value pattern if we use gap base 2 since all its values are positive. This has the following pattern;

h	GB2	Difference
1	7	6
5	11	6
7	13	6
11	17	6
13	19	6
17	23	6
19	25	6

Table 55

Hence if we know h, then the ending prime in gap base 2 will be given by $(h+6)$. For example, if $h = 71$, then the ending prime in gap base 2 is (71+6) = 77. Therefore the sieve will produce only numbers that end in 5 for gap base 4 and 7 for gap base 2, that is a (5,7) group. This may also explain the inherent group condition of choosing one pair per group, that is, (9,7) does not qualify because we cannot have the condition $(h+6)$ maintained.

According to the definition, there is also a specific pattern between elements of the subsets for a given value of h. That is, if for a given h, then $f(x) = p$ and $f(x) = q$ are elements of the function, then;

$$q - p = 2h$$

This tells us that the sieving process is looking for related primes and pseudoprimes on the basis of $h = 6x \pm 1$. Therefore in a given set we only consider elements that are $2h$, and the interesting observation is that all such numbers in an h-subset end with the same digit! For example, if we consider $h = 41$, then we know the ending prime will be $41 + 6 = 47$, hence we will have a (5,7) pattern. Now the difference between any two numbers in this set will be $2 \times 41 = 82$. Take for example the two numbers;

$$p = 470{,}184{,}984{,}535 \qquad q = 470{,}184{,}984{,}617$$

The difference between the two numbers is 82, and the pattern is (5,7).

Gap Base 4	Gap Base 2
$6^n - h = 6x - 1$ Hence $x = (6^n + (1-h))/6$	$6^n + h = 6x + 1$ Hence $x = (6^n + (h-1))/6$

Table 56

What is intriguing is the fact that we have two different sieve definitions but the behavior is similar in having a consistent structure though different in result. That is, we have specific end digit behavior in each case. This makes the two systems to be related in some way since they are dealing with the same class of numbers.

6.10 The 6-Power Set

Classification is a powerful tool that propelled discovery after discovery in chemistry and physics. It is based on drawing conclusions from observed patterns, and is here presented as a technique called pattern methodology. By introducing the concept of gap base system it enhanced the way the integers are understood in terms of primes and pseudoprimes. This is further developed in this book to have a system called the universal number distribution theorem (UNDT), which develops four other sets called Class 1A consisting of odd numbers, Class 2A, Class 1B and Class 2B consisting of even numbers. Whilst the derivation of these sets will be explained later, they are applied here to make deeper observations on integer patterns and distribution behavior.

Consider the set $6^n \pm h$. In terms of the discovery process, this became an inspiration to consider other sets in the same format and highlight some of their properties. It is evident that the exercise can be carried out wider and deeper, hence only some sets are used to illustrate the research.

Definition

Let there be a sieve $f(m,n) = (6m)^n \pm h$*, where;*

a) $1 \leq m \leq \infty$*, and* $1 \leq n \leq \infty$
b) $h = 6x \pm 1, h > 0$*, and* $1 \leq x \leq \infty$
c) A specific sieve equation is derived by keeping m *and* h *constant while varying* n*.*

This type of sieve is referred to as a 6-power sieve because it is in multiples of six. Hence $6^n \pm h$ results from $m = 1$. The resulting sets are denoted according to the value of $6m$ as $S(6m)$, for example $S(6)$, $S(12)$ and so on. Each set will be found to have its own particular characteristics. The aim is to look for a particular rule that defines a pattern for the set.

6.10.1 The Set S(6)

This set is defined as $f(1,n) = (6)^n \pm h$, where for a particular set we keep h constant, and get the next set in the same group by varying h. For example, let $h = 1$, then we have the set $f(1,n) = (6)^n \pm 1$ where $(6)^n - 1$ define gap base 4 and $(6)^n + 1$ gives gap base 2. Note that Class 1B, 2B and 2A denote even numbers, P a prime number and Ps a pseudoprime.

h	n	GB2	Class	Class	GB4	Class	Class
1	1	7			5		
1	2	37	1A	GB2	35	1A	GB4
1	3	217	2B	GB4	215	2B	1A
1	4	1,297	2A	GB2	1,295	2A	Ps,GB4
1	5	7,777	Ps, GB4	GB4	7,775	Ps, GB4	1A
1	6	46,657	1B	1A	46,655	1B	Ps,GB2
1	7	279,937	1A	GB2	279,935	1A	GB4
1	8	1,679,617	2B	GB4	1,679,615	2B	1A
1	9	10,077,697	2A	GB2	10,077,695	2A	Ps,GB4
1	10	60,466,177	P, GB4	GB4	60,466,175	P, GB4	1A
1	11	362,797,057	1B	1A	362,797,055	1B	Ps,GB2

Table 57

The rule for this set is found by observing that for a given h, it has a repeating group of numbers on the right where such numbers are always 5 in number. Once these numbers are determined, then the left numbers are derived. These are called polynomial roots, and they are mostly defined by a left and right behavior. In the pattern development, the values for $n = 1$ are ignored as these do not seem to be part of a pattern. Perhaps there is some meaning to them that still has to be determined. Hence the pattern is considered from $n = 2$.

We can change the value of h and choose one in gap base 2 and another in gap base 4, that is $h = 7$ and $h = 11$. The same pattern is still observed though a different set of values occur. The classification of the roots is also consistent for a given set.

h	n	GB2	Class	Class	GB4	Class	Class
GB2							
7	1	13			-1		
7	2	43	1A	P, GB2	29	1A	P, GB4
7	3	223	2B	P, GB4	209	2B	1A
7	4	1303	GB2	1A	1289	2A	P, GB4
7	5	7783	GB4	P, GB4	7769	GB4	1A
7	6	46663	1B	1A	46649	1B	Ps, GB2
GB4							
11	1	17			-5		
11	2	47	1A	P, GB4	25	1A	Ps, GB2
11	3	227	2B	1A	205	2B	P, GB4
11	4	1307	GB2	P, GB2	1285	2A	Ps, GB2
11	5	7787	GB4	1A	7765	GB4	Ps, GB4
11	6	46667	1B	P, GB2	46645	1B	1A

Table 58

Now there are two groups of data; one for GB2 and one for GB4. Hence let us consider data for gap base 2, where this is also aggregated in terms of the gap base of h. This gives us the following table.

GB2 Group

h is GB2					
h = 1	37	17	97	77	57
h = 7	43	23	03	83	63
h = 13	49	29	09	89	69
h = 19	55	35	15	95	75
h = 25	61	41	21	01	81
h is GB4					
h=5	41	21	01	81	61
h= 11	47	27	07	87	67
h = 17	53	33	13	93	73
h = 23	59	39	19	99	79
h = 29	65	45	25	05	85

Table 59

It is noted that the pattern formation is consistent and in steps of 6 from the previous h. This type of data could be applied in cryptography to develop code keys. Now we consider the next group for GB4.

GB4 Group

h is GB2					
h = 1	35	15	95	75	55
h = 7	29	09	89	69	49
h = 13	23	03	83	63	43
h = 19	17	97	77	57	37
h = 25	11	91	71	51	31

h is GB4

h = 5	31	11	91	71	51
h = 11	25	05	85	65	45
h = 17	19	99	79	59	39
h = 23	13	93	73	53	33
h = 29	07	87	67	47	27

Table 60

The same observation is made here, that is, the sets are consistent and the difference to the next row is given by a subtraction of 6. In GB2 group it was an addition, in this group it is a subtraction. Now the most interesting observation is made with regard to the structure of this set system. It is indeed amazing as shown by the diagram.

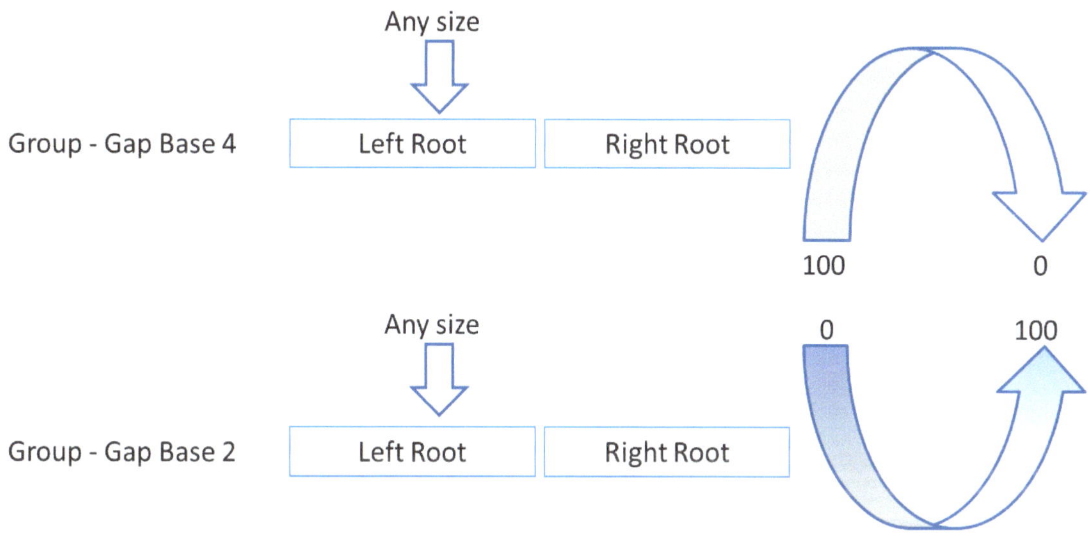

Figure 45

All the right roots range between 0 and 100 for any value of h, almost like a modulo operation while the left root can take any size. This then gives a very simple rule for determining the left root for the S(6) set. Therefore if we consider the GB2 group we note that the numbers between 0 and 100 are derived through a truncation process. Choosing the group that starts with 81 on the fourth column, we add 6 until we stop at 345. This gives us table A.

Table A

81	111	141	171	201	231	261	291	321
87	117	147	177	207	237	267	297	327
93	123	153	183	213	243	273	303	333
99	129	159	189	219	249	279	309	339
105	135	165	195	225	255	285	315	345
	141		201		261		321	

Transformation process: *truncate the two right digits*

Table B

81	11	41	71	01	31	61	91	21
87	17	47	77	07	37	67	97	27
93	23	53	83	13	43	73	03	33
99	29	59	89	19	49	79	09	39
05	35	65	95	25	55	85	15	45
	41		01		61		21	

Table 61

Then we transform this table to table B by truncating the last two digits on the right, and this gives us our five number pattern arrangement for $f(1,n) = (6)^n \pm h$. We ignore the grey colored numbers to get the set of five to correspond accurately. A similar reasoning applies for the GB4 group, only this time we go in reverse instead. For example in this group we have the (71, 65, 59, 53, 47) at the fourth column. If from table A we take the numbers from 141 to 201, and we truncate the last two digits on the right, then starting from 71 backwards derives the intended numbers that have been shown in color as {71, 65, 59, 53, 47}.

Table A

141	171
147	177
153	183
159	189
165	195
	201

Transforms to

Table B

41	71
47	77
53	83
59	89
65	95
	01

Table 62

There is no evident pattern for deriving the left root.

6.10.2 The Set S(12)

That is, $m = 2$, hence we have $f(2,n) = (12)^n \pm h$. This gives the following pattern, keeping m and n constant.

GB2	GB4	h	
3,5,9,7	1,3,7,5	1	GB2
1	9		
GB2	GB4		
3,5,9,7	1,3,7,5	7	GB2
1	9		
GB2	GB4		
5,7,1,9	1,5,3,9	13	GB2
3	7		
GB2	GB4		
5,9,7,3	1,3,7,5	19	GB2
1	9		
GB2	GB4		
9,3,1,7	7,9,3,1	25	GB2
5	5		
GB2	GB4		
3,7,5,1	3,5,9,7	31	GB2
9	1		
GB2	GB4		
7,1,9,5	9,1,5,3	37	GB2
3	7		
GB2	GB4		
1,5,3,9	5,7,1,9	43	GB2
7	3		
GB2	GB4		
5,9,7,3	1,3,7,5	49	GB2
1	9		

GB4	GB2	h	
7,9,3,1	7,9,3,1	5	GB4
5	5		
GB4	GB2		
3,5,9,7	1,3,7,5	11	GB4
1	9		
GB4	GB2		
7,1,9,5	9,1,5,3	17	GB4
3	7		
GB4	GB2		
1,5,3,9	5,7,1,9	23	GB4
7	3		
GB4	GB2		
5,9,7,3	1,3,7,5	29	GB4
1	9		
GB4	GB2		
9,3,1,7	7,9,3,1	35	GB4
5	5		
GB4	GB2		
3,7,5,1	3,5,9,7	41	GB4
9	1		
GB4	GB2		
7,1,9,5	9,1,5,3	47	GB4
3	7		
GB4	GB2		
1,5,3,9	5,7,1,9	53	GB4
7	3		

Table 63

Re-arranging in terms of missing digits such as (1,9) does not seem to yield any particular pattern. However, in this case we do not have the demonstration of the H_{30} set. This is shown by the fact that if we take $h = 7$, then the next set should be $h = 37$, but the pattern and missing digits are a different configuration.

GB2	GB4	h	
3,5,9,7	1,3,7,5	1	GB2
1	9		
3,5,9,7	1,3,7,5	7	GB2
1	9		
5,9,7,3	1,3,7,5	19	GB2
1	9		
5,9,7,3	1,3,7,5	49	GB2
1	9		
GB2	GB4		
5,7,1,9	1,5,3,9	13	GB2
3	7		
7,1,9,5	9,1,5,3	37	GB2
3	7		
7,9,1,5	9,5,1,3	67	GB2
3	7		

GB4	GB2		
3,5,9,7	1,3,7,5	11	GB4
1	9		
5,9,7,3	1,3,7,5	29	GB4
1	9		
5,9,7,3	1,7,3,5	59	GB4
1	9		

Table 64

Similarly, there is no evidence of a clear and consistent pattern here if we vary n and keep m and h constant.

h	n	GB2	GB4
1	1	13	11
1	2	145	143
1	3	1,729	1,727
1	4	20,737	20,735
1	5	248,833	248,831
1	6	2,985,985	2,985,983
1	7	35,831,809	35,831,807
1	8	429,981,697	429,981,695
1	9	5,159,780,353	5,159,780,351
1	10	61,917,364,225	61,917,364,223

Table 65

Other values of h do not show any interesting observable pattern. So this set has no fireworks except for the fact that we have the end-digit pattern (3, 5, 9, 7) for GB2 and (1, 3, 7, 5) for GB4 for $h = 1$, and so on.

6.10.3 The Set S(18)

The equation here is given as $f(3,n) = (18)^n - h$ for GB4 and $f(3,n) = (18)^n + h$ for GB2. The table shows that m is fixed at $m = 3$, h is fixed at $h = 1$, but what varies is n. This is in-line with the definition.

h	n	GB2	Class	Class	GB4	Class	Class
1	1	19	P/Ps	P/Ps	17	P/Ps	P/Ps
1	2	325	P/Ps	P/Ps	323	P/Ps	P/Ps
1	3	5,833	1B	1A	5,831	1B	P
1	4	104,977	2B	P	104,975	2B	1A
1	5	1,889,569	2B	Ps	1,889,567	2B	1A
1	6	34,012,225	1B	1A	34,012,223	1B	Ps
1	7	612,220,033	2A	Ps	612,220,031	2A	Ps
1	8	11,019,960,577	1A	Ps	11,019,960,575	1A	Ps
1	9	198,359,290,369	1A	Ps	198,359,290,367	1A	Ps
1	10	3,570,467,226,625	2A	Ps	3,570,467,226,623	2A	Ps

Table 66

The first major observation is the occurrence of elements for $n = 1$ and 2. Thereafter the pattern changes from $n = 3$ to infinity. For the set S(6) it was only $n = 1$. These two numbers are always either prime or pseudoprime for any value of h. The abbreviation P/Ps means the number is a prime or a pseudoprime. The un-answered question here is why this type of pattern, that the first two numbers are always prime or pseudoprime. There could be some reason.

However, we do have a specific pattern, there are polynomial roots for $f(3,n) = (18)^n - h$. We have the same left and right pattern, but this time we start with the left root in order to identify the right root. Each number can be split into two characteristics. For example the number 5831 is composed of 58 and 31. These are referred to as called polynomial roots of 5831. The first root does not conform to $f(x) = 6x \pm 1$ whilst the second root conforms.

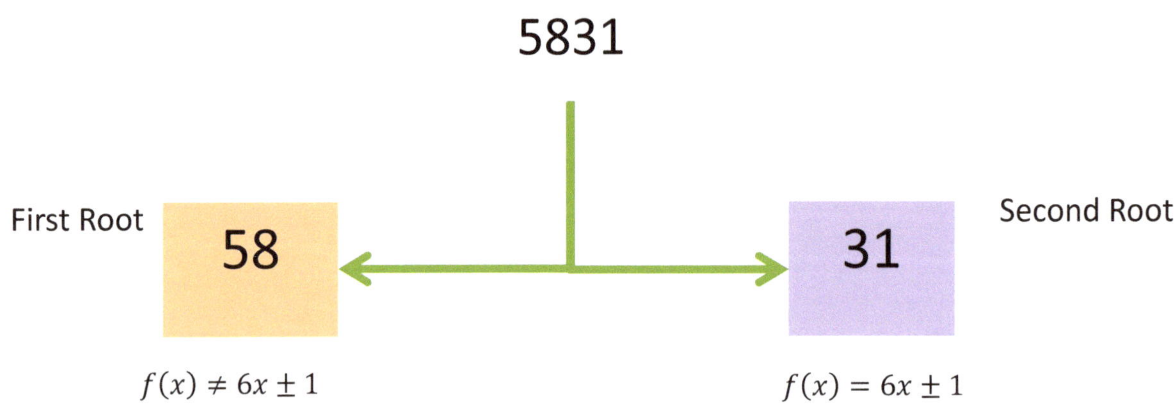

Figure 46

This gives us the following possibilities in terms of roots for a given number. Since the left root is a composite, it implies that other classes will apply.

Left Root	Right Root
Class 1A	Class 1A
Class 1A	Prime or Pseudoprime in GB2 or GB4
Class 2A	Class 1A
Class 2A	Prime or Pseudoprime in GB2 or GB4
Class 1B	Class 1A
Class 1B	Prime or Pseudoprime in GB2 or GB4
Class 2B	Class 1A
Class 2B	Prime or Pseudoprime in GB2 or GB4

Table 67

In particular, h has no influence on the gap base of the prime or pseudoprime. The arrangement of the right roots appears to be random; it can be a prime or pseudoprime that is in gap base 2 or 4, or it can be an element of Class 1A.

Gap Base 4

Class 1B → 5,831 ← p/ps

Class 2B → 104,975 ← Class 1A

Class 2B → 1,889,567 ← Class 1A

Gap Base 2

Class 1B → 5,833 ← Class 1A

Class 2B → 104,977 ← p/ps

Class 2B → 1,889,569 ← p/ps

Figure 47

The rate at which the composite left root is growing is much smaller than the right root. If we consider the number of digits, whether for gap base 2 or gap base 4, then starting from $n = 3$, it is observed that the number of digits R_d for the right root is given by;

$$R_d = n - 1$$

The remaining number of digits then should be for the left root. In terms of generality, the number of digits will differ depend on the value of h for the same value of n. The pattern of roots will still be there for different values of h, but the h-factor will slightly change the count value.

	GB4		
n	$f(3,n) = (18)^n - h$	Left Root	Right Root
1	17	0	2
2	323	0	3
3	5,831	2	2
4	104,975	3	3
5	1,889,567	3	4
6	34,012,223	3	5
7	612,220,031	3	6
8	11,019,960,575	4	7
9	198,359,290,367	4	8
10	3,570,467,226,623	4	9
11	64,268,410,079,231	4	10

Table 68

The number of elements for the right root occur in steps of four, hence a group may be defined where $n = 4g$, and g is the group number. Then the number of digits for the left root is given by L_d where;

$$L_d = g + 2$$

Therefore if we are considering the fifth group, then then number of digits is (5+2)=7. The value of n is (5x4) = 20. Therefore the number of digits for the right root is R_d = 20-1 = 19. Hence the total number of expected digits is (7+19) = 26. Therefore as a general formula, the expected number of digits is N_d where;

$$N_d = (5n + 4)/4$$

This assumes that n is taken in multiples of 4.

6.10.4 The Set S(24)

The set is given by $f(4,n) = (24)^n \pm h$. The $S(24)$ set has the following pattern for given values of h that is consistent. A value of $h = 29$ is chosen at random, and a two number pattern cycle is observed. That is 05 and 53 for GB2 and 47 and 95 for GB4.

h	n	GB2	Left Root	GB4	Left Root
29	1	53		(5)	
29	2	605	2A	547	GB4
29	3	13,853	2A	13,795	GB4
29	4	331,805	2A	331,747	GB4
29	5	7,962,653	2A	7,962,595	GB4
29	6	191,103,005	2A	191,102,947	GB4
29	7	4,586,471,453	2A	4,586,471,395	GB4
29	8	110,075,314,205	2A	110,075,314,147	GB4
29	9	2,641,807,540,253	2A	2,641,807,540,195	GB4
29	10	63,403,380,965,405	2A	63,403,380,965,347	GB4

Table 69

A general pattern is shown for other values of h. The pattern has a left and right polynomial root system, where the left root is always of class 2A for GB2 and GB4 for gap base 4 for $h = 29$. Numbers produced by this function will alternate in successive pairs, for example 25 and 77, 29 and 81, and so on. Hence the right root is used to determine the left root. Only positive values are considered in order to preserve consistency.

	GB2 Group		GB4 Group	
	1st	2nd	1st	2nd
h = 1	25	77	23	75
h = 5	29	81	19	71
h = 7	31	83	17	69
h = 11	35	87	13	65
h = 13	37	89	11	63
h = 17	41	93	7	59
h = 19	43	95	5	57
h = 23	47	99	1	53
h = 25	49	01	51	99
h = 29	53	05	47	95

Table 70

These can actually be re-arranged as follows.

h is in GB2

	GB2 Group - Increasing		GB4 Group - Decreasing	
	1st	*2nd*	*1st*	*2nd*
h = 1	25	77	23	75
h = 7	31	83	17	69
h = 13	37	89	11	63
h = 19	43	95	5	57
h = 25	49	01	51	99
h = 31	55	7	45	93

h is in GB4

	GB2 Group - Increasing		GB4 Group - Decreasing	
	1st	*2nd*	*1st*	*2nd*
h = 5	29	81	19	71
h = 11	35	87	13	65
h = 17	41	93	7	59
h = 23	47	99	1	53
h = 29	53	5	47	95
h = 35	59	11	41	89

Table 71

From $h = 1$ to $h = 23$, the classification of the left root is consistently the same but changes from $h = 25$. However, the two number pattern for the right root remains consistent.

h	n	GB2		GB4	
1	1	25		23	
1	2	577	GB4	575	GB4
1	3	13,825	2A	13,823	2A
1	4	331,777	GB4	331,775	GB4
1	5	7,962,625	2A	7,962,623	2A
1	6	191,102,977	GB4	191,102,975	GB4
1	7	4,586,471,425	2A	4,586,471,423	2A

Table 72

The general pattern for end-digits is given below.

h	GB4	GB2	h	GB4	GB2
1	(3,5)	(5,7)	5	(9,1)	(9,1)
11	(3,5)	(5,7)	25	(9,1)	(9,1)
31	(3,5)	(5,7)	35	(9,1)	(9,1)
41	(3,5)	(5,7)	55	(9,1)	(9,1)
61	(3,5)	(5,7)	65	(9,1)	(9,1)
71	(3,5)	(5,7)	95	(9,1)	(9,1)
91	(3,5)	(5,7)	7	(7,9)	(1,3)
13	(1,3)	(7,9)	17	(7,9)	(1,3)
23	(1,3)	(7,9)	37	(9,7)	(1,3)
43	(1,3)	(7,9)	67	(9,7)	(1,3)
73	(1,3)	(7,9)	77	(9,7)	(1,3)
83	(1,3)	(7,9)	97	(9,7)	(1,3)
			19	(5,7)	(3,5)
			29	(5,7)	(3,5)
			49	(5,7)	(3,5)
			59	(5,7)	(3,5)
			79	(5,7)	(3,5)
			89	(5,7)	(3,5)

Table 73

Values of h that end with the same digit have the same pattern for their digit end as well depending on the gap base. For example if h ends with a 3, then any gap base 4 element will end with a (1,3) or a (3,1) pattern and a gap base 2 element will end in a (7,9) or (9,7) digit pattern. This is also a different result from the 30-set, where the ending pattern was related to values of h that had a difference of 30.

Now when we test whether the values of $S(24)$ have roots, we find that we cannot establish a consistent pattern such that the left has a root satisfying the condition $f(x) \neq 6x \pm 1$ and the right component $f(x) = 6x \pm 1$.

6.10.5 The Set S(30)

The set $S(18)$ demonstrated a remarkable property where it had a root construction of its elements. It has the same root construction system in gap base 2 and gap base 4. However, for the set $S(30)$ there is a different scenario. Gap base 2 has a different root construction system from gap base 4. The other difference is that the construction for set $S(30)$ is more constrained, there is a more regular pattern in the construction compared to set $S(18)$. The table shows the pattern up $h = 25$. The first root is always class 2B.

Base	h	First Root	Second Root
GB2	1	Class 2B	Class 1A
GB4	5	Class 2B	Ps - GB4
GB2	7	Class 2B	Class 1A
GB4	11	Class 2B	P/Ps GB4
GB2	13	Class 2B	Class 1A
GB4	17	Class 2B	P/Ps GB4
GB2	19	Class 2B	Class 1A
GB4	23	Class 2B	P/Ps GB4
GB2	25	Class 2B	Class 1A

Table 74

For gap base 4, the value of h has an influence on the type of root that we get. This is a very different outcome from set $S(18)$, where the value of h had no regular predictable influence.

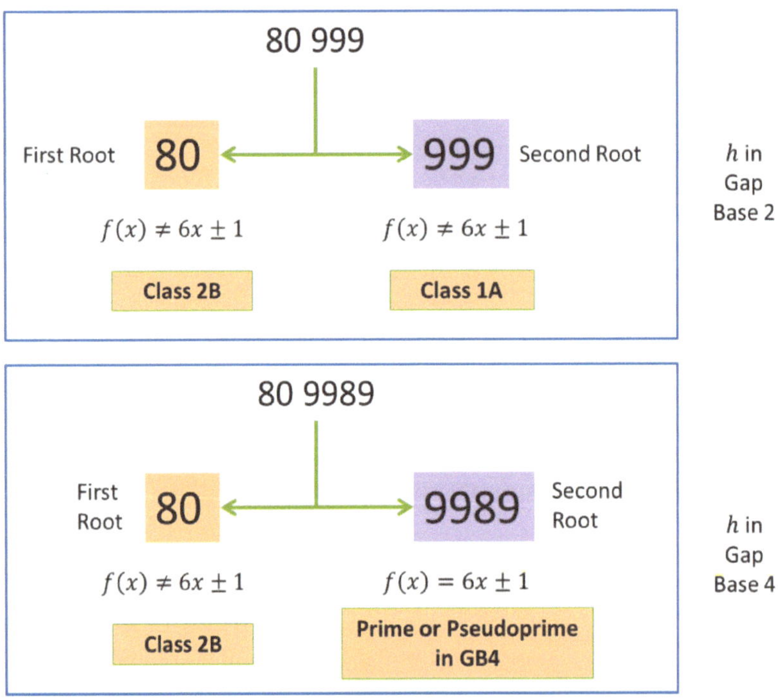

Figure 48

For example for h being gap base 2, then the root will be an element of class 1A only, no other class is involved. When h is gap base 4, then the right root will either be a prime or a pseudoprime in gap base 4 only. The result of this influence in set $S(30)$ is that it appears more structured.

The other point of interest about the set $S(30)$ is the fact that it groups certain types of prime numbers and pseudoprime together. For example, $h = 1$ derives the primes or pseudoprimes having a string of nines, for example such a number can be 656,099,999,999. It also relates numbers with a string of zeroes and a prime or pseudoprime at the end, such as 656,100,000,001.

An example is given that illustrates the pattern for $h = 101$. The common property between set $S(18)$ and set $S(30)$ is the fact that there is a consistent pattern such that the left root satisfies the condition $f(x) \neq 6x \pm 1$ and the right root the condition $f(x) = 6x \pm 1$.

			GB4	GB2
m=3	h	n		
30	101	1	(71)	131
30	101	2	799	1,001
30	101	3	26,899	27,101
30	101	4	809,899	810,101
30	101	5	24,299,899	24,300,101
30	101	6	728,999,899	729,000,101
30	101	7	21,869,999,899	21,870,000,101
30	101	8	656,099,999,899	656,100,000,101
30	101	9	19,682,999,999,899	19,683,000,000,101
30	101	10	590,489,999,999,899	590,490,000,000,101

Table 75

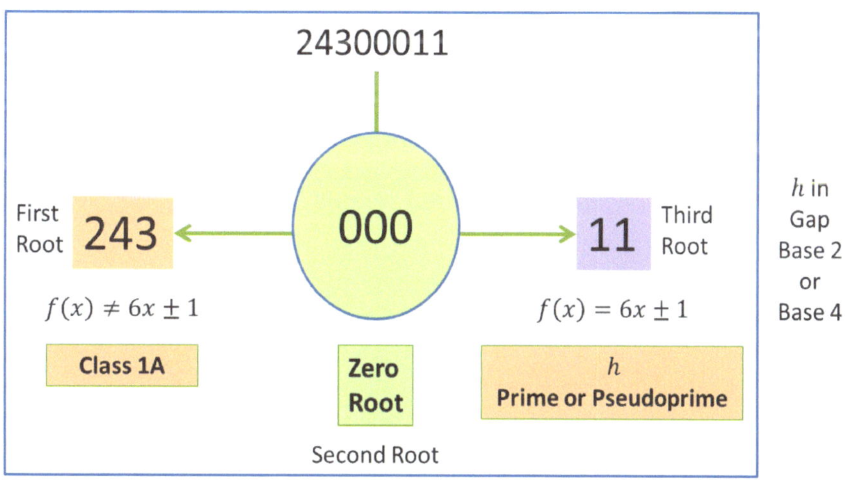

Figure 49

For the h in gap base two, there is something that is also new. Instead of two roots, there are now three roots; class 1A, the zero root, and h. This arises from the fact that we are using 30^n, and this gives us powers of 3^n. When $h = 1$, the number of zeroes is given by $(n - 1)$. The third root is always defined by h, whether it is gap base 2 or 4. Therefore in a similar manner, h has an influence on the third root, it is always either a prime or pseudoprime whose value is h.

6.10.6 The Set S(36)

This is the last set considered simply because it is a variation of the set S(6) and S(24). It has the following structure, a left and polynomial right root. The right root is used to determine the left root. The distinct features of the set are found from the relativity and the difference. The difference follows exactly the same pattern for gap base 2 and 4. The relativity pattern is also exactly the same; a small number followed by a big number. Unlike S(24), the left root has different classes.

h	n	GB2	**Relativity**	**Difference**	GB4	**Relativity**	**Difference**
1	1	37	*Small*	77-37 = 40	35	*Small*	75-35 = 40
1	2	1,297	*Big*	97-37 = 60	1,295	*Big*	95-35 = 60
1	3	46,657	*Small*	97-57 = 40	46,655	*Small*	95-55 = 40
1	4	1,679,617	*Small*	57-17 = 40	1,679,615	*Small*	55-15 = 40
1	5	60,466,177	*Big*	77-17 = 60	60,466,175	*Big*	75-15 = 60
1	6	2,176,782,337	*Small*	77-37 = 40	2,176,782,335	*Small*	75-35 = 40
1	7	78,364,164,097	*Big*	97-37 = 60	78,364,164,095	*Big*	95-35 = 60
1	8	2,821,109,907,457	*Small*	97-57 = 40	2,821,109,907,455	*Small*	95-55 = 40
1	9	101,559,956,668,417	*Small*	57-17 = 40	101,559,956,668,415	*Small*	55-15 = 40
1	10	3,656,158,440,062,977	*Big*	77-17 = 60	3,656,158,440,062,975	*Big*	75-15 = 60

Table 76

How does one derive these numbers? The cycle is also based on 100, and the transformation rule through truncation of the last two digits. Hence 60 is the truncation cycle generator.

GB2

	Original Set	Truncated Result
Add	37	37
60	97	97
60	157	57
60	217	17
60	277	77
60	337	37
60	397	97
60	457	57
60	517	17
60	577	77

GB4

	Original set	Truncated Result
Add	35	35
60	95	95
60	155	55
60	215	15
60	275	75
60	335	35
60	395	95
60	455	55
60	515	15
60	575	75

Table 77

This then derives the above cycle demonstrated by the set.

6.11 Cyclic Truncation Transforms

One of the aims of the above exhaustive exercise has been to demonstrate the usefulness of the gap base system in classifying prime numbers as well as in unearthing very interesting properties. It has illustrated in a striking manner the relationship between prime numbers and even the pseudoprimes, where this would not have been possible if the framework was not applied.

However, a fitting conclusion would be the development of truncation transforms. This is a phenomena that has been observed in the prime number patterns, especially the 6-power set. This is also a simple yet very interesting number pattern, and it is an intrigue as to how it gets involved with the 6-power set, as well as other sets.

We start with a very simple set, the set of counting numbers. Counting can be in the form of a pattern, for example using odd numbers or even numbers. Counting can also be cyclic, for example if we count in multiples of ten. Therefore we develop a model that links counting and cycle properties together through a transformation system.

Consequently, if we start counting from 15, we call 15 the cycle initiator, and if we are adding 1, then 1 is the cycle generator. This will obviously generate the counting numbers from 15. Now if we truncate the right digit, then the first element will 5, and this cycle will end with 4 to start another similar cycle. This is where the term cycle truncation comes from. This is called a single transform because it consists of a single digit only.

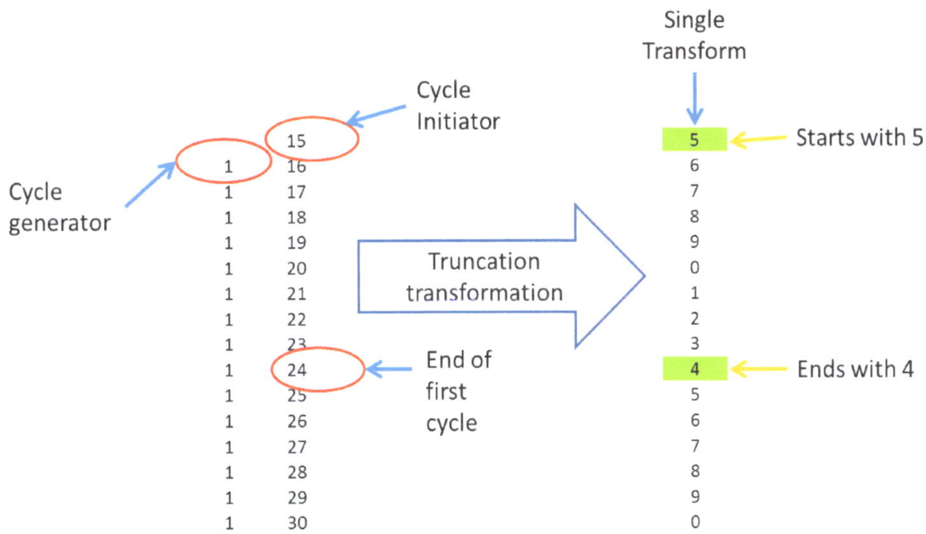

Figure 50

The set derived from the cycle generator is called the original set. The set that result from the transformed elements is called a cycle transform set. We also note the right digit of the generated numbers; these consist of 1, then 2, then 3 and so on. This is to be expected but is noted as part of the concept development.

Definition

For a given truncation, the elements of the cycle transform set are always unique.

Now let us have 7 as our generator, and we keep the same initiator. The transform set is formed from that, and the cycle is established. However, we now go a step further and define derivative sets. The set D_1 is defined to consist of elements that define the start of the cycle, and the beginning of the other cycles that follow. Since our set ends at 127, this implies we have only two elements 15 and 85. The set D_2 consists of elements that show regularity of pattern on the right digit, for example (2,2), (5,5), (7,7) and so on while the remaining elements define D_3. That is, D_3 is a complementary set.

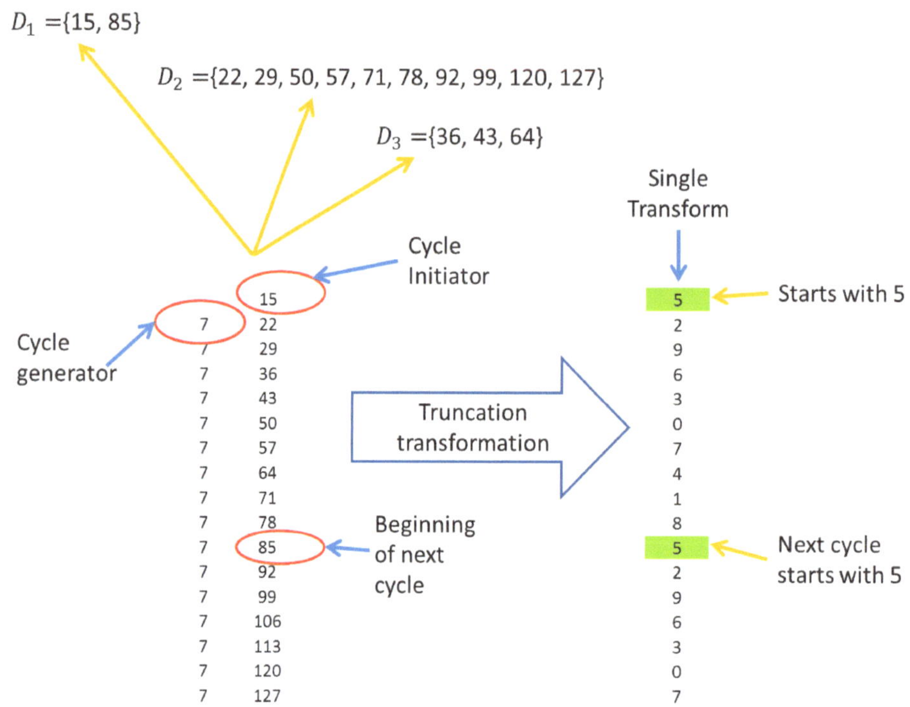

Figure 51

Let us consider using 2 and 6 as cycle generators and using 73 and 31 respectively as cycle initiators. Then we realize that the process generates elements of the set D_t, which consists of primes and pseudoprimes. The new development is that in both cases the cycle depth is now 5 instead of 10. This occurs when the cycle generator is 2 or 6 or a multiple thereof. When we had generators that were not a multiple of 2 or 6, then the cycle was longer at 10 digits.

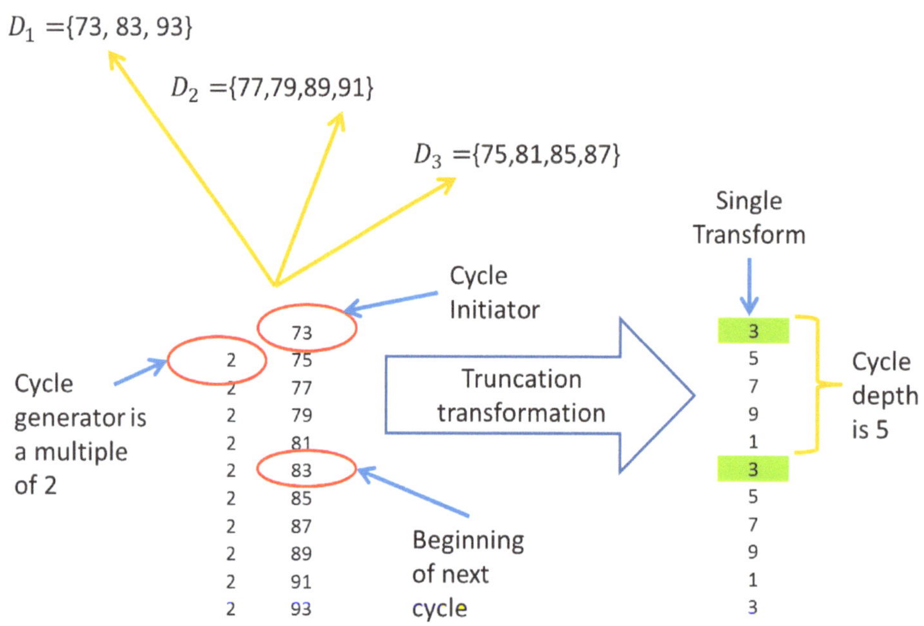

Figure 52

All along we have been discussing a single digit transformation without reference to primes or pseudoprimes, but we can define the sets D_i taking recognition of this. For example the set $D_2 = \{77,79,89,91\} \subset D_t$.

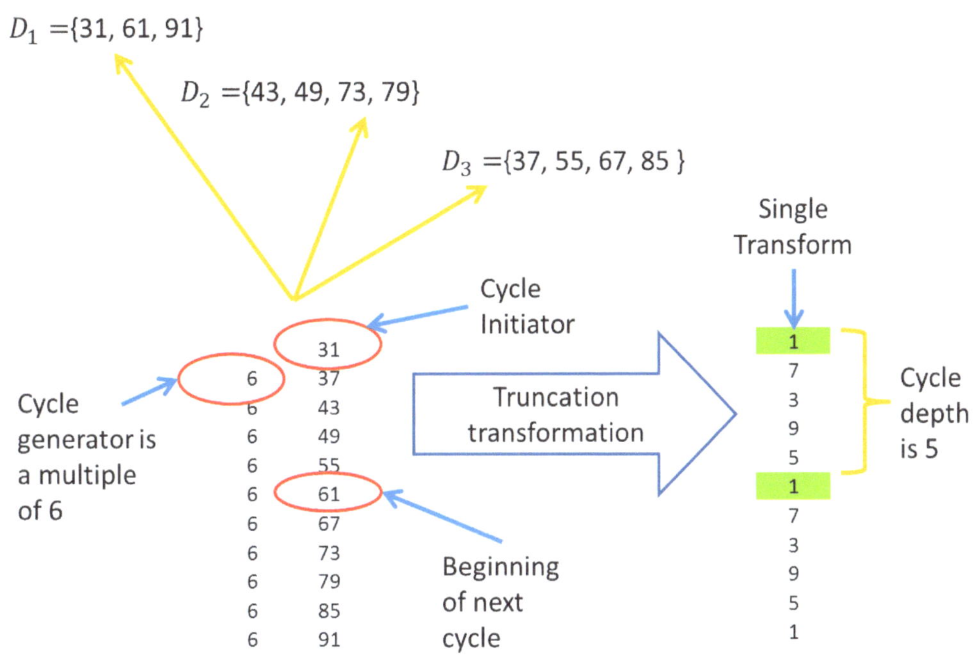

Figure 53

However in the set with generator 6, the occurrence of primes and pseudoprimes has been ignored, and instead $D_2 = \{43,49,73,79\}$ because we are looking at the pattern of the left digit as basis for grouping. The consistent approach would be to consider the elements of D_t to define D_2.

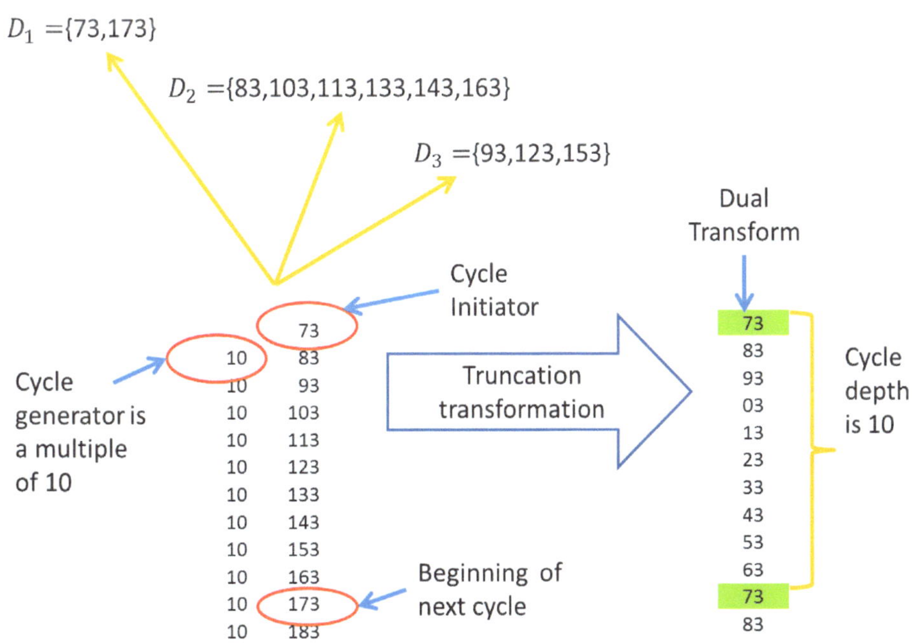

Figure 54

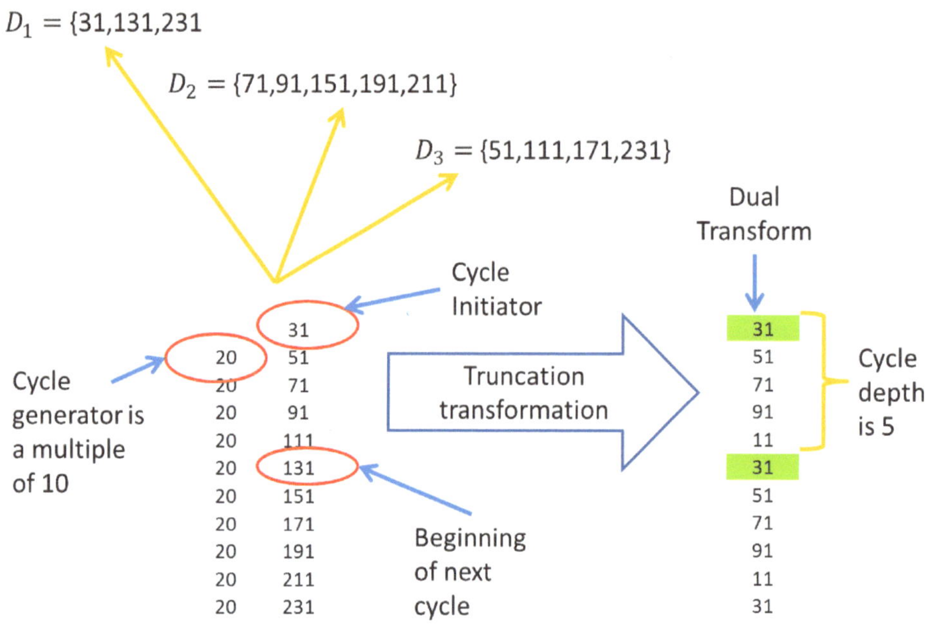

Figure 55

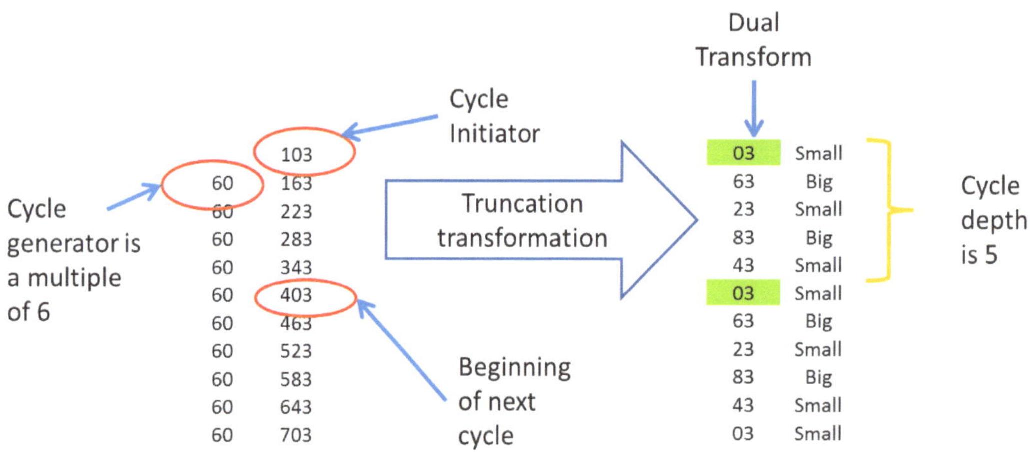

Figure 56

Having given some examples, a summary is made that defines this behavior as a pattern. All odd numbers have a cycle length of 10, while even numbers in the cycle pattern of 2-4-6-8 have a cycle length of 5. Multiples of 5 have a cycle length of 2, and 100 of 1. No other pattern exists for the truncation transform except as single and dual.

	Generator Number	Single/Dual	Cycle length
Odd	1, 3, 7, 9, 11, 13, 15, ...	S	10
Even	2, 4, 6, 8, 12,14, 16, 18, 22....	S	5
Multiples of 5	5, 15, 25, 35, 45, 55...	S	2
	10, 30,40,70,90	D	10
	20, 50, 60, 80	D	5
	100	D	1

Table 78

That is, the numbers repeat as a cycle from 0 to 10, or from 1 to 100, exactly the same behavior that was observed for the 6-power set. This is true even the generators are more than 100. No particular explanations are evident for the above construct, that is, why a particular cycle length is observed.

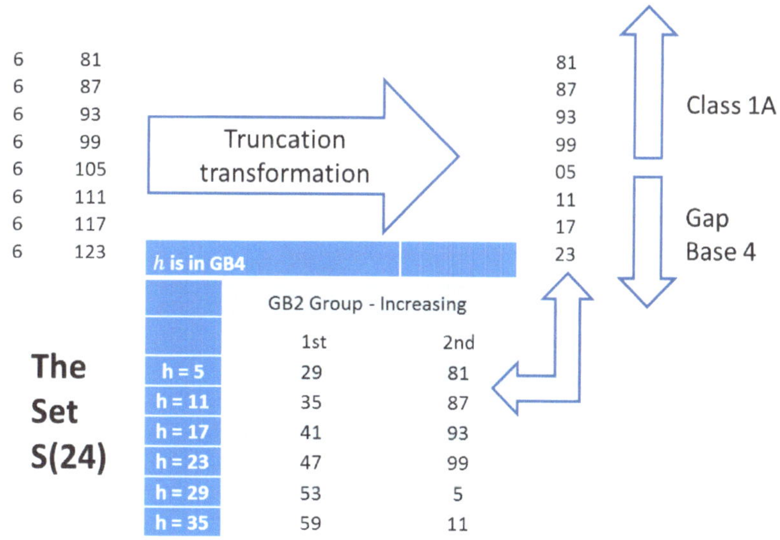

Figure 57

6.12 Cyclic Transform In Gap Base Structure

Cyclic transforms are evident in the formation of prime numbers. If one puts the prime numbers in sequence, then the pattern for such transforms is not evident. Gap base 4 numbers from 26 161 001 to 26 161 595 are chosen, and a distinct pattern emerges. Each number in a column has a root defined by gap base 2, class 1A and gap base 4 number between 1 to 100 from the right side.

	GB4	GB4	GB4	GB4	GB4	GB4
	Gap Base 2	Class 1A	Gap Base 4	Gap Base 2	Class 1A	Gap Base 4
1	26161001	26161103	26161205	26161301	26161403	26161505
2	26161007	26161109	26161211	26161307	26161409	26161511
3	26161013	26161115	26161217	26161313	26161415	26161517
4	26161019	26161121	26161223	26161319	26161421	26161523
5	26161025	26161127	26161229	26161325	26161427	26161529
6	26161031	26161133	26161235	26161331	26161433	26161535
7	26161037	26161139	26161241	26161337	26161439	26161541
8	26161043	26161145	26161247	26161343	26161445	26161547
9	26161049	26161151	26161253	26161349	26161451	26161553
10	26161055	26161157	26161259	26161355	26161457	26161559
11	26161061	26161163	26161265	26161361	26161463	26161565
12	26161067	26161169	26161271	26161367	26161469	26161571
13	26161073	26161175	26161277	26161373	26161475	26161577
14	26161079	26161181	26161283	26161379	26161481	26161583
15	26161085	26161187	26161289	26161385	26161487	26161589
16	26161091	26161193	26161295	26161391	26161493	26161595
17	26161097	26161199		26161397	26161499	

Table 79

The pattern is consistent for any number range that is selected. Now if we assume two digit numbers less than 100, then 1 is represented as 01, hence the remaining number on the left defines another root. The combination of the left and right root is called a cyclic transform structure. For example we can say $[261610] + [01] = 26161001$, where $[261610]$ is a root on the left and $[01]$ a root on the right.

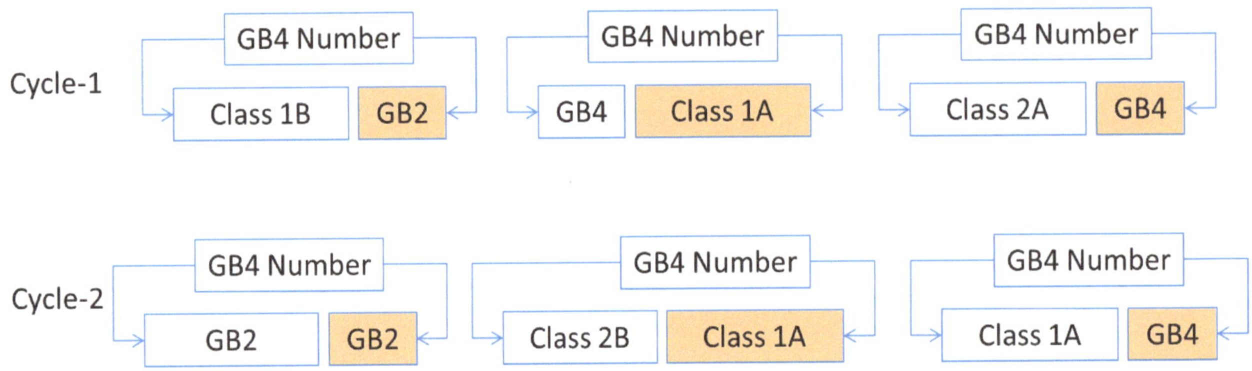

Figure 58

If the numbers of the left root are classified, then a consistent pattern is also observed. The construction of the gap base 4 number has a two cycle system that repeats itself. The *GB2-CLASS 1A-GB4* structure is common to both cycles, but what is interesting is that the cycle is built on the basis of all the UNDT classes being used at least once with the underlying structure. Amazing!

Theorem

Let there be a number W_g, then W_g consists of right root m and left root n, denoted $W_g = [n] + [m]$ where;

a) $m \in D_t,\ 1 \leq m \leq 97$
b) $m \in Class\ 1A,\ 3 \leq m \leq 99$
c) $n \in D_t,\ 1 \leq n \leq \infty$
d) $n \in D_e,\ 2 \leq n \leq \infty$
e) $n \in Class\ 2A,\ 6 \leq n \leq \infty$

The cyclic transform structure explains the behavior of patterns that have been derived and demonstrated in this chapter. The usefulness of the theorem lies in the fact that we can easily construct a number at random that can either be composite, prime or pseudoprime using roots. For example, the following numbers are generated at random, and tested for type.

$[n]$	Type	$[m]$	Type	$[n] + [m]$	Type
60000	Class 2A	23	GB4	6000023	Prime
1788093	Class 1A	77	GB4	17809377	Composite
550302	Class 2A	37	GB2	55030237	Pseudoprime
611123	GB4	45	Class 1A	61112345	Pseudoprime

Table 80

Now if $W > 0$ and $W = W_g \pm 6i$, where $1 \leq i \leq \infty$ this will generate a set that is either Class 1A (that is, odd composites) or the set D_t which implies a set consisting of primes and pseudoprimes.

6.13 Population Theory and Probability of Occurrence

If we consider the cyclic transform in terms of one digit, then the fundamental structure may be defined in the following manner: 5-1-7-3-9. The pattern is derived by observing that the difference between the next of next number is always 2. That is 7-5 = 2 or 3-1=2. Now except for the first prime number 5, the rest of the numbers that end with 5 are pseudoprimes for positive numbers. Therefore this implies that the probability for a prime number occurring is actually defined by a space $[a, b] = [a, a + 18]$ in GB2 or GB4 where a corresponds to the end-digit 1. Assuming the that we have $6x \pm 1$ to generate GB4 and GB2, then for any value of x, we know that $x = 5t - 4$ in GB4 or $x = 5t - 1$ in GB2 will give us a pseudoprime that ends with 5, where $1 \leq t$.

GB4	GB2
1	3
7	9
3	5
9	1
5	7
1	3
7	9
3	5
9	1
5	7
1	3
7	9
3	5
9	1

GB4	GB2
1	
7	
3	
9	1
	7
	3
	9

GB4	GB2
	1
	7
1	3
7	9
3	
9	

Table 81

Therefore the space $[a, b]$ will always lie between t and $(t + 1)$ that defines where a possible prime occurrence will take place.

Definition

In the prime space $[a, b] = [a, a + 18]$ for the consecutive end-digit 1-7-3-9, this maps a prime number population space where either a prime number or pseudoprime occurs.

This prime space is fixed, and therefore the implication is that the prime numbers and pseudoprimes are fighting to occupy positions within it. There are four positions in gap base 2 and another four in gap base 4, making a total of eight positions to fight for. The space is special because it is also responsible for defining the gap development in the distribution of the prime numbers. Because of this, one of its major consequences is to raise the question whether prime numbers occur at infinity. Accordingly, the space is said to establish the Fixed Population Theory of the Prime Number, abbreviated as FPT-PN.

The probability of any location in the population space getting a prime number or pseudoprime is 50%, which implies the probability of either of them is 6.25% for gap base 2 or 4.

	GB2	GB4
Probability of prime in GB2 = 0.5x0.5x0.5x0.5= 0.0625	1	3
same	7	9
Probability of prime in GB4 = 0.5x0.5x0.5= 0.125	3	5
same	9	1

Table 82

To calculate the probability of occurrence in both of them, we have to note the occurrence of the pseudoprime ending with 5. Therefore for a space $[a, b]$ in GB2 or GB4 for the configuration 1-7-3-9, the multiple of 5 occurs once.

Hence the probability that a number will be prime in GB2 and GB4 in space $[a, b]$ is 0.125x0.0625 = 0.0078125, that is 0.78125%. This gives us the probability generally of a prime number to occur. Since the probability is influenced by growth of the gap between the prime numbers due to pressure from the pseudoprimes, the correct statement of the probability is;

$$\textit{Probability(Prime Occurrence)} \leq 0.78125\%$$

The assumption is that the probability is progressively decreasing as we go to infinity due the "fight". The graphic below shows such a pattern at the beginning, using GB4 to start the 1-7-3-9 cycle.

Gap Base 2	Gap Base 4
7	11
13	17
19	23
25	29
31	35

Figure 59

The same result will be obtained for the pseudoprime, which means the two types of numbers have an equal probability of occurring. It must be remembered though the prime and pseudoprime are "fighting" to fill in these four slots at equal probability in GB2 or GB4 and that in reality the pseudoprimes gain ascendance as x increases towards infinity because of their generation through multiplication.

The Algebraic Sieve

One of the first application of the classification system concerned primality testing. This was an obvious development from the fact that the sieve presented a systematic group of numbers. But it also led to the conceptualization of the prime space concept which provided further understanding of distribution behavior.

OUTCOME

Further developments in the algebraic sieve and prime space.

7 The Algebraic Sieve Concept

The paradigm shift in concept development in this theory came from the definition of the gap base system followed by the concept of an algebraic sieve. Without the gap base system, the algebraic sieve would not have much meaning and application; it is the combination of these two concepts that created potency. The concept of the algebraic sieve is formed through the multi-valued function technique which expresses the framework for the fundamental gap theorem.

This develops a structure where we have two sets $6x - 1$ and $6x + 1$ being related to x in a fixed manner. The relationship generates two types of numbers; the prime numbers and the set of pseudoprimes. As a result, the grid is a conceptual construct in which the prime numbers exist, but they do not exist alone. This is the other major assumption, that prime numbers in this grid co-exists with the pseudoprimes that are also related to the prime generation process rather than being a side event. The above assumptions have the important implication that,

a. all the prime numbers now have a specific position in the grid
b. each specific position has a known value that can be found in the grid, and
c. the grid consists of primes and a set of pseudoprimes.

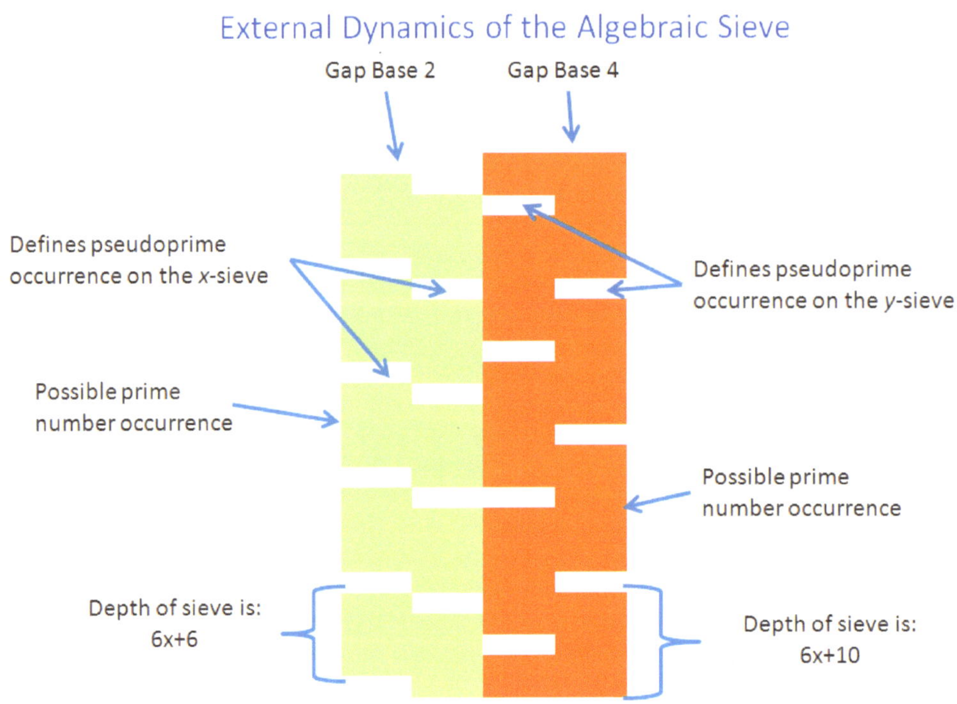

Figure 60

Hence, we predict a prime in terms of its position in the grid rather than describing it in terms of trying to define an nth prime. That is, we are not looking for the nth prime, rather we are looking for the value of a given prime at the nth position of the grid. Therefore, when we can establish a relationship between these two, we effectively for the basis of a theory to predict where we can find a prime number on the grid and the value of that prime number.

The sieve as a construct is assumed to be a representation of a pseudoprime occurrence in gap base 2 or gap base 4, where such an occurrence is controlled by the sieve depth. Conceptually, using the fundamental gap theorem, then;

a. the first part of the sieve is called the x-sieve, and
b. the second part of the sieve is called the y-sieve.
c. Combining the two of them gives us the xy-sieve.

We consider an example for the basis of defining a sieve structure. The following shows shaded factors of 5 and 11. Factors of 5 occur on both the x-sieve and the y-sieve, the same being true for factors of 11.

Key		Factors of 5		Key			Other	
		Factors of 11					Repeat	
x	GB 4	GB2		x	GB4	GB 4	GB2	GB2
1	5	7		1		5	7	
2	11	13		2		11	13	
3	17	19		3		17	19	
4	23	25		4		23	25	
5	29	31		5		29	31	
6	35	37		6		35	37	
7	41	43		7		41	43	
8	47	49		8		47	49	
9	53	55		9		53	55	
10	59	61		10		59	61	
11	65	67		11		65	67	
12	71	73		12		71	73	
13	77	79		13		77	79	
14	83	85		14		83	85	
15	89	91		15		89	91	
16	95	97		16		95	97	
17	101	103		17		101	103	
18	107	109		18		107	109	
19	113	115		19		113	115	
20	119	121		20		119	121	
21	125	127		21		125	127	
22	131	133		22		131	133	
23	137	139		23		137	139	
24	143	145		24		143	145	

Table 83

When we divide by 5, 55 will be shaded, and when we divide by 11 it will be shaded again. The red color shows multiple shading for either gap base 2 or 4. The second part of the graphic shows all possible shadings using the bluish color from other factors such as 13, 17, 19 and so on. The red shows repeat from 13, 17, 19, 23 in the second part of the graphic. The un-shaded parts that remain are then prime numbers.

What now remains is to describe the process in an algebraic manner. Several equation formats can be used to define the sieve. For example in this case for demonstration we apply the equations $y = 6x + 7$ and $y = 6x + 11$ where these generate gap base 2 and 4 respectively. Note that:

1. Each prime number and pseudoprime defines an xy-sieve; hence, there are an infinite number of sieves. As a result, each sieve is defined by $f(x) = (y_1, y_2)$.
2. Let there be a sieve defined by y_1, then the sieving process will be defined by a sieve step of y_1, the same being true for y_2.
3. Each sieve step defines a sieve depth.

For example, when dividing by five, we call that a 5-sieve; by seven, a 7-sieve; by twenty five, a 25-sieve. In terms of the external properties of the xy-sieve, we observe the following, assuming the equations $y_1 = 6x + 7$ and $y_2 = 6x + 11$:

a. The primes are either in gap base 2 or gap base 4, that is, the x-sieve or the y-sieve.
b. The sieve identifies areas at which a "hit" will occur for a pseudoprime—it is called a pseudo-hit. This corresponds to non-occurrence of a prime number.
c. For natural primes, prime generation varies from $-1 \leq x \leq \infty$, and as x varies, it defines a unique sieve through sieve depth. The sieve is considered for positive primes only but can do negatives as well.
d. Normally, prime numbers for the same x can be said to define the xy-sieve pair. Therefore (7, 11) form the same pair, and (5, 7) form a cross pair on the sieve if they have a different value of x.
e. The next pseudo-hit will always occur at (sieve depth + 1) from the previous hit.
f. Each sieve depth for a given x defines a possible occurrence of a prime number. The sieve depth has the same definition whether we have a cross pair or not.

The power of the algebraic sieve is this: *It knows algebraically where the pseudoprimes are, and it knocks them out to remain with the prime numbers*. And because it knows, it does not take long to do the operation.

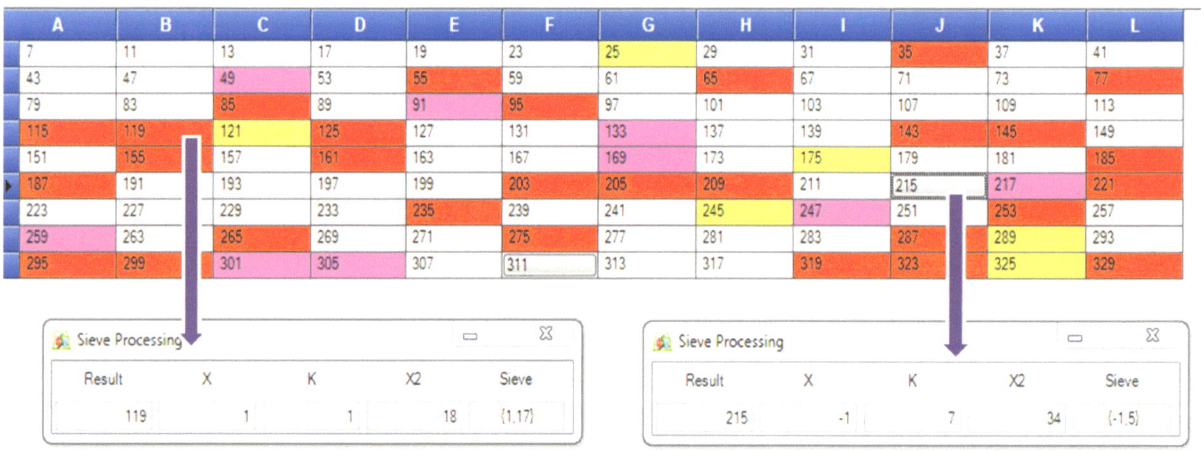

Figure 61

An example is made from the 7-sieve which starts in gap base 2, hence it is an x-sieve. The function used is $f(x) = 6x + 1$ for gap base 2 to illustrate that it does not matter what function we apply. While x is held constant, k varies to identify the spots that define a hit in the sieve. But there are multiples of 7 on both x and y sieves, so it also looks for hits on the y-sieve, where this is referred to as the x-sieve on y.

Where Hit Occurs	Gap Base 2	Gap Base 4
Gap Base 2 "Hit"	$x - sieve$	$x - sieve\ on\ y$
Gap Base 4 "Hit"	$y - sieve\ on\ x$	$y - sieve$

Table 84

		x-sieve	GB-2	x-sieve on y	GB-4
		x+(6x+1)k	Hit	(-6x+1-1)+(6x+1)k	Hit
x	k	x1	f(x1)	x2	f(x2)
1	0	1	7	-1	-7
1	1	8	49	6	35
1	2	15	91	13	77

1	3	22	133	20	119
1	4	29	175	27	161
1	5	36	217	34	203
1	6	43	259	41	245
1	7	50	301	48	287
1	8	57	343	55	329
1	9	64	385	62	371
1	10	71	427	69	413

Table 85

The operational assumptions and equations for a particular sieve are as follows, one sieve acting in gap base 2 and gap base 4.

1. There exists on the sieve a point $(x_1, y_{1,} y_2)$, where y_1, y_2 is a prime or pseudoprime, and $-\infty \leq x_1 \leq \infty$, that is, x_1 can take any position on the sieve, where y_1 and y_2 defines a sieve.
2. Hence, for a given y_1 operating in gap base 2, then
    a. $x_1 = \frac{1}{6}(y_1 - 7)$ for the x-sieve
    b. $x_2 = \frac{1}{6}(-y_1 - 11)$ for the x-sieve on the y-sieve
3. Hence, for a given y_2 operating in gap base 4, then
    a. $x_3 = \frac{1}{6}(y_2 - 11)$ for the y-sieve
    b. $x_4 = \frac{1}{6}(-y_2 - 7)$ for the y-sieve on the x-sieve
4. For k an integer, then a "hit" in the sieve occurs whenever;
    a. $x = x_1 + ky_1$, at $(x, f(x))$, where $-\infty \leq k \leq \infty$, from initial point $(a_{1,} a_2)$ on the x-sieve.
    b. $x = x_2 + ky_1$, at $(x, f(x))$, where $-\infty \leq k \leq \infty$, from initial point $(a_{3,} a_4)$ on the y-sieve.
    c. $x = x_3 + ky_2$, at $(x, f(x))$, where $-\infty \leq k \leq \infty$ from initial point $(b_{1,} b_2)$ on the y-sieve.
    d. $x = x_4 + ky_2$, at $(x, f(x))$, where $-\infty \leq k \leq \infty$ from initial point $(b_{3,} b_4)$ on the x-sieve.

Hence, for example, if we want to sieve all spots that are divisible by the prime number 13 in the x-sieve, where the x-sieve uses $f(x) = 6x + 7$, then;

a. $x_1 = \frac{1}{6}(13 - 7) = 1$ for the x-sieve, and $f(x) = 13$ giving us the point $(a_{1,} a_2) = (1, 13)$ as our initial point.
b. Now $x = x_1 + ky_1$, hence $x = 1 + 13k, -\infty \leq k \leq \infty$.
c. Therefore, the equation for the x-sieve for 13 is;
   $(1 + 13k, f(1 + 13k)), -\infty \leq k \leq \infty$, where $f(1 + 13k) = 6(1 + 13k) + 7$.

Now for the x-sieve on the y-sieve (meaning that we are still using 13), then;

a. $x_2 = \frac{1}{6}(-13 - 11) = -4$ for the y-sieve giving us the point $(a_{3,} a_4) = (-4, -13)$ as our initial point.
b. Now $x = x_2 + ky_1$, hence $x = -4 + 13k, -\infty \leq k \leq \infty$
c. Therefore, the equation for the y-sieve for 13 is
   $(-4 + 13k, f(-4 + 13k)), -\infty \leq k \leq \infty$, where $f(-4 + 13k) = 6(-4 + 13k) + 11$.

SIEVE	-4	-13		Look for factors of		13			
X₂ =	1	1	1	1	1	1	1	1	1
k =	1	2	3	4	5	6	7	8	9
x=	9	22	35	48	61	74	87	100	113
$f(x)$ =	65	143	221	299	377	455	533	611	689

Table 86

This will give all points from negative to positive infinity that are not occupied by a prime number and are pseudoprimes that are multiples of 13 on the x-sieve and the x-sieve acting on the y-sieve. Therefore $f(x)$ defines a "pseudo hit" where $x = -4 + 13k$.

For $x = 1$, then $y_2 = 17$. We can find the sieve starting in gap base 4. For the y-sieve:

a. $x_3 = \frac{1}{6}(17 - 11) = 1$ for the y-sieve giving us the point $(b_1, b_2) = (1,17)$ as our initial point.

b. Now $x = x_3 + ky_2$, hence $x = 1 + 17k, -\infty \leq k \leq \infty$

c. Therefore, the equation for the y-sieve for 17 is

$(1 + 17k, f(1 + 17k)), -\infty \leq k \leq \infty.$

Now for the y-sieve on the x-sieve, then,

a. $x_4 = \frac{1}{6}(-17 - 7) = -4$ for the y-sieve on the x-sieve, and $f(x) = 17$ giving us the point $(-4, 17)$ as our initial point.

b. Now $x = x_4 + ky_1$, hence $x = -4 + 17k, -\infty \leq k \leq \infty$.

c. Therefore, the equation for the y-sieve on the x-sieve for 17 is

$$(-4 + 17k, f(-4 + 17k)), -\infty \leq k \leq \infty.$$

SIEVE	1	17		Look for factors of		17
X₃=1	1	1	1	1	1	1
k =	1	2	3	4	5	6
x=	18	35	52	69	86	103
$f(x)$ =	119	221	323	425	527	629

Table 87

Hence, the general sieve equation is given as,

1. $(x_i + y_1k, f(x_i + y_1k)), -\infty \leq k \leq \infty, i = 1,2$
2. $(x_j + y_2k, f(x_j + y_2k)), -\infty \leq k \leq \infty, j = 3,4$

Note when $k = 0$, we have $(x_1, f(x_1))$, the starting point of the sieve. We can make the sieve to be more useful by choosing the starting point and then let k vary over a narrow range rather than an infinite range. That is $k_1 \leq k \leq k_2$ for a given range $[a, b]$. Hence, for example, if we want to sieve between 6000 and 8000, we set k_1 and k_2 accordingly. This approach creates efficiency in identifying the primes.

Hence, considering only integer values, we may expand the above equation as;

1. $(a_1, a_2) = (x + (6x + 7)k, 6\{x + (6x + 7)k\} + 7)$ for the x-sieve

$$a_1 = x + (6x + 7)k$$

$$a_2 = 6a_1 + 7$$

2. $(a_{3,}a_4) = (\frac{1}{6}(-f(x) - 11) + f(x)k, -f(x) - 11 + 6(f(x)k) + 11)$ for the x-sieve acting on the y-sieve. Note $f(x) = 6x + 7$ as this is a gap base 2 operation.

$$a_3 = \frac{1}{6}(-f(x) - 11) + f(x)k$$
$$a_4 = -f(x) - 11 + 6(f(x)k) + 11$$

3. $(b_{1,}b_2) = (x + (6x + 11)k, 6\{x + (6x + 11)k\} + 11)$ for the y-sieve.

$$b_1 = x + (6x + 11)k$$
$$b_2 = 6b_1 + 11$$

4. $(b_{3,}b_4) = (\frac{1}{6}(-f(x) - 7) + f(x)k, -f(x) - 7 + 6(f(x)k) + 7)$ where this is for the y-sieve acting on the x-sieve. Note $f(x) = 6x + 11$ as this is a gap base 4 operation.

$$b_3 = \frac{1}{6}(-f(x) - 7) + f(x)k$$
$$b_4 = -f(x) - 7 + 6(f(x)k) + 7$$

As noted, the way the sieve works is to keep x constant while varying the value of k. A simple example is made for k = 0 to 2, and x is constant.

	x	1	k	0
The sieve is		13		
	1	13	Gap Base 2, *x*-sieve	
	-4	-13	Gap Base 2, *x*-sieve on y-sieve	

	x	1	k	1
The sieve is		13		
	14	91	Gap Base 2, *x*-sieve	
	9	65	Gap Base 2, *x*-sieve on y-sieve	

	x	1	k	2
The sieve is		13		
	27	169	Gap Base 2, *x*-sieve	
	22	143	Gap Base 2, *x*-sieve on y-sieve	

Table 88

For $k = 0$, this defines the initial sieve. In gap base 2 we sieve using (1,13) while in gap base 4 we use (-4, -13). By varying k, we get the sieve coordinates for each hit, for example $k = 1$ gives us (14,91) and (9,65) while $k = 2$ gives us (27,169) and (22,143).

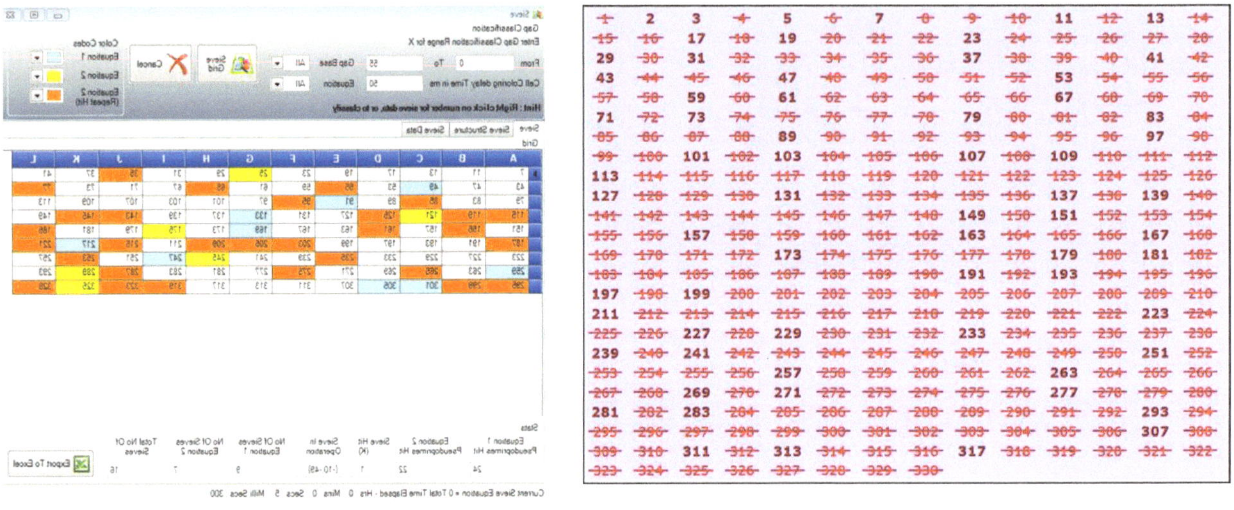

Figure 62. The algebraic sieve in comparison to an internet demonstration of sieve of Erostathenes.

The advantage of the algebraic sieve is its efficiency, handling less numbers with precision.

The Algebraic Prime Space

Outcomes

Law	Consequence and attribute of law
The Prime Gap Law	c) *The gap base structure has a probabilistic influence on the distribution of gap formation between prime numbers.* d) *The average natural prime gap increases in proportion to $n\{f(x_a) \cdot f(x_b)\}$ where n is the number of pseudoprimes in range $[a, b]$ at position x.* **Consequence and attribute:** The gap base structure creates a sieve that determines position in terms of 2, 4, 6. A prime or pseudoprime must fall into this structure without fail.
The Law of Pseudoprimes	*Let $W = f(x_a) \cdot f(x_b)$, where $f(x_a) = 6x_a \pm 1$ and $f(x_b) = 6x_b \pm 1$, then W is a pseudoprime.* **Consequence and attribute:** That pseudoprimes owe their initial existence to the prime numbers. But once established, pseudoprimes dominate the primes and whatever position they miss is occupied by a prime number. Because of the attribute of product, they self-generate.
Gap Population Law	*For a given prime space $[a, b]$ the population of smaller gaps diminishes towards zero as x increases to infinity.* **Consequence and attribute:** The consequence is that primes are not infinite, and that the smaller gaps are the first casualties to go out of the scene.
Pseudoprime Gap Law	*The average natural gap to the next pseudoprime decreases tending towards 3 as $f(x) = 6x \pm 1$ increases.* **Consequence and attribute:** All the above laws make this law to exist. The exact pseudoprime difference is 2 between gap bases, 4 diagonally and 6 within a gap base. But once 2 is established, 4 and 6 are automatic.
Prime Gap Structure Law	*The natural prime number gap increases in multiples of six that are randomized by the inter-gap base relationship.* **Consequence and attribute:** That is, prime numbers are generated in an orderly manner, but are randomized by the gap base structure.
Fixed Space Distribution Law	*There exists a self-determining fixed space [a,b] consisting of prime and pseudoprime numbers.* **Consequence and attribute**: Makes the distribution to be non-linear and set-based.
The Co-existence Theorem	*For every (x, y_1, y_2), then (y_1, y_2) are either a pseudoprime or a prime number.* **Consequence and attribute**: The theorem defines the algebraic sieve.
Law of Integer Distribution	*For the range $[a, b]$, then $f(b) - f(a) = n_1 + n_2 + n_3$ where n_1 is the number of primes, n_2 the number of pseudoprimes, n_3 the number of composites and $f(b) - f(a)$is a constant.* **Consequence and attribute:** There is a deterministic distribution of numbers such that this defines a space $[a, b]$.

Table 89

8 Distribution and Gap Dynamics

There are three main ways of studying the prime numbers, the fourth is of interest;

a) The distribution of the prime numbers
b) The gap between the prime numbers
c) The symmetric patterns of prime numbers
d) The truncation properties of the prime numbers

So much of the mystery of prime numbers is in their distribution, and the gap between the numbers. However, there is virtually no study of prime numbers in relation to other numbers. Therefore the algebraic sieve provides a new framework that we call the UNDT to unearth more hidden information. Since we have a specific construct, this means specific relationships in terms of laws can be defined to describe these behaviors. If a dimension consists of 1000 prime numbers, then we notice that in general, the distribution of primes per 1000 goes down.

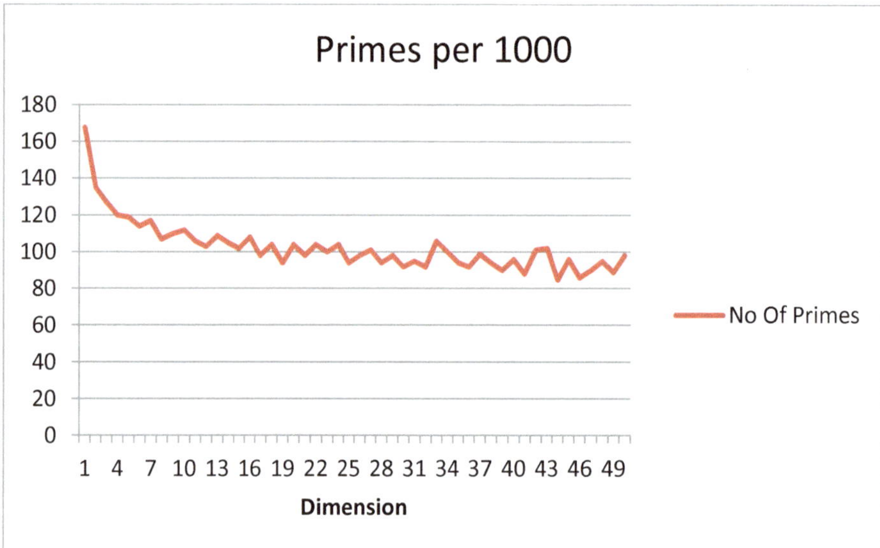

Figure 63

The interesting thing about the sieve construct is that pseudoprimes are actually initially generated by prime numbers. Hence without the first prime numbers, there would be no pseudoprimes yet at the same time pseudoprimes determine where further prime number occurrence takes place. Three laws are defined that are collectively known as the Law of Primes because of the way they are interrelated. The laws are derived as a result of the definition of the algebraic sieve.

The Law of Primes is a combination of four laws defined by:

a) *The Law of Pseudoprimes*. This is how pseudoprimes are formed and defined relative to the prime numbers.
b) *The Prime Gap Law*. This law defines how the prime gap forms.
c) *The Co-existence Theorem*. This defines the positional relationship between prime numbers and pseudoprimes.
d) *The Fixed Space Distribution Law*. This states that prime numbers exist in a specific space that is fixed such that the distribution is dependent on the pseudoprimes.

Most of what is derived with respect to the theory on integers is described by these three laws. For example, the framework for factorization is a consequence of the law of primes. Later the law of primes is expressed in an algebraic format to include the set of composite numbers.

The Law of Pseudoprimes

Let $W = f(x_a) \cdot f(x_b)$ *, where* $f(x_a) = 6x_a \pm 1$ *and* $f(x_b) = 6x_b \pm 1$*, then W is a pseudoprime.*

Therefore $f(x) = 6x \pm 1$ and $-\infty \leq x \leq \infty$, hence all pseudoprimes are defined by $f(x_a) \cdot f(x_b)$.From this law we further define a specific type of factor called the ditation factor.

Pair Theorem

For W a pseudoprime, then $f(x_a)$ *and* $f(x_b)$ *are ditation factors of W if and only if* $W = f(x_a) \cdot f(x_b)$.

For example $W = 1519 = 31 \cdot 49$. The ditation factors of 1519 are obviously 31, 49 but also 7 and 217 since these are elements of D_t. These are the biggest factors in the context of $f(x) = 6x \pm 1$. However, 49 consists of prime factors 7 and 7, hence the prime factors of 1519 are 7, 7, 31. The critical thing about ditation factors is that they are always a pair.

$f(x_a)$	$f(x_b)$
7	217
31	49

Table 90

Consequently they are also called factor pairs, and they generate the so called factor pair theory for factorization.

If pseudoprimes did not exist, the prime numbers would have an official gap of 2 and 4. They would occur as a regular pattern in this manner. Therefore the above law of the pseudoprimes is directly responsible for the generation of irregular gaps between the prime numbers. All prime gaps are in multiples of six in a given gap base.

$f(x_i)$	$f(x_j)$	Product Result	Comment
Prime	Prime	Pseudoprime	Prime Dependent
Prime	Pseudoprime	Pseudoprime	Prime Dependent
Pseudoprime	Prime	Pseudoprime	Prime Dependent
Pseudoprime	Pseudoprime	Pseudoprime	Self-generating

Table 91

A new pseudoprime can either be generated by any two prime numbers, or the combination of a prime and a pseudoprime. If the product is formed by two pseudoprimes, then it is said to be self-generating, that is, it does not need a new prime number to be defined. For example 25 is a prime dependent pseudoprime, but 625 is a self-generated pseudoprime since $625 = 25 \cdot 25$. Any pseudoprime that is defined by a new prime number is prime dependent.

Consequently 625 is the first self-generating pseudoprime. It is evident that as prime-dependent pseudoprimes increase, so does the probability of occurrence of a self-generating pseudoprime. For example the pseudoprime 875 is the next self-generating pseudoprime after 625, and the gap is not so large. The list of self-generating primes from zero to 10 000 is as follows, provided the last pseudoprime we consider is 175. More pseudoprimes will derive a higher number, hence this is just for illustration.

Similarly, the list does not include prime dependent pseudoprimes. In each one thousand the number of pseudoprimes is counted, that is those that are self-generating. Altogether there are 1054 pseudoprimes in gap base 2 and 1050 pseudoprimes in gap base 4 from zero to 10 000, and of these only 63 are self-generating. If we assume the first prime number is 7 (sieve generator $6x + 7$ and $6x + 11$), then there are 611 primes in gap base 2, and 615 in gap base 4. Because this is a fixed structure, for both gap bases it adds up to 1665 each, which implies a total of 3330 positions on the algebraic sieve.

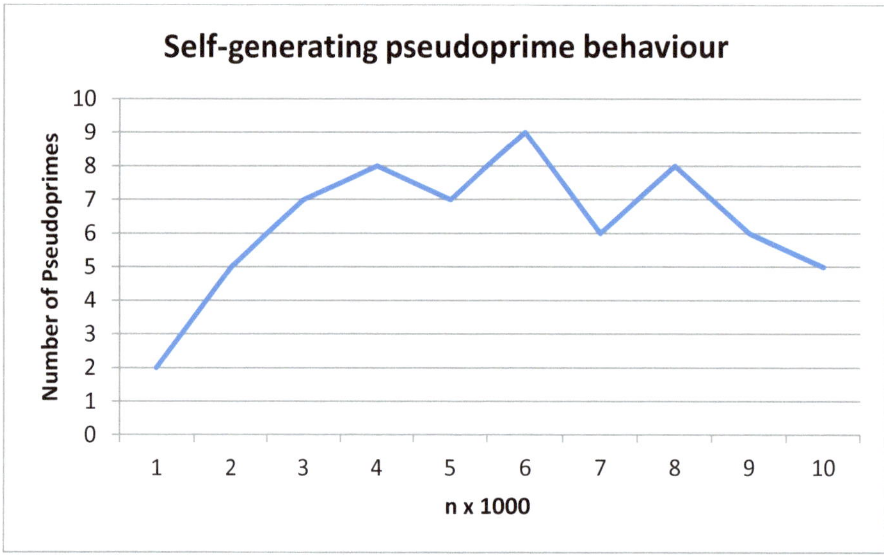

Figure 64

Self-generating pseudoprimes from 0-10 000

	625	1225	2125	3025	4025	5005	6125	7105	8075	9025
	875	1375	2275	3125	4165	5075	6175	7225	8125	9163
		1625	2375	3185	4225	5225	6325	7315	8525	9425
		1715	2401	3325	4235	5425	6517	7475	8575	9625
		1925	2695	3575	4375	5425	6545	7595	8645	9775
			2875	3625	4655	5525	6875	7735	8855	
			2975	3773	4675	5635		7889		
				3875		5831		7975		
						5929				
Count	2	5	7	8	7	9	6	8	6	5

Table 92

It can be noted that multiples of five dominate the group of self-generating pseudoprimes.

8.1 Gap Types

Prime numbers are studied in terms of their gaps. The same applies to the pseudoprimes. When prime numbers are arranged in their natural order, they define the natural gap, the same being true for pseudoprimes. The major difference in gap theory here is the UNDT framework, because prime numbers now exist in relation to pseudoprimes.

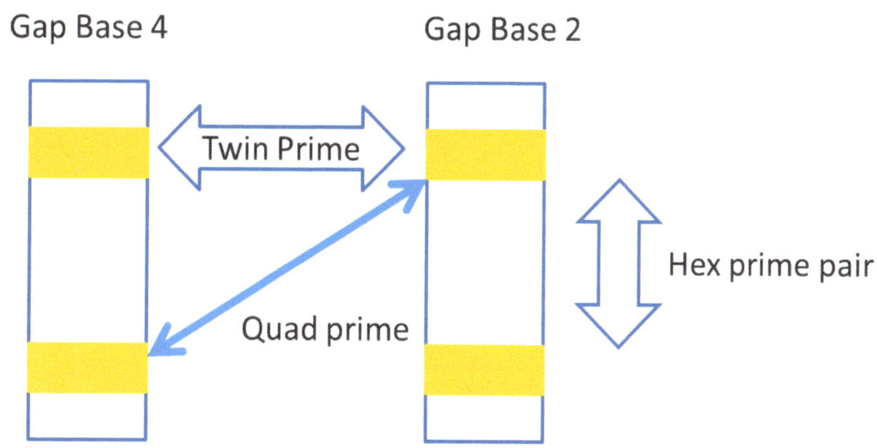

Figure 65

In the UNDT framework, there are three types of gaps that are fundamental.

1. The twin difference. This is a difference of two, and results in a twin prime pair or twin pseudoprime pair. This difference is between gap base 2 and 4 for the same value of x.
2. The Quad difference. This is a difference of four, and results in a quad prime pair or quad prime pseudoprime. The difference is between gap base 4 and 2 and occurs diagonally for x and $(x + 1)$.
3. The hex difference. This is a difference of six, and results in a hex prime pair or hex pseudoprime pair. This difference occurs within a given gap base for x and $(x + 1)$.

It is therefore not surprising that the UNDT framework is based upon these differences as a gap base. Whilst these may be defined in terms of either being prime numbers or pseudoprimes, it is also possible to define the differences as a combination of both prime and pseudoprime.

8.101 Gap Distribution

The study of prime numbers outside the context of the UNDT framework is then referred to as a linear analysis of the primes because it is one dimensional in approach. It is also non-dynamic because it ignores the role of the pseudoprime in the distributive behavior of the prime number. The UNDT framework presents a specific structure in which gap development takes place as observed from the gap graphic. That is, prime gaps don't just form; they have to fit into the structure of the gap base system. One may then assume that it implies it is more difficult for a gap 2 to form then a gap 4, or that gap 6 should be the easiest. It is the easiest because it is within the gap base whereas the other two are occurring between gap bases which decrease the probability of their occurrence.

A statistical analysis bears out the influence of the gap base structure on the distribution of gaps. A concept of dimension is defined as $x = 1$ has prime numbers from 1 to 100 000, where this is effectively a prime space $[a, b]$. Since $x = 1$ is 1 to 100 000, so $x = 4$ is values from 400 000 to 499 999.

GB-2 Stats				GB-4 Stats			
Gap 2 Primes	761	Pseudo Primes	12827	Total Gap 4	746	Pseudo Primes	12831
Gap 6 Primes	634	Composites	83334	Total Gap 6	650	Composites	83334
Over Gap 6 Primes	2446	Largest Gap	90	Total Over Gap 6	2441	Largest Gap	114
Total Primes	3841	Constant Value	16668	Total Primes	3837	Constant Value	16668

Figure 66

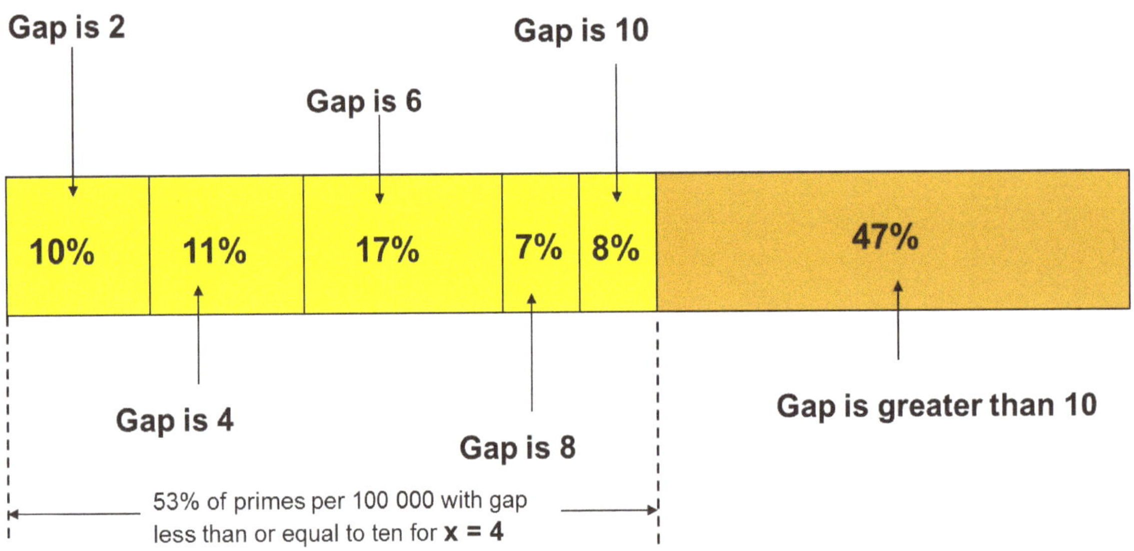

Figure 67

A similar analysis can be carried out but using the pseudoprimes this time, in order to make a comparison in terms of behavior. Gap 6 has a higher probability of occurrence at 17% =(634+650)/7678, while the diagonal gap 4 does better than gap 2 at 11% and 10% respectively. The distribution also indicates that for gaps in a given range, then 2, 4, 6 up to 10 occur most frequently compared to gaps above gap 10 at 47%. This is also probably due the structure of the gap base system.

Hence in the UNDT framework there are two things that affect the prime gap, it is the structure and the fact that the gap is proportional to the number of pseudoprimes in the respective gap bases.

The Prime Gap Law

a) *The gap base structure has a probabilistic influence on the distribution of gap formation between prime numbers.*
b) *The average natural prime gap increases in proportion to* $n\{f(x_a)\cdot f(x_b)\}$ *where* n *is the number of pseudoprimes in range* $[a,b]$ *at position* x.

It is interesting to question whether there is a shift to the right or to the left in terms of gap distribution for more than 10 and less than 10. For example by the time we get to x = 9990, we have the following result.

	Gap 2	Gap 4	Gap 6	Gap 8	Gap 10	Gap >10
x=9990	6%	6%	12%	5%	7%	64.12%

Table 93

A very surprising behavior indeed, as it means the larger gaps are actually crowding the smaller gaps on average. Gap 2 and 4 are almost put under pressure at the same rate since their values are the same. This leads to a very important observation that larger gaps begin to dominate as x increases towards infinity, hence the population of the other gaps less than ten goes down. Definitely there is a progressive shift towards the left.

The table makes a direct comparison that is given in the graph. The value of x is multiplied by 100 000. At x =1, the gap greater than 10 occupies a much smaller space.

X	Gap>10	*x*	Gap>10
1	36.67%	70	54.83%
2	43.02%	100	55.82%
3	44.64%	300	56.77%
4	46.66%	500	59.22%
5	47.13%	800	59.52%
6	47.72%	9990	64.12%

Table 94

The gap greater than 10 occupancy is increasing steadily with x, giving the effect of "the big gap pressure" on smaller prime gaps. The blue line shows the jump as x becomes very large.

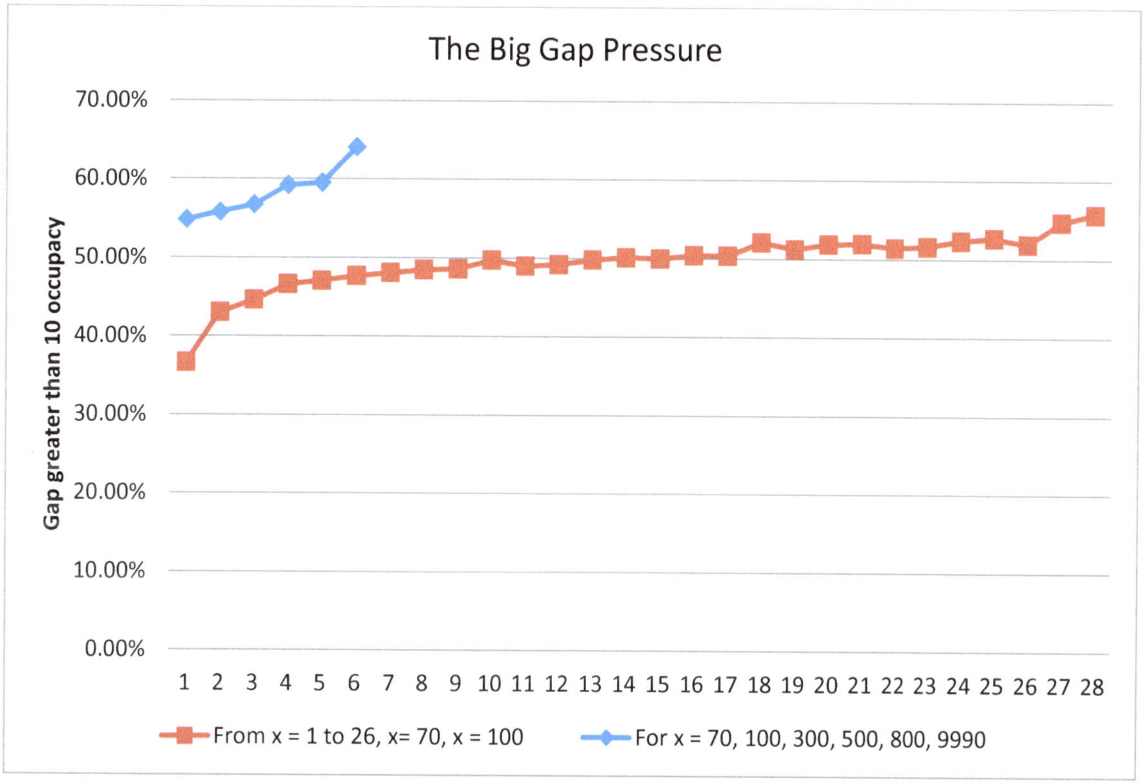

Figure 68

This observation is then stated as a gap distribution law. The choice of gaps ten and less was really arbitrary, one could have perhaps gone up to twelve which is double six. However, the trend will be the same.

Gap Population Law

For a given prime space $[a, b]$ *in* D_t *the population of smaller gaps diminishes towards zero as* x *increases to infinity.*

The law does not necessarily say the population will be zero, but it is on a downward trend towards zero.

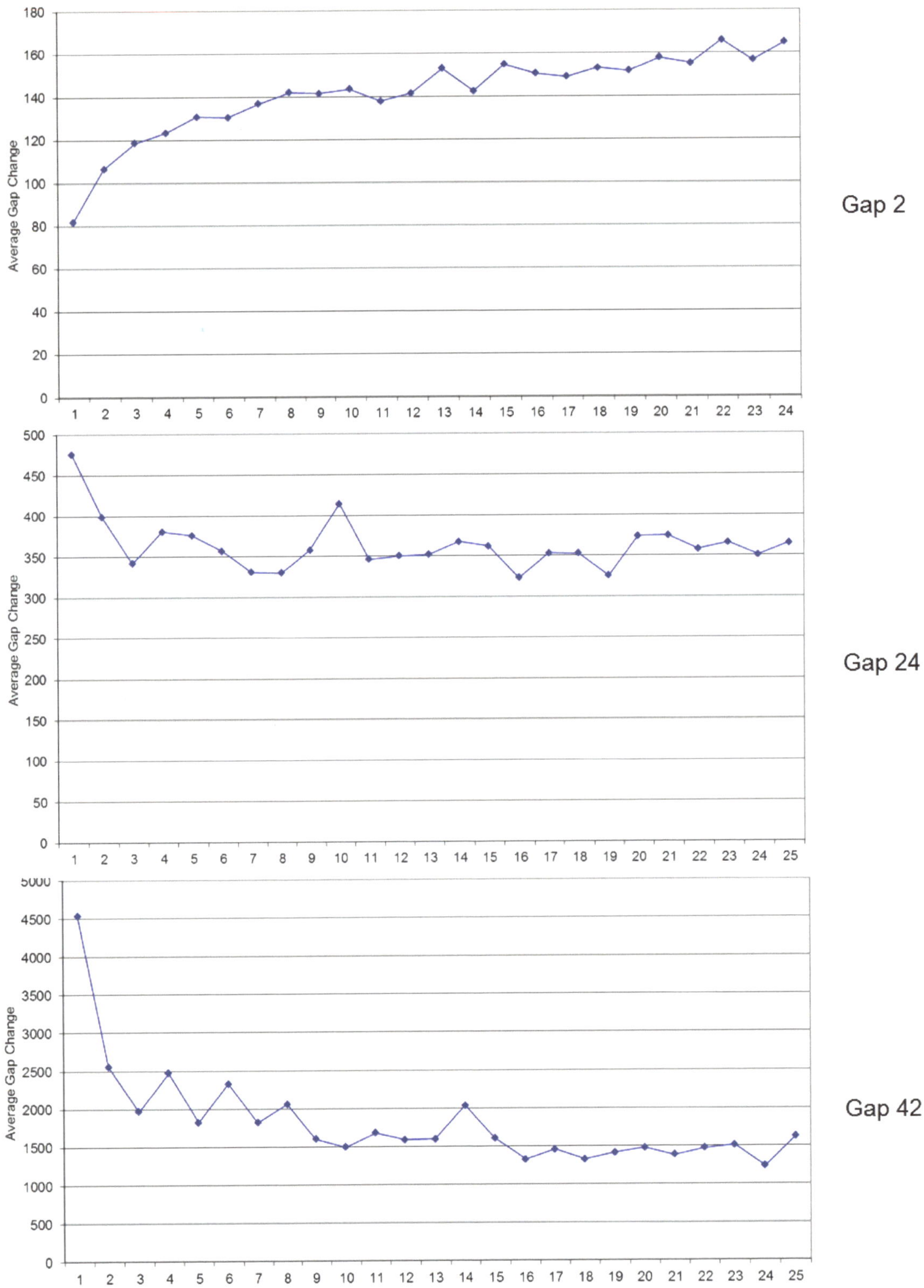

Figure 69

The gap population law explains why the rate of change of gaps is different as shown by gap 2, gap 24 and gap 42. Gap 2 is said to have a positive acceleration, gap 24 a steady acceleration and gap 42 a negative acceleration. This means it takes longer to find the next gap 2, but gap 24 is going at the same pace while gap 42 it takes a shorter and shorter time to find the next gap 42 as x increases. These graphs demonstrate the gap population law at work.

Now what has been considered are characteristic of prime numbers, not pseudoprimes. We can also look at pseudoprimes in a given space, where we assume the space is changing position with respect to x. The generation of pseudoprimes is defined by $f(x_a) \cdot f(x_b)$. Consequently, a different law can also be stated in terms of the pseudoprimes and the average gap to the next pseudoprime.

The minimum gap between any two pseudoprimes is 6 within a given gap base, the same being true for the prime numbers. The minimum difference between gap base 2 and gap base 4 pseudoprime is 2, while diagonally between the bases the minimum is 4. Therefore it implies that as we increase the number of pseudoprimes, the average difference between pseudoprimes will fluctuate between 2 and 4 so as to tend towards an asymptotic straight line of gap 2 while the prime number gap will be increasing.

Pseudoprime Gap Law

The average natural gap to the next pseudoprime decreases tending towards 2 as $f(x) = 6x \pm 1$ *increases.*

This means the pseudoprimes increase as x increases, therefore the prime gap law and the pseudoprime gap law are operating in opposite direction with respect to each other. One is maximizing the gap while the other is minimizing the gap. The two laws emphasize the fact that the two numbers are in the same space fighting for position.

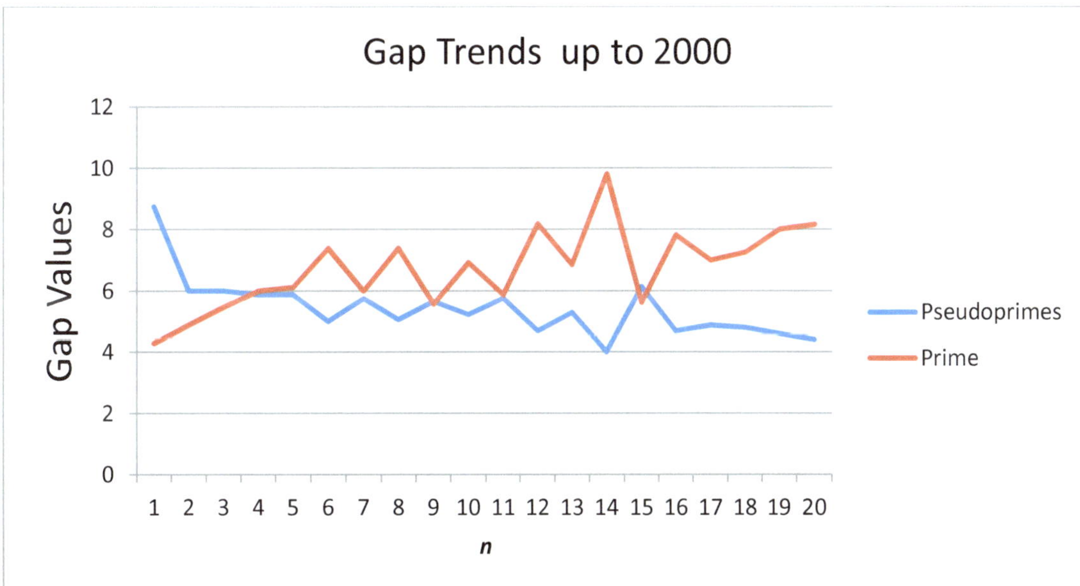

Figure 70

Actually for any pseudoprime values that follow each other, then $W_1 - W_2 = 2$ and $W_2 - W_3 = 4$ where W_1 is in gap base 4, W_2 in gap base 2, and W_3 in gap base 4. Hence $W_1 - W_3 = 6$. This can form a sequence;

$$\sum [W_1, W_2, W_3]_1, [W_1, W_2, W_3]_2 \cdots [W_1, W_2, W_3]_n$$
$$n = x_2 - (x_1 + 1)$$

In other words, we The sequence implies that pseudoprime numbers follow each other in occurrence, where this tant to break up the gap into two and four format in order to see the structure. Typically we know that 6 = 4+2, where this explans why gap 6 fits in either gap base or gap base4. Hence a large gap is made of a large sequence. For example if we have the prime 8467 and the next prime as 8501 with a gap of 34 where $x_1 = 1411, x_2 = 1417$. Hence $n = 1417 - (1411 + 1) = 5$. Therefore the gap is 2(5)+ 4(5+1) = 34.

<table>
<tr><th>Start GB2</th><th>Start GB4</th><th>End GB4</th><th>End GB2</th></tr>
<tr><td colspan="3">$2n + 4(n + 1)$</td><td></td></tr>
<tr><td></td><td colspan="2">$2n + 4n$</td><td></td></tr>
<tr><td colspan="4">$2(n + 1) + 4(n + 1)$</td></tr>
<tr><td></td><td colspan="3">$2(n + 2) + 4(n + 1)$</td></tr>
</table>

Table 95

a. For gap AB, where $x_1 = 56187, x_2 = 56192$. Hence $n = 56192 - (56187 + 1) = 4$. Therefore the gap structure 2(4+2)+4(4+1)=30.
b. For gap BC, where $x_1 = 56192, x_2 = 56198$. Hence $n = 56198 - (56192 + 1) = 5$. Therefore the gap structure 2(5+1)+4(5+1)=38.
c. For gap CD, where $x_1 = 56198, x_2 = 56203$. Hence $n = 56203 - (56198 + 1) = 4$. Therefore the gap structure 2(4)+4(4+1)=28.
d. For gap AD, where $x_1 = 56187, x_2 = 56203$. Hence $n = 56203 - (56187 + 1) = 4$. Therefore the gap structure 2(16)+4(16)=96.

	x	GB4	GB2	
	56186	337115	337117	
A	56187	**337121**	337123	
	56188	337127	337129	
	56189	337133	337135	
	56190	337139	337141	
	56191	337145	337147	
	56192	337151	**337153**	B
	56193	337157	337159	
	56194	337163	337165	
	56195	337169	337171	
	56196	337175	337177	
	56197	337181	337183	
	56198	337187	**337189**	C
	56199	337193	337195	
	56200	337199	337201	
	56201	337205	337207	
	56202	337211	337213	
D	56203	**337217**	337219	

Table 96

AD gives a gap that is in gap base 2 only, while AB gives a gap that is between GB4 and GB2. On the other hand BC gives us a GB2 gap only and CD gives us a gap between GB2 and GB4.

It is also of interest to find out the behavior of the gaps in the respective gap bases. Two graphs are made where we take the maximum gap in a given range. The range is set to 100 for the first graph and 1000 for the next graph. It appears that the larger gaps tend to occur in gap base 4, hence smaller gaps are expected to occur in gap base 2. This is actually confirmed by the third graph. Infact AD is an example of how large gaps can be in gap base 4.

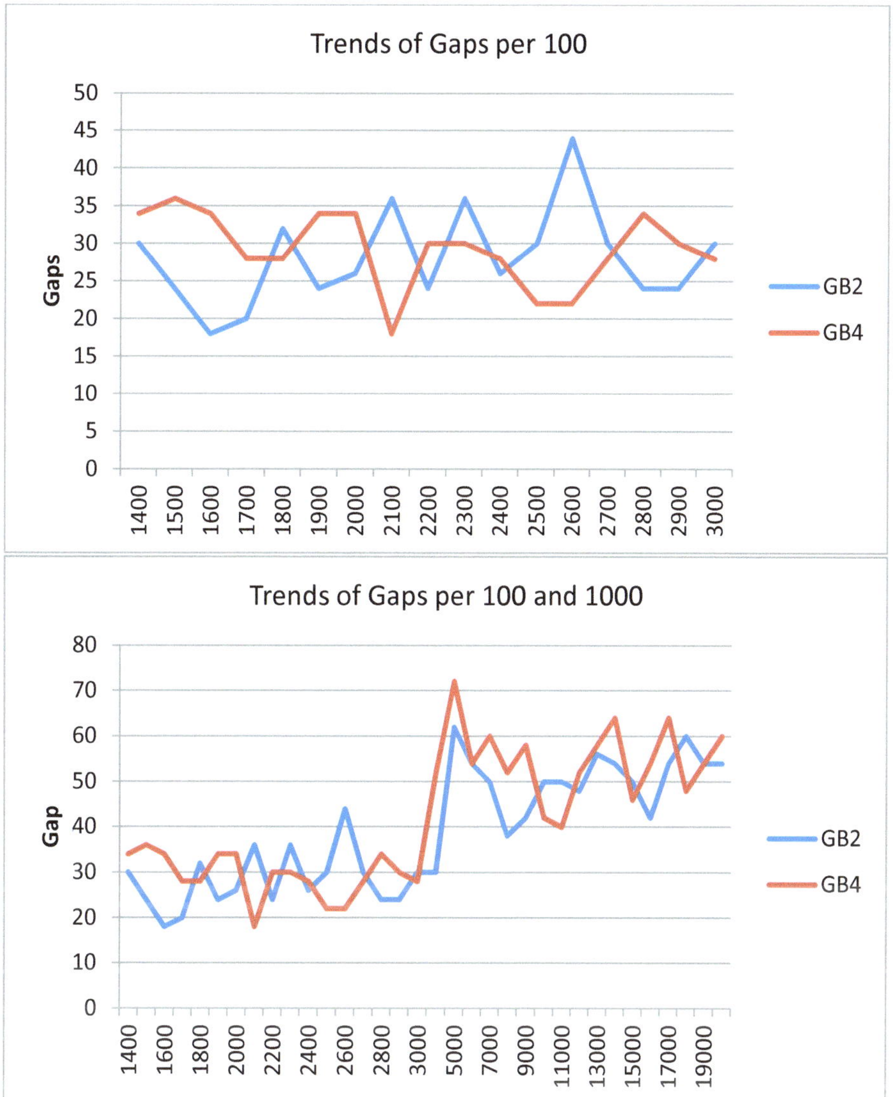

Figure 71

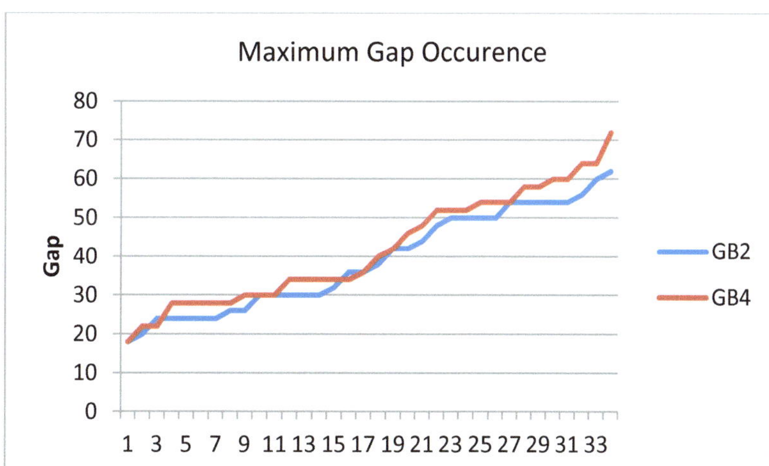

Figure 72

8.102 The Lead and Diagonal Gaps

Let us consider again the gap base and the gaps that occur. All prime numbers have a base gap, where this is defined as the gap between consecutive prime numbers in the same gap base. The base gap gives us a basic approach to understanding how the prime gap builds up through the following law. The building block for prime gaps is 2, 4 and 6 which is a sum of 2 and 4. The "weed" effect is used to described the fact that prime numbers tend to appear anywhere unpredictably. The interest is what actually causes the "weed" effect behavior.

Prime Gap Structure Law

The natural prime number gap increases in multiples of six that are randomized by the inter-gap base relationship.

In other words, it is having the two gap bases that causes the prime numbers to be random, otherwise the behavior would have been constant. This is illustrated by the table that indicates the various gap scenarios that can be defined to understand the distribution in terms of the gap base structure. It is clear that the gap in each base is in multiples of six, and that it is when the prime numbers are in consecutive order that the random process is generated. The natural prime defines the prime numbers in their consecutive form without the gap base structure, where this is said to derive the natural prime gap as a difference between a given prime number and the previous one.

The lead gap is defined as the difference between the a gap base 2 number with a gap base 4 number in the same row, while a diagonal gap is defined as the difference of gap base 2 number with a previous gap base 4 number.

Gap Base 2	Base Gap	Gap Base 4	Base Gap	Lead Gap	Diagonal Gap	Natural Primes	Natural Gap	Natural Primes	Natural Gap
7		11		-4	2	7			
13	6	17	6	-4	2	11	4	101	4
19	6	23	6	-4	8	13	2	103	2
31	12	29	6	2	8	17	4	107	4
37	6	41	12	-4	2	19	2	109	2
43	6	47	6	-4	14	23	4	113	4
61	18	53	6	8	14	29	6	127	14
67	6	59	6	8	14	31	2	131	4
73	6	71	12	2	8	37	6	137	6
79	6	83	12	-4	14	41	4	139	2
97	18	89	6	8	14	43	2	149	10
103	6	101	12	2	8	47	4	151	2
109	6	107	6	2	20	53	6	157	6
127	18	113	6	14	26	59	6	163	6
139	12	131	18	8	20	61	2	167	4
151	12	137	6	14	20	67	6	173	6
157	6	149	12	8	14	71	4	179	6
163	6	167	18	-4	14	73	2	181	2
181	18	173	6	8	20	79	6	191	10
193	12	179	6	14	20	83	4	193	2
199	6	191	12	8	20	89	6	197	4
211	12	197	6	14		97	8	199	2

Table 97

The lead gap shows the relationship between the gap behavior between the two gap bases in terms of increases in magnitude. The diagonal gap measures the same relationship but reveals a very interesting observation besides the fact that it increases faster than the lead gap. That is, from gap base 4 to gap base 2, the gap between any two such prime numbers is generally greater. This indicates a structural difference between gap base 2 and gap base 4. However, the natural gap increases at a slower rate.

A concept of gap acceleration is used to compare how the gap is changing in the given variables. Taking prime numbers from 7 to 211, then we have the following values for gap acceleration, which is defined simply as the average rate of change of the gap.

GB2 - Base Gap	GB4 - Base Gap	Lead Gap	Diagonal gap	Natural Gap
9.714286	8.85714	4.181818	13.42857	4.571429

Table 98

The indication is that at this range, gap base 2 is accelerating slightly faster than gap base 4 in its base gap distribution. The lead gap is the slowest and the diagonal gap is the fastest. It is expected that the lead gap should increase faster than the natural gap, so at larger values this will change. It is low because it began with negative values. Whilst the lead gap compares generation of prime numbers between the two gap bases, the diagonal gap shows how the two gap bases relate to each other in terms of gap. Hence from the diagonal gap, there is an increasing widening of the gap between the two gap bases. This difference in gap base behavior from prime gaps shows the significance of the structure in understanding distribution and behavior. Even in finding prime factors, the speed of finding factors is not the same.

A similar exercise can be done for the pseudoprimes up to 211, and the following result is obtained. The measurement of change in the average natural gap for the pseudoprime is small, and does not throw much light on what is happening. However, when we use the different types of gap definitions such as lead gap, we are able to see trends much more clearly. It is evident that in terms of the natural gap, the pseudoprime has the same type of structure as the prime numbers, indicating how closely related the numbers are.

Gap Base 2	Base Gap	Gap Base 4	Base Gap	Lead Gap	Diagonal Gap
25		35		-10	
49	24	65	30	-16	14
55	6	77	12	-22	-10
85	30	95	18	-10	8
91	6	119	24	-28	-4
115	24	125	6	-10	-4
121	6	143	18	-22	-4
133	12	155	12	-22	-10
145	12	161	6	-16	-10
169	24	185	24	-16	8
175	6	203	18	-28	-10
187	12	209	6	-22	-16
205	18				

Psdprimes	Natural Gap
25	
35	10
49	14
55	6
65	10
77	12
85	8
91	6
95	4

Psdprimes	Natural Gap
115	20
119	4
121	2
125	4
133	8
143	10
145	2
155	10
161	6
169	8
175	6
185	10
187	2
203	16
205	2
209	4

Table 99

The lead gap for the pseudoprimes is increasing in the negative showing that gap base 4 generates pseudoprimes faster than gap base 2. However, the gap between gap base 4 and gap base 2 pseudoprime is narrowing slightly as the pseudoprimes increase as indicated by the diagonal gap. The average gap is found for each column and the results are shown as follows.

GB2 – Base Gap	GB4 – Base Gap	Lead Gap	Diagonal Gap	Natural Gap
15	15.81818	-18.5	-3.45455	7.9375

Table 100

Gap base 4 is having a slightly faster gap acceleration compared to gap base 2 for the base gap, which is not surprising and corresponds to the behavior of the prime gap acceleration. However, note that the acceleration for the pseudoprimes is higher than for the prime numbers, which means the prime numbers are under pressure from the pseudoprimes. In the lead gap, there is a negative acceleration on the pseudoprimes, which explains why the prime gap acceleration is comparatively low. In terms of generation of pseudoprimes, there is a marked difference between the two gap bases as shown by the negative values. The value of the diagonal acceleration is lower than the lead gap acceleration, showing that for the pseudoprimes the gap bases are coming closer together. This is an opposite behavior from the prime numbers. The natural gap acceleration of the pseudoprimes is higher than the prime gap shows more activity than the prime numbers. The average natural gap for prime numbers will continue to increase while the pseudoprime average natural gap will tend towards 3.

As a result, using this approach, we can see how two different types of numbers that are related are behaving in their distribution in the same prime space $[a, b]$.

8.2 The Co-existence Theorem

The above laws illustrate a very close mutual relationship between the set of prime numbers and pseudoprimes where this can be expressed as a theorem. The laws imply that we cannot separate the two types of numbers, especially in the context of studying the distribution behavior of prime numbers.

The Co-existence Theorem

For every (x, y_1, y_2), then (y_1, y_2) are either a pseudoprime or a prime number.

The theorem provides an exact relationship between primes and pseudoprimes. Its first application is the development of the primality test using the algebraic sieve. An example of the law of primes is illustrated as in the table for $f(-1) = 5$. Note that the difference in each case is $6 \cdot 5 = 30$, which implies all these positions are taken up as a result of the existence of the prime factor 5.

Constant at $f(-1) = 5$	Gap Base 2 $f(-1) \cdot (6x + 7), -1 \leq Int(x) \leq \infty$	Gap Base 4 $f(-1) \cdot (6x + 11), -1 \leq Int(x) \leq \infty$
5	5 × 1 = 5	5 × 5 = 25
5	5 × 7 = 35	5 × 11= 55
5	5 × 13 = 65	5 × 17 = 85
5	5 × 25= 125	5 × 23 = 115
5	5 × 31= 155	5 × 29 = 145
5	*Continues to infinity*	*Continues to infinity*

Table 101

The prime number 7 would result in $6 \cdot 7 = 42$, that is a gap of 42. Therefore, as $f(x_a) \cdot f(x_b)$ increases, the number of primes is a given range decreases, and the gap between them increases.

GB-4	GB-2
5	7
11	13
17	19
23	25
29	31
35	37
41	43
47	49

Prime 5&7	Product
5,5	25
5,7	35
7,7	49

Table 102

Let us consider $y = 6x + 7$ and $y = 6x + 11$ as our sieve generator, then for $x = 0$, we have the primes $(5,7)$ as our first result. We could consider (1,5) as our first pair, but since 1 is an identity, (5,7) as the first pair does not create any contradiction. Consequently, they are the first primes to generate any pseudoprime, and these pseudoprimes are 25,35 and 49. This is a multi-value function, that is, x gives us two results. We take the larger value to determine the prime boundary of the pseudoprime generation, in this case 7. The prime boundary is therefore $7^2 = 49$, hence we can find the number of pseudoprimes generated for this prime boundary as 3.

Pseudoprime Generation with prime boundary 169.

1	5	7
2	11	13
3	17	19
4	23	25
5	29	31
6	35	37
7	41	43
8	47	49
9	53	55
10	59	61
11	65	67
12	71	73
13	77	79
14	83	85
15	89	91
16	95	97
17	101	103
18	107	109
19	113	115
20	119	121
21	125	127
22	131	133
23	137	139
24	143	145
25	149	151
26	155	157
27	161	163
28	167	169

5,5	25
5,7	35
7,7	49
5,11	55
5,13	65
7,11	77
5,17	85
5,19	95
7,17	119
7,19	133
5,23	115
5,25	125
7,23	161
7,25	175
5,29	145
5,31	155

Primes	CODE
5,7	
11,13	
5,7 & 11,13	

Table 103

After (5,7) we have (11,13) as the next pair. Therefore $13^2 = 169$ as our prime number boundary. An alternative method is to use a given range $[a, b]$ where $a = 0$ as a means to find out the number of primes and pseudoprimes, and demonstrate that the pseudoprimes are increasing relative to the prime numbers. This is a unique feature of the algebraic sieve.

X Range	Prime Boundary	Square of Prime	Primes Gap Base 2	Primes Gap Base 4	Total Primes	Pseudoprimes Gap Base 2	Pseudoprimes Gap Base 4	Total Pseudoprimes	Pseudoprime - prime
[0,5]	5	25	5	5	10	1	1	2	-8
[0,8]	7	49	6	8	14	3	1	4	-10
[0,28]	13	169	18	20	38	11	9	20	-18
[0,60]	19	361	34	37	71	27	24	51	-20
[0,104]	25	625	55	58	113	50	47	97	-16
[0,160]	31	961	78	83	161	83	78	161	0
[0,228]	37	1369	105	113	218	124	116	240	22
[0,308]	43	1849	139	142	281	170	167	337	56
[0,400]	49	2401	175	180	355	226	221	447	92
[0,504]	55	3025	209	223	432	296	282	578	146
[0,620]	61	3721	257	261	518	364	360	724	206
[0,748]	67	4489	298	310	608	451	439	890	282

Table 104

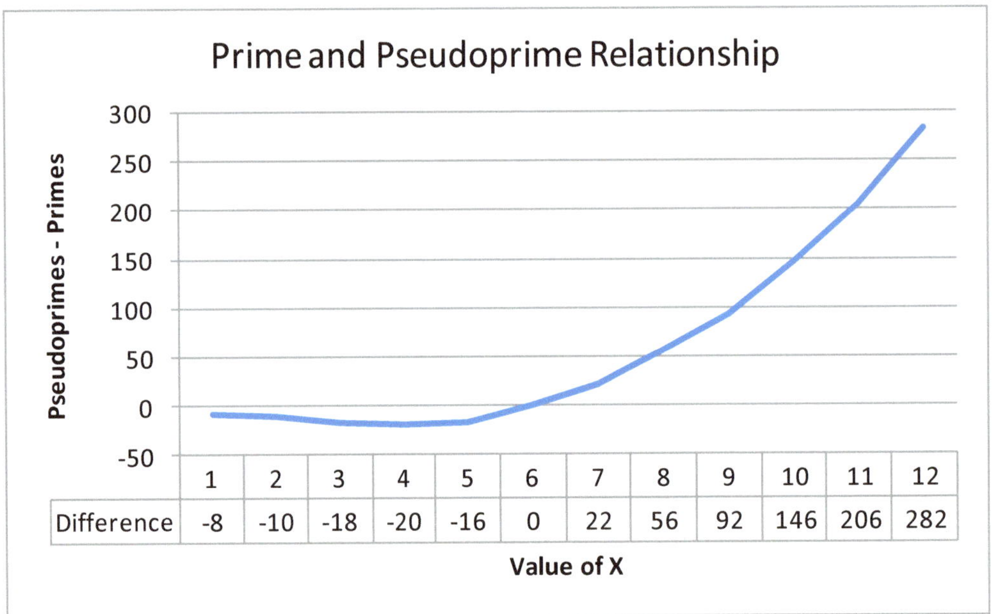

Figure 73

The difference relationship in terms of a trend line is approximated as;

$$Difference = 4.2318x^2 - 30.258x + 28.455$$

This gives us the approximate rate of change of $8.46x - 30.26$. This is a guess approximation, and will differ depending of the range chosen. It serves to illustrate the fact that pseudoprimes are increasing faster with respect to prime numbers. Later such a difference is defined in terms of a given space at position x.

8.3 Fixed Space Distribution Law

Often the major interest as indicated is always on the distribution of prime numbers. However, the focus now is a distribution of both primes and pseudoprimes since it is very clear that they influence each other. The distribution of prime numbers and pseudoprimes is considered in terms of the gap base.

Consider a distribution from *x=20 000* to *x=25 000*. We can look for primes in gap base 2 and gap base 4. For example, we have 277 primes of gap 2 in gap base 2, and 274 primes of gap 4 in gap base 4.

	GB2	GB4	
Gap 2	277	274	Gap 4
Gap 6	244	228	
Over Gap 6	752	774	
Total primes	1273	1276	
Pseudoprime	3728	3725	
Space	5001	5001	

Table 105

The notable observation is the fact that the distribution between the gap bases is always more or less the same as shown by the graph. The total number of pseudoprimes is also more or less the same between the gap bases. This is indeed a surprising behavior for a space where we assume a random behavior. Secondly, the distribution always occurs over a fixed space that is defined by the sum of the primes and pseudoprimes. This is a major difference in the analysis of prime number distribution; the fact that it is always considered over a fixed space.

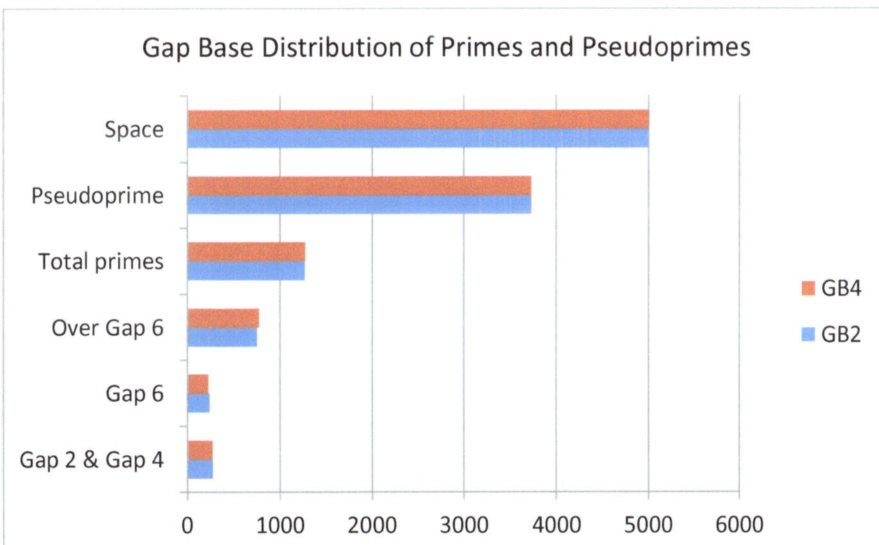

Figure 74

This is almost like the law of primes that also informs us about the distribution of prime numbers with respect to pseudoprimes. The difference is the emphasis that the primes and pseudoprimes are always considered to exist in a fixed space. This implies that whenever we study them, our conclusions should be based on the fact that they exist in a fixed space.

Fixed Space Distribution Law

There exists a self-determining fixed space [a,b] consisting of prime and pseudoprime numbers.

This law can be illustrated by taking values from any point and choosing a fixed range. This example consider from x = 20 000 to x = 25 0000 up to 60 000.

x	20000	25000	25000	30000	30000	35000	35000	40000
Gap Base	GB2	GB4	GB2	GB4	GB2	GB4	GB2	GB4
Gap 2	277	274	276	271	276	267	250	262
Gap 6	244	228	221	225	221	212	195	206
Over Gap 6	752	774	743	759	747	742	752	751
Total primes	1273	1276	1240	1255	1244	1221	1197	1219
Pseudoprime	3728	3725	3761	3746	3757	3780	3804	3782
Space	5001	5001	5001	5001	5001	5001	5001	5001

x	40000	45000	45000	50000	50000	55000	55000	60000
Gap Base	GB2	GB4	GB2	GB4	GB2	GB4	GB2	GB4
Gap 2	250	252	241	237	279	246	229	254
Gap 6	207	239	204	206	194	224	197	197
Over Gap 6	754	720	729	738	738	726	743	734
Total primes	1211	1211	1174	1181	1211	1196	1169	1185
Pseudoprime	3790	3790	3827	3820	3790	3805	3832	3816
Space	5001	5001	5001	5001	5001	5001	5001	5001

Table 106

A graphical presentation is as shown. Note the fixed space that it is equal and the distribution pattern. This is represented by the outer circle. The inner circle demonstrates the distribution with respect to prime and pseudoprimes, and the relative totals. The pattern repeats as GB2-GB4. It is evident that the pattern is more or less the same in terms of distribution.

The circle gives an interesting presentation of the relationship arising from the distribution and behavior because this is a fixed space. Each of the inner circle strips start from the inside to show relationship.

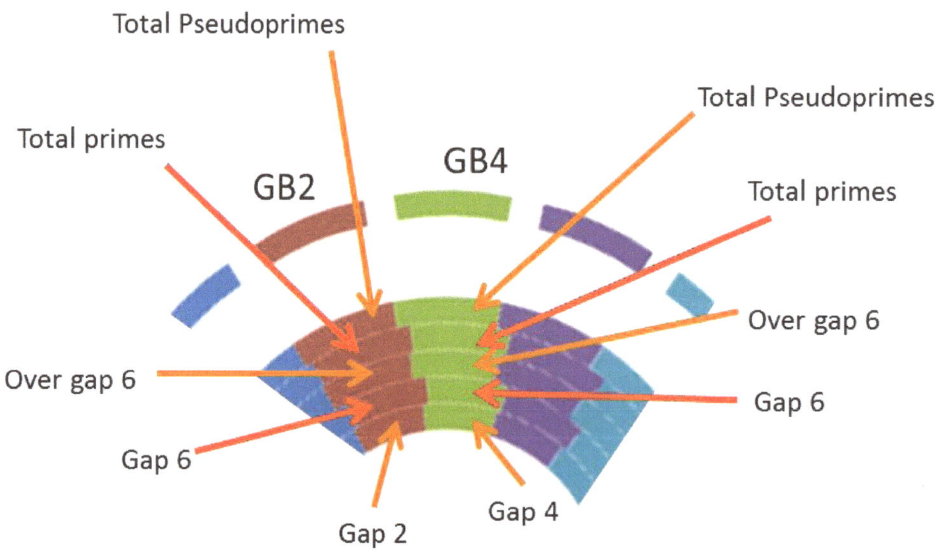

Figure 75

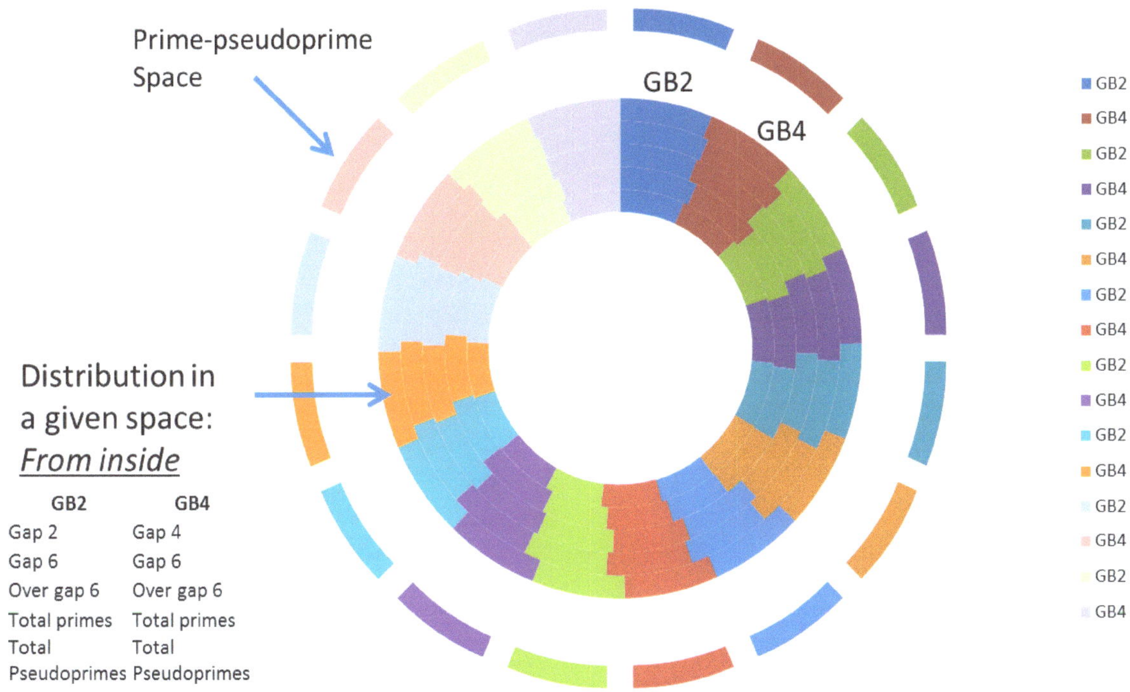

Figure 76

This law affects interpretation and application of prime number behavior, and consequently the distribution.

8.4 The Algebraic Prime Space

Using the Law of Primes, the prime numbers are directly connected to the pseudoprimes in terms of their occurrence. The two numbers seemingly fight for the same space because they are connected algebraically, while the composite is independent from the prime number. The prime number behavior in this context can be studied by using a concept of a fixed space, which is an application of the fixed space distribution law.

Now for gap base 2, $f(a) = 6a + k_1$ and for gap base 4 we have $f(b) = 6b + k_2$, where k_1 and k_2 determine the equations used to generate the gap base, such that $k_1 < k_2$ and they are either prime or pseudoprime. An example is $6x + 7$ for gap base 2 and $6x + 11$ for gap base 4. We have $k_1 > k_2$ when we include negative values, such as $6x \pm 1$ where $k_1 = 1,\ k_2 = -1$. Furthermore, we assume that $a < b$, then;

$$f(b) - f(a) = 6(b - a) + k_2 - k_1$$

Therefore for the algebraic prime space, we assume that $f(a)$ always starts in gap base 2 and that $f(b)$ is in gap base 4 in order to simplify our assumptions. Alternatively, one can start at gap base 4 and end at gap base 2 as shown in the above table.

Now let $a = x_a$ and $b = x_b$ then the number of positions in the sieve from x_a to x_b is found as follows. From $x_b - x_a$ we get the number of sieve positions less 1, hence we add 1, then multiply by 2 since the model has gap base 2 and 4. Therefore;

$$2(x_b - x_a + 1) = c$$

Prime
Pseudoprime

x	sieve	
23	145	149
24	151	155
25	157	161
26	163	167
27	169	173
28	175	179
29	181	185
30	187	191
31	193	197
32	199	203
33	205	209
34	211	215
35	217	221
36	223	227

15 primes
13 pseudoprimes
c = 28 = (36-23+1)*2

x	sieve	
40	247	251
41	253	257
42	259	263
43	265	269
44	271	275
45	277	281
46	283	287
47	289	293
48	295	299
49	301	305
50	307	311
51	313	317
52	319	323
53	325	329

15 pseudoprimes
13 primes
c = 28 = (53-40+1)*2

Table 107

Now c will always be constant for the chosen values of x_a to x_b provided this difference is kept constant, therefore this is said to construct a notion of a space $[a, b]$. This space can occupy a position x on the sieve, and x can be variable. For example in the above table, $x_a = 23$ and $x_b = 36$ therefore position $x = 23$. Now we can have the same space defined at $x = 40$, such that c is the same value, what differs is the distribution of the primes and pseudoprimes. Hence it is like we are sliding down a given region on the sieve.

Definition

For a range $[a, b]$ let there be an algebraic space $[a, b, x]$, where such a range is constant from $(b - a)$ but varies its position in x.

We can then consider the co-existence theorem in a space$[a, b, x]$ which states that any position is either occupied by a prime number or a pseudoprime. Since the space is a fixed structure at a given x, then at any point in the prime number distribution, the sum of the primes and pseudoprimes is a constant for the given range difference.

That is, if the number of primes is n_1 and the number of pseudoprimes is n_2, then we have;

$$n_1 + n_2 = c$$

As already noted c is a constant for the range difference $(b - a)$. Similarly;

$$2(x_b - x_a + 1) = n_1 + n_2$$

Therefore, in a given range difference, the probability of getting a prime is $\frac{n_1}{c}$ and a pseudoprime $\frac{n_2}{c}$. We note that $n_1 < n_2$ for a reasonably large range. Now as x increases, the value of n_2 also increases. This then demonstrates that the generation of pseudoprimes affects the distribution of the prime numbers in an inverse relation.

A pseudoprime therefore is not independent of the prime number in terms of its behavior and distribution as shown by the constant c. In studying numbers, this implies three forms of numbers; primes, pseudoprimes, and composites.

Now $f(b) - f(a) = 6(b-a) + k_2 - k_1$, which means that this is also a constant if a and b are fixed. Hence we have;

$$6(b-a) + k_2 - k_1 - 2(b-a+1) = 2(2b-2a-1) + k_2 - k_1$$

Hence let;

$$n_3 = 2(2b-2a-1) + k_2 - k_1$$

a) $k_2 - k_1 = 2$, if we start with gap base 4
b) $k_2 - k_1 = 4$, if we start with gap base 2

Then this is also constant for a given space $[a,b]$. It turns out that this constant is the number of composites in $[a,b]$. Then it follows that, for a given range we have the following distributive behavior that summarizes the law of integer distribution.

Law of Integer Distribution

For the range $[a,b]$ then $f(b) - f(a) = n_1 + n_2 + n_3$ where n_1 is the number of primes, n_2 the number of pseudoprimes, n_3 the number of composites and $f(b) - f(a)$is a constant.

That is the sum of primes, pseudoprimes and composites is defined by the given range. Now we know that $n_1 + n_2 = c$ where c is a constant. This value c is the same for both gap base 2 and gap base 4, therefore n_3 is the same for both gap bases. Similarly, for a given range $[a,b]$, then it means $f(b) - f(a)$ is also a constant.

Consider [a,b] = [10,24]. Let $k_1 = 7,\ k_2 = 11$, hence we have equations $6x + 7$ for gap base 2 and $6x + 11$ for gap base 4. Then the total primes and pseudoprimes will always be 30 for this range, that is;

$$2(x_b - x_a + 1) = 2(14+1) = 30$$

For the composites, then;

$$2(2b-2a-1) + k_2 - k_1 = 2(48-20-1) + 11 - 7 = 58$$

However, in each gap base we will have 58/2 = 29 composites. The total constant for the space is 30+58 = 88, where this is $f(b) - f(a) = 6x_b + 11 - (6x_a + 7)$.

$$f(b) - f(a) = 155 - 67 = 88$$

Range	Primes	Pseudoprimes	Total	Composites	Constant Sum
[10,24]	18	12	30	58	88
[19,33]	16	14	30	58	88
[60,74]	15	15	30	58	88
[560,574]	8	22	30	58	88
[1560,1574]	12	18	30	58	88
[11560,11574]	9	21	30	58	88
[211560,211574]	6	24	30	58	88

Table 108

The dynamics of the prime space are indeed very interesting, the fact that the position x of the space can be anywhere on the algebraic sieve but some descriptions remain constant.

For a given range [a,b];

- Let the number of primes in gap base 2 be k
- Let the number of primes in gap base 4 be l

then in terms of the distribution of primes, it has been observed that $k \approx l$.

Hence we can always estimate the number of primes in another gap base since they are approximately equal through $0.5n_1$. For example for [10,24], [19,33], and [1560,1574] the number of primes are equal in each gap base. This leads to the observation that one can search for such algebraic prime spaces. That is, find two prime spaces with the same number of primes.

Range	Primes Gap Base 2	Primes Gap Base 4	Pseudoprimes Gap Base 2	Pseudoprimes Gap Base 4
[10,24]	9	9	6	6
[19,33]	8	8	7	7
[60,74]	8	7	7	8
[560,574]	3	5	12	10
[1560,1574]	6	6	9	9
[11560,11574]	4	5	11	10
[211560,211574]	3	3	12	12

Table 109

It is quite interesting to note that c is constant whether you are in gap base 2 or 4. For example for [560,574] we have (3+12) in gap base 2 and (5+10) in gap base 4.

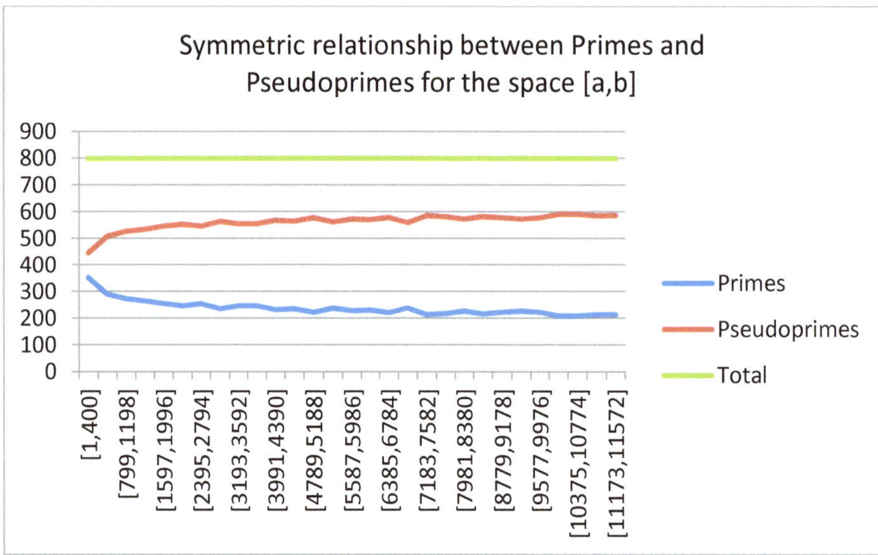

Figure 77

We also note that $n_1 + n_2 = c$ produces a symmetric relationship between primes and pseudoprimes in terms of distribution and behavior. An example is given for $[a, b] = [1,400]$ where $c = 800$.

This pattern will be generally true for any space, and clearly shows that pseudoprimes increase while primes decrease. Another approach is to designate each space in terms of n as a position, for example $n = 1$ would represent [1,400], $n = 2$ for [799,1198] and so on.

8.5 The Prime Number Probability Index

In the algebraic space, the way prime numbers, pseudoprimes and composites are distributed is directly related to our choice of x_a andx_b. The algebraic space also derives a probability space because our variables c and n_3 are fixed at any value of x for a given $[a, b]$. This derives the following probabilities of occurrence for a given a range $[a, b]$.

$$P(prime) = \frac{n_1}{f(b) - f(a)}$$

$$P(pseudoprime) = \frac{n_2}{f(b) - f(a)}$$

$$P(composite) = \frac{n_3}{f(b) - f(a)}$$

This approach gives us an alternative way of investigating the density of primes. This is typically given as an approximation $\pi(x) = \frac{x}{Log\, x}$, called the prime number theorem, and the notation $\pi(x)$ denotes the number of primes less than or equal to x. The range $[a, b]$ gives us an opportunity to study the prime density over a given range. Consequently, an abstract measure called the P-Index is defined that is related to the probability of occurrence of a prime, a pseudoprime, and a composite.

Definition

Let the probability of occurrence of prime, pseudoprime and composite in the range [a,b] be T*, then the P-index is given by* $10000T$.

This will make the index to vary between zero and two or three hundred. For example if we define a range $x = 10$, and $x = 24$, that is $[a, b] = [10,24]$, then $f(b) - f(a) = 88$. The table shows the total primes, pseudoprimes and composites, and the P-Index. An example of a calculation is given for the space [19,33].

Space [19,33]	Occurrence	Probability
Prob(Prime)	16/88	0.181818182
Prob(Pseudoprime)	14/88	0.159090909
Prob(Composite)	58/88	0.659090909
Prob of Occurrence	Product	0.019064613
P-Index	x10 000	190.6461307

Table 110

Note that the range is kept constant, and the total number of primes plus pseudoprimes and composites remains constant. However, the P-Index is not constant because of the variation of the primes in the given range. Therefore it accurately reflects the dynamics of density behavior of the prime numbers relative to the composites.

Range	P-Index
[10,24]	183.8373
[19,33]	190.6461
[60,74]	191.4972
[560,574]	149.7934
[1560,1574]	183.8373
[11560,11574]	160.8577
[211560,211574]	122.5582

Table 111

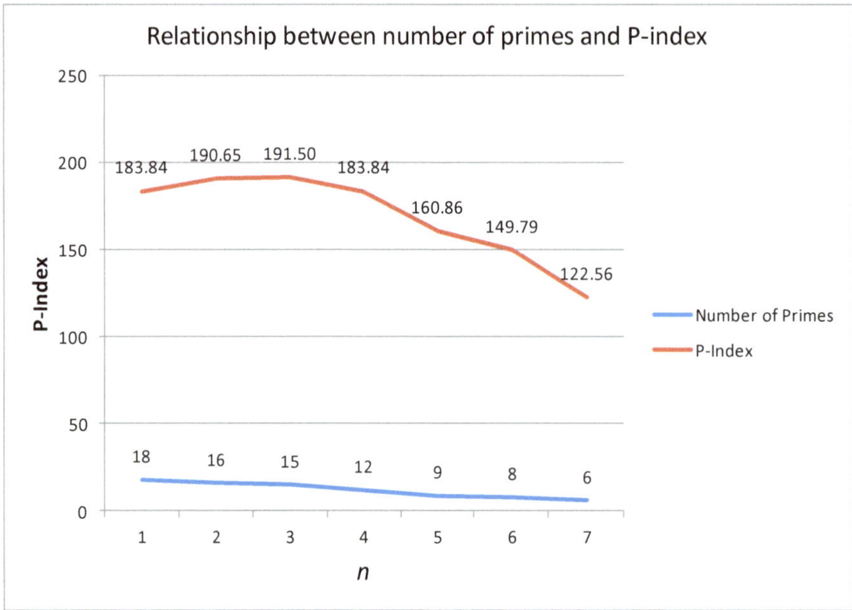

Figure 78

The P-index is a downward trend that follows the behavior and distribution of the prime numbers.

8.6 Prime Space Search

We note that the space $[a, b]$ occurs at position x, and it can start at any position. For example we can have [10,24] and then [15,29]. In that case we can define in general the algebraic prime space $[a, b, x]$ starting at any position x, and the value of x can be random. Alternatively, instead of being random, we can define x to be regular from a given initial point. This means the value of a or b can be calculated for regular intervals. For a given $[a, b]$, one can search for a certain number of primes and pseudoprimes, where the result is called a prime space. It is called a prime space because we try to look for the closest value of primes in the given range $[a, b]$, and in some way this makes $[a, b]$ a prime space because of a specific position and condition.

The notation we use for prime space is $[Space: n_1, n_2]$, where n_1 is the number of primes and n_2 is the number of pseudoprimes. When we search for the prime space, we can find our exact specification for the space, hence the find is referred to as an exact space. However, it can happen that we do not find an exact space but instead get something close. This is referred to as a nearby prime space.

Therefore a prime space is found in an algebraic space $[a, b, x]$;

$$f(b) - f(a) = [Space: n_1, n_2] + n_3$$

In order to define the search, we use the concept of a cumulative space. Such a space always starts at $x = 1$, that is $a = 1$. Secondly, we note that there are many equations that give the algebraic sieve. In this case we assume that the generator is $f(x) = 6x + 7$ and $f(x) = 6x + 11$. Therefore $x = 0$ gives 7 and 11, but $x = 1$ gives 13 and 17. Hence $a = 1$ starts counting primes from 13 and 17. If we use $f(x) = 6x \pm 1$, then the start point will be different, and the algebraic space will count accordingly but the basic equation format remains the same.

Our first space therefore is defined as $[1, b]$ and $(b-1)$gives the depth of the space. In general this is defined with the notation $Cum[a, b]$, where $2(b-a+1) = n_1 + n_2$. The start point for the next space will then be a sum of $(b-1)$ to each component of [a,b]. Hence;

$$[a, b] + [b-1, b-1] = [1, b] + [b-1, b-1]$$

This simplifies to $[b, 2b-1]$. If we add the third space $(n = 3)$, then we have;

$$[b, 2b-1] + [b-1, b-1] = [2b-1, 3b-2]$$

Now in $2b-1, 2 = 3-1 = n-1$, and $1 = 3-2 = n-2$. Hence

$$2b-1 = (n-1)b-(n-2)$$

Similarly for $3b = nb$, and $2 = 3-1 = n-1$. Hence $3b-2 = nb-(n-2)$. Therefore, in general for the cumulative space:

$$Cum[a, b] = [1, b] + \sum[(n-1)b-(n-2), nb-(n-1)], \text{ where } 2 \leq n \leq \infty$$

The value of n starts at 2 since $[1, b]$ is already pre-defined. This value of n counts the number of spaces until the find is made. The variable part is $[(n-1)b-(n-2), nb-(n-1)]$, and this is what we use to find a given space. The tool that we use to search for the prime space is $c = n_1 + n_2$ as this defines the prime space. The depth of the space cannot be more than c, therefore for the cumulative space, we set $b = n_1 + n_2$. Consequently this is called a fixed cumulative space since $b = c$.

For example consider $[Space{:}\ 150,300]$, this means the space has 150 primes and 300 pseudoprimes, hence $a = 1$ and $b = 450$. When you perform a search to see is such a space exists, then we find $[a, b] = [1348, 1797]$ as the exact space that suits the given condition. In other words in this range we will find 150 prime numbers and 300 pseudoprimes. The search takes place in either gap base 2 or gap base 4. We can actually calculate the number of composites beforehand using the equation. That is,

$$4(450) - 4(1) + 4 - 2 = n_3$$

$$n_3 = 1798$$

The number of composites will be constant at any value of $[a, b]$, and therefore [1348,1797] will have the same value of composites as [1,450]. Similarly for example $n = 3$ gives;

$$Cum\ [a, b] = [(n-1)b-(n-2), nb-(n-1)]$$

$$Cum\ [a, b] = [(3-1)450-(3-2), 3(450) - (3-1)]$$

$$Cum\ [a, b] = [899, 1348]$$

Therefore we can express as summation $Cum[a, b] = [1{,}450] + [450{,}899] + [899{,}1348]$.

Results Summary

	Result
N	4
Gap 2 Primes	41
Gap 6 Primes	31
Over Gap 6 Primes	78
Total	150
Prime Space	[1,150]
Nearby Space	[1348,1797]
Exact Space	Yes
Composites	2244
Pseudos	300

Ranges

n	a	b
1	1	450
2	450	899
3	899	1348
4	1348	1797

Figure 79

But if you seek $[Space\colon 150{,}350]$, you will get a nearby space with 149 primes and 351 pseudoprimes. What this means is that you cannot get the space you are actually looking for, but a nearby space is the closest you can come to considering the given conditions. Whereas it took $n = 4$ to get the exact space, it took $n = 7$ to get a nearby space. Note that we did not change the number of primes, only the pseudoprime value was changed. Similarly for example $n = 5$ gives, $b = 500$;

$$Cum\ [a,b] = [(n-1)b-(n-2), nb-(n-1)]$$

$$Cum\ [a,b] = [(5-1)500-(5-2), 5(500) - (5-1)]$$

$$Cum\ [a,b] = [1997, 2496]$$

Results Summary

	Result
N	7
Gap 2 Primes	42
Gap 6 Primes	33
Over Gap 6 Primes	74
Total	149
Prime Space	[1,149]
Nearby Space	[2995,3494]
Exact Space	No
Composites	2494
Pseudos	351

Ranges

n	a	b
1	1	500
2	500	999
3	999	1498
4	1498	1997
5	1997	2496
6	2496	2995
7	2995	3494

Information

Switch to Gap Classification Tab for detail.

Figure 80

This shows the inter-relationship between the primes and pseudoprimes in terms of the prime space search.

8.7 Prime Space as Count Function

There are several ways to define a prime count, most of which are approximating functions. However, the concept of prime space makes a more precise approach towards a prime count function through the relationship that we have in the three types of numbers, that is; $f(b) - f(a) = [Space{:}\, n_1, n_2] + n_3$.

Solving the space component we get;

$$[Space{:}\, n_1, n_2] = f(b) - f(a) - n_3$$

Consequently, if we specify the algebraic space function $f(b) - f(a)$, and specify the value of composites n_3 that we want, then we know that the total prime and pseudoprime count will be $[Space{:}\, n_1, n_2]$. For example consider $a = 56$ and $b = 120$. Then $f(b) = 6(120) + 11 = 731$, and $f(a) = 6(56) + 7 = 347$. Hence $f(b) - f(a) = 347$. Therefore we know that $[Space{:}\, n_1, n_2] = 65$ exactly. This means we can actually define another version of the prime space called the variable cumulative space as shown.

Definition

The variable cumulative space where $1 \leq n \leq \infty$ is given by;

$$Cum[a, b] = [1, nb]$$

Since a is constant, this means we define a new space each time through nb by varying n. For example, consider $b = 100$, then we observe that composite numbers increase linearly as we increase b, the same being true for prime numbers and pseudoprimes.

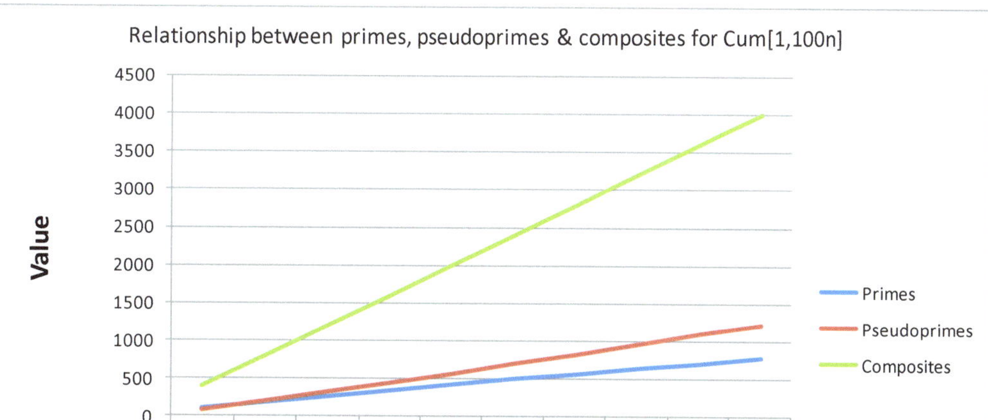

	1	2	3	4	5	6	7	8	9	10
Primes	106	192	275	353	427	499	571	642	708	780
Pseudoprimes	94	208	325	447	573	701	829	958	1092	1220
Composites	398	798	1198	1598	1998	2398	2798	3198	3598	3998

Figure 81

However, the P-index shows that there is a difference even though the relationship appears linear. The other point of interest about the cumulative space is the fact that linearity means approximating descriptive equations can be used to define relationship between the three types of numbers for a given space. For example, for $[a, b] = [1,100n]$ $1 \leq n \leq \infty$, the trend lines for Estimate-1 equations are;

1. $Composites = 400n - 2$
2. $Pseudoprimes = 125.82n - 47.333$
3. $Primes = 74.18n + 47.333$

Therefore adding we get $Primes + Pseudoprimes = 200n$. Hence for $n = 1000$, then the approximate count is as follows;

a) $Composites = 399\,988$
b) $Pseudoprimes = 125{,}773$
c) $Primes = 74{,}227$

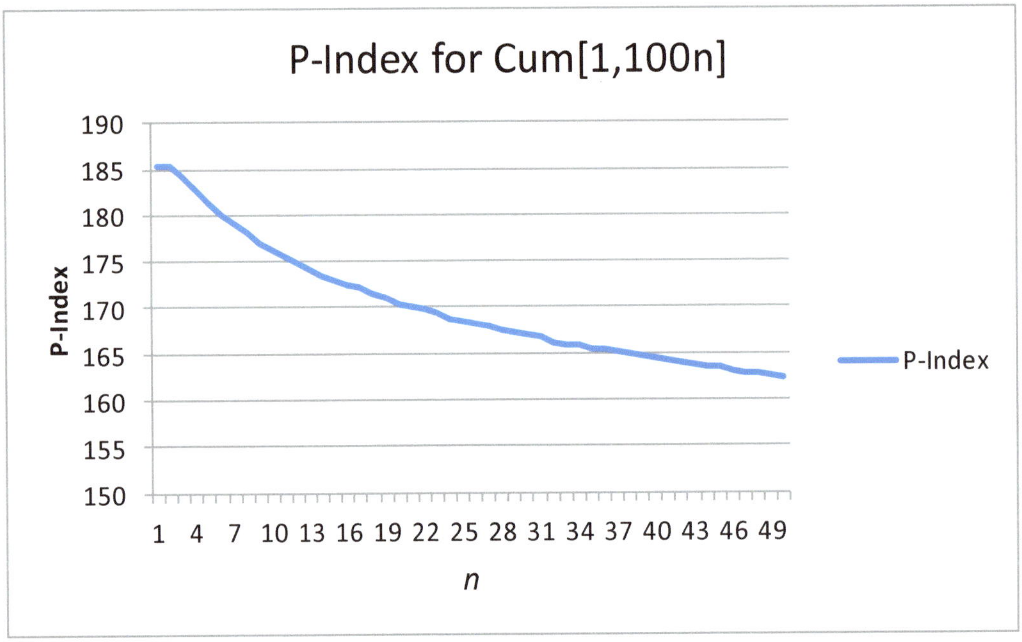

Figure 82

Increasing n to 20 gives us Estimate-2 equations, which is a better approximation for the count. That is;

1. $Composites = 1000n - 12$
2. $Pseudoprimes = 130.82n - 73.858$
3. $Primes = 69.185n + 73.858$

We note that the composite count function remains unchanged, and similarly, the $Primes + Pseudoprimes = 200.01n$, which is technically the same as the previous one. This is to be expected since this is also a constant $n_1 + n_2$. Let $n = 1000$, then we get the approximations;

1. $Composites = 399\,988$
2. $Pseudoprimes = 130{,}746$
3. $Primes = 69{,}259$

We assume that the more accurate value is 69,259 for the prime numbers compared to 74,227.

Our interest is drawn to the sum $n_1 + n_2$ in relation to b. If $b = 100$, then we have $n_1 + n_2 = 200n$, if we substitute $a = 1$ and $b = 100n$ in $2(x_b - x_a + 1)$. That is $n_1 + n_2 = 200\,000$ for $n = 1000$. If $b = 200$, then $n_1 + n_2 = 400n$. That is generally we can assume for a given space $Cum[a, b] = [1, nb]$, then the sum of the primes and pseudoprimes in that space will be $n_1 + n_2 = 2nb$.

Hence the sum of the primes and pseudoprimes in the space $[1, nb]$ is increasing at a constant rate of two hundred. For $b = 100$, the number of composites is given by $n_3 = 400n - 2$. When $b = 200$, then $n_3 = 800n - 2$. Hence for the composites, for the space $Cum[a, b] = [1, nb]$, then the total number of composites will be;

$$n_3 = 4nb - 2$$

Now $f(b) - f(a) = n_1 + n_2 + n_3$, therefore $f(b) - f(a) = 6nb - 2$. All of this indicates the deterministic properties observed in the distribution of prime numbers. The composite also increase at a constant rate, and the same obviously applies to the total sum.

n	Range	Primes	Pseudoprimes	Total	Composites
1	[1,100]	106	94	200	398
2	[1,200]	192	208	400	798
3	[1,300]	275	325	600	1198
4	[1,400]	353	447	800	1598
5	[1,500]	427	573	1000	1998
6	[1,600]	499	701	1200	2398
7	[1,700]	571	829	1400	2798
8	[1,800]	642	958	1600	3198
9	[1,900]	708	1092	1800	3598
10	[1,1000]	780	1220	2000	3998
11	[1,1100]	849	1351	2200	4398
12	[1,1200]	916	1484	2400	4798

Table 112

Note that the number of primes and pseudoprimes is growing at a fixed rate of two hundred while the composites at a fixed rate of four hundred. This shows that the algebraic sieve is precisely structured.

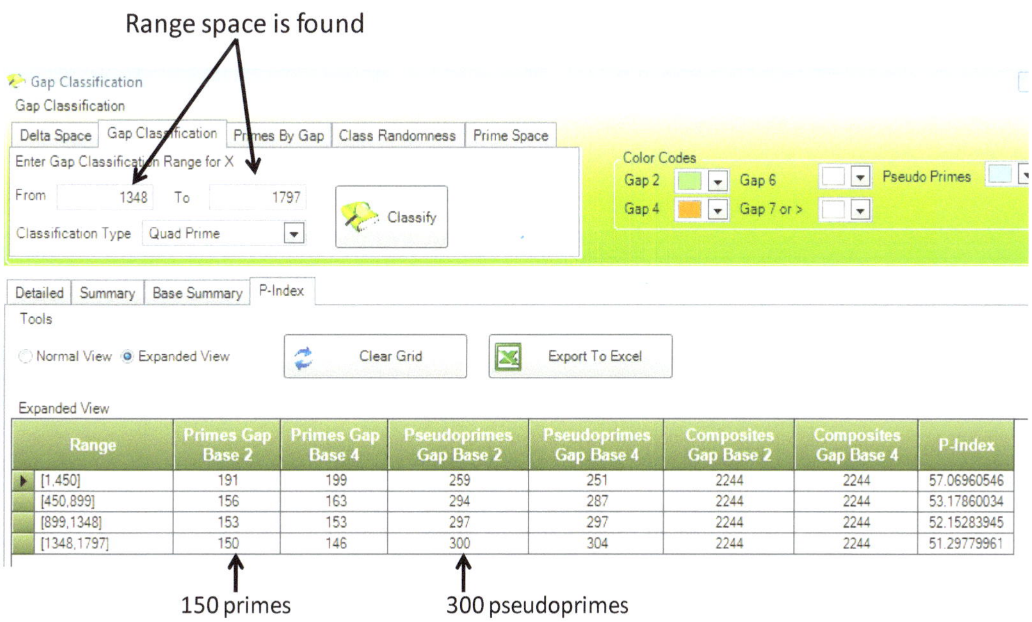

Range	Primes Gap Base 2	Primes Gap Base 4	Pseudoprimes Gap Base 2	Pseudoprimes Gap Base 4	Composites Gap Base 2	Composites Gap Base 4	P-Index
[1,450]	191	199	259	251	2244	2244	57.06960546
[450,899]	156	163	294	287	2244	2244	53.17860034
[899,1348]	153	153	297	297	2244	2244	52.15283945
[1348,1797]	150	146	300	304	2244	2244	51.29779961

Figure 83

We may not know the position of value of the next prime number, hence the random nature of primes, but the distribution happens in a measured and precise manner. Hence the variable cumulative space brings out this attribute of the set of integers in general.

8.7.1 Comparing the Prime Count Functions

The prime counting function is generally defined as $\pi(x)$ which gives us the number of primes less than or equal to for a given number x. It is referred to as the prime number theorem, and the number of primes is given by $\pi(x) \approx x/Ln(x)$. There are many prime number counting functions that are a better approximation than this, but this is a standard function that is normally used as a starting point. In the cumulative prime space, n_1 gives us the prime count. Therefore we can compare three prime count functions for this particular example.

- $\pi(x) = x/Ln(x)$, according to prime number theorem (PNT)
- $n_1(10) = 74.18n + 47.3$, for variable cumulative space $[a, b], n = 10$
- $n_1(20) = 69.185n + 73.858$, for variable cumulative space $[a, b], n = 20$

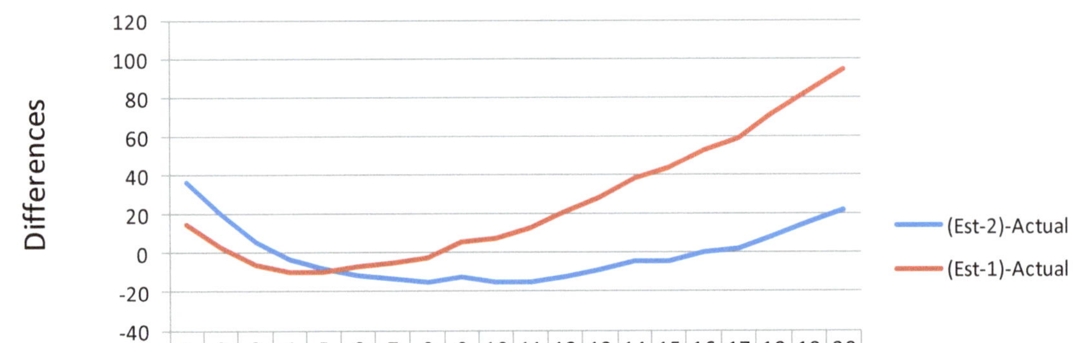

n	1	2	3	4	5	6	7	8	9	10	11	12	13	14	15	16	17	18	19	20
(Est-2)-Actual	37	20	6	-3	-8	-11	-13	-15	-12	-15	-15	-12	-9	-4	-4	0	2	9	16	22
(Est-1)-Actual	15	3	-6	-10	-10	-7	-5	-2	6	8	13	21	29	39	44	53	59	72	84	95

Figure 84

For $n = 10$, we have estimate-1 equation and $n = 20$ estimate-2 equation. We can then find the difference between the actual number of primes and the estimate in order to establish the behavior of the count function. The first estimate equation is on the higher side, while the second estimate equation is much closer to actual values. However, both estimates are closer or better than the one provided by the prime number theorem as shown by the graph. It is evident that the higher the value of n, the better the estimate equation that we can derive.

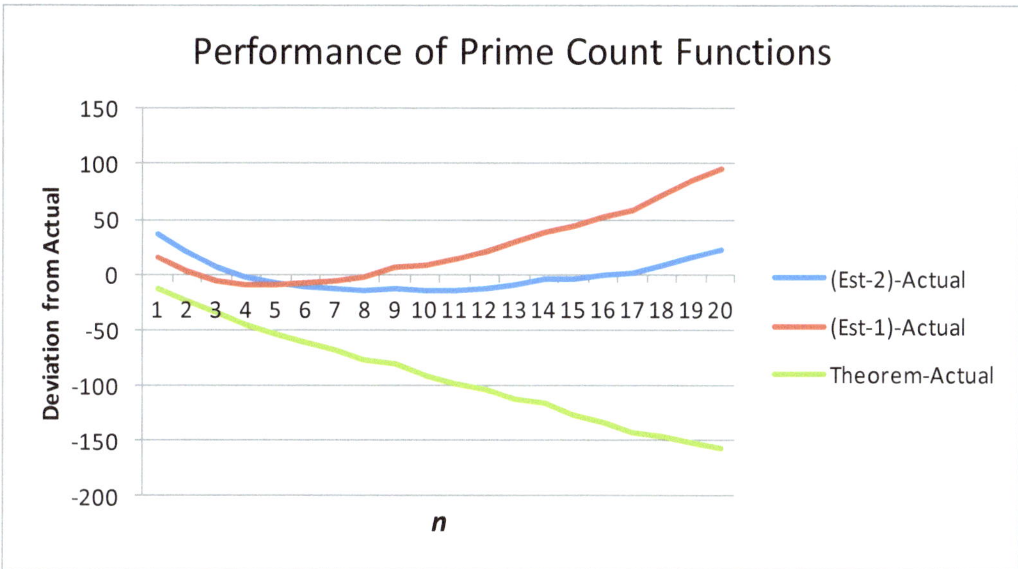

Figure 85

We can write the prime number theorem equation in terms of a trend line, and compare the standard deviation for each value of n in multiples of ten. Thus we have five equations as follows.

Description	Equation Trend line		Standard Deviation
P NT	$56.659n + 117.52$		93.41493
Estimate-1	$n_1(10) = 74.18n + 47.3$	0.6376	165.3042
Estimate-2	$n_1(20) = 69.185n + 73.858$	1.0675	94.97094
Estimate-3	$n_1(30) = 66.386n + 98.251$	1.48	58.27074
Estimate-4	$n_1(40) = 64.501n + 119.73$	1.8563	38.42198
Estimate-5	$n_1(50) = 63.051n + 143.27$	2.2723	32.07532

Table 113

We can consider a general approach by noting the following.

a) As n increases, the standard deviation goes down, implying a better accuracy for the prime count function.
b) The count is poor for lower values of but more accurate at higher values if n is large, the converse being true.
c) We will derive different equations when we vary the value of b.

It is of interest to compare the two types of cumulative prime spaces that have been developed.

a) In the fixed cumulative space, the number of primes with n decreases, and stops when a find or nearby find has been made.
b) In the variable cumulative space, the number of primes increases to infinity with n.
c) In both cases, the P-index decreases because the number of primes per unit of space is thinning out.

These are two opposing characteristics help to highlight the behavior of prime numbers. They are increasing to infinity, yet at the same time their occurrence per unit of space is decreasing.

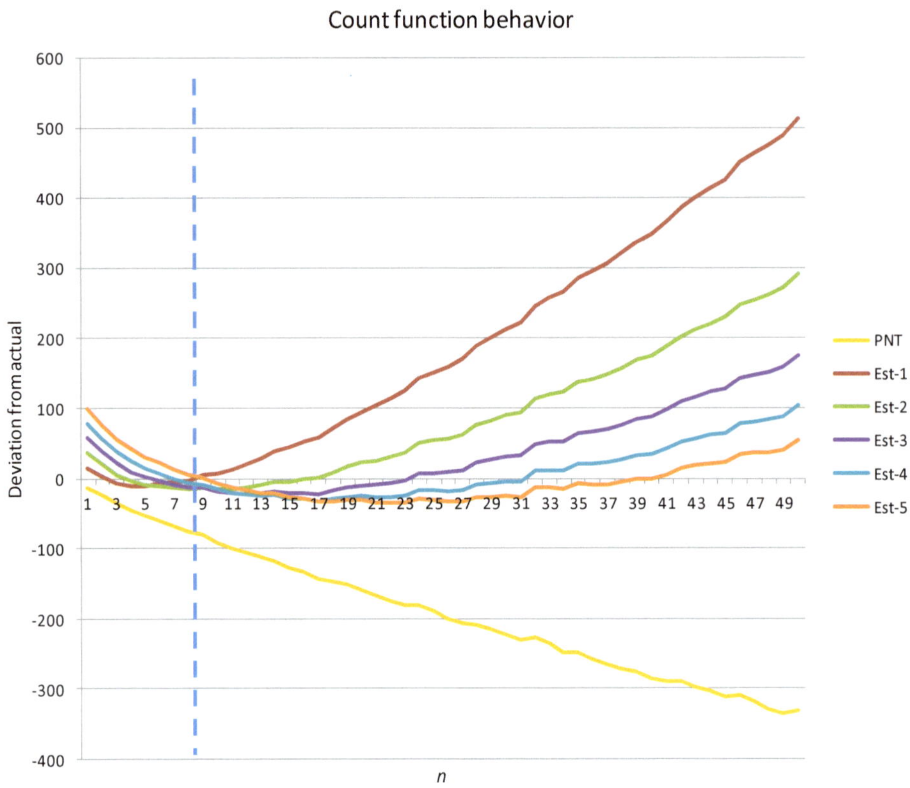
Count function behavior
Deviation from actual
n
600
500
400
300
200
100
0
-100
-200
-300
-400
1 3 5 7 9 11 13 15 17 19 21 23 25 27 29 31 33 35 37 39 41 43 45 47 49
PNT
Est-1
Est-2
Est-3
Est-4
Est-5

Figure 86

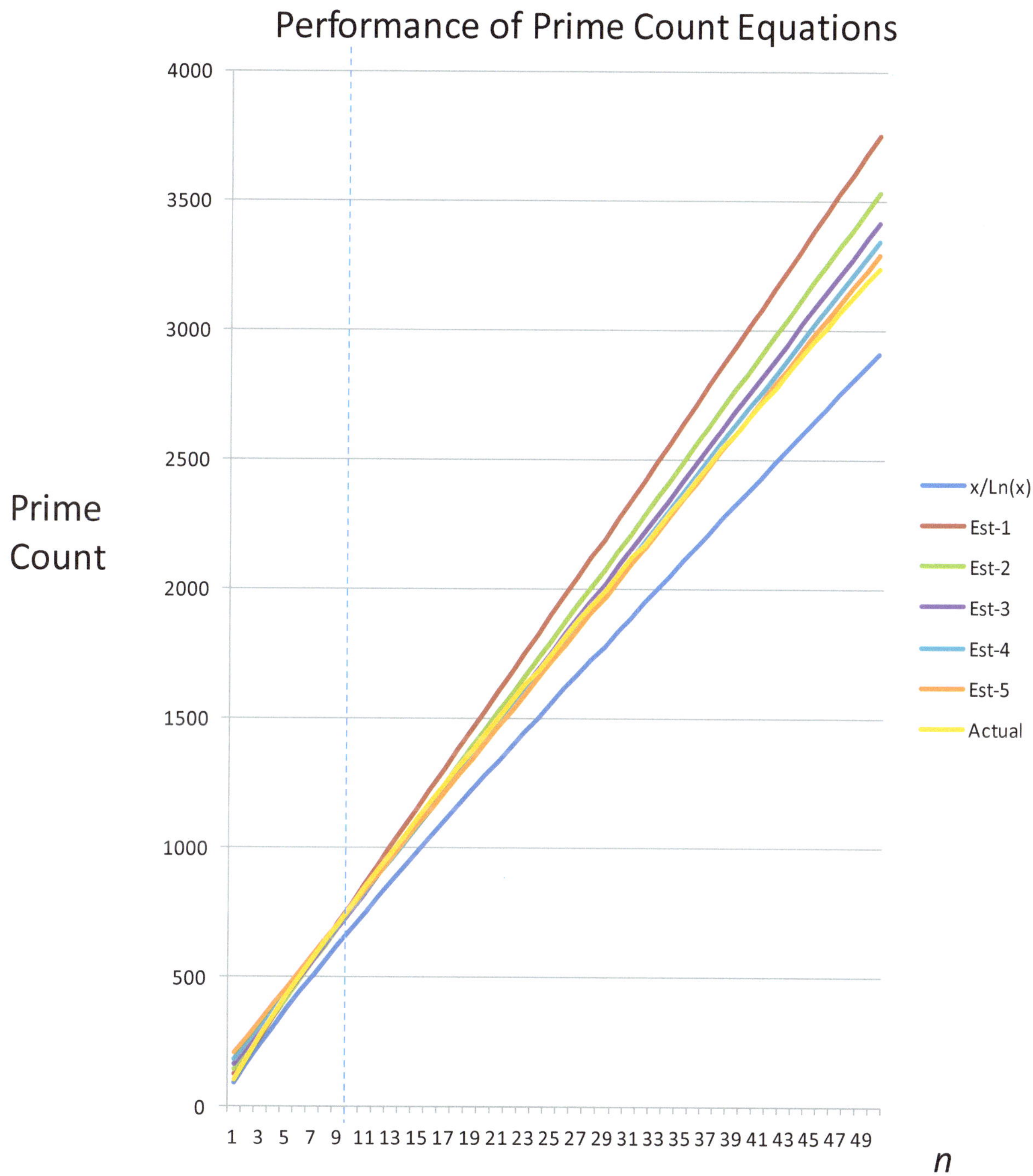

Figure 87

8.8 The Difference Gap

For the prime space $[a, b, x]$, then we know that $f(b) - f(a) = n_1 + n_2 + n_3$ and $n_1 + n_2 = c$. However, we can consider the difference between the primes and pseudoprimes, that is;

$$d = n_2 - n_1$$

where this is a prime space difference gap. For example consider the space $[1,600,x]$, then the following curve appears as a standard form. It is a trend line determined as $d_{gap} = 461.14x^{0.0679}$.

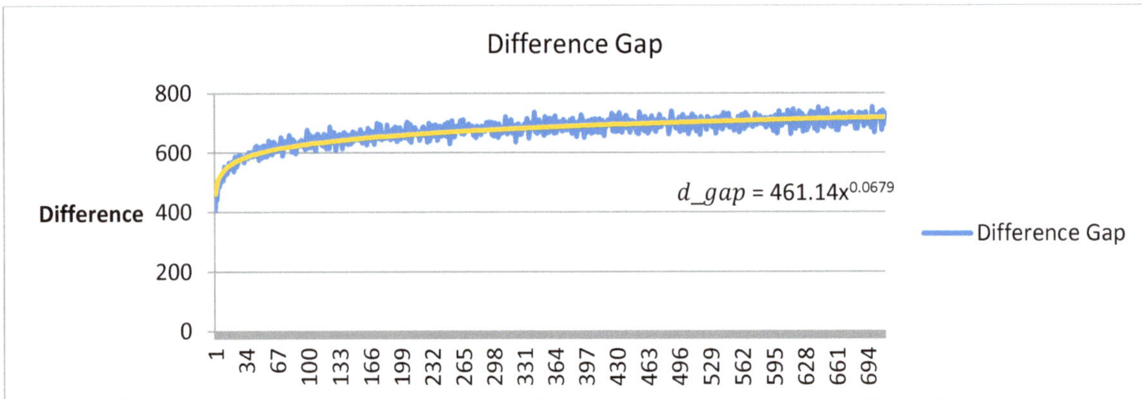

Figure 88

Now the maximum value of this difference is $d_{gap} = c$ when $n_1 = 0$ since $n_1 + n_2 = c$. However, the curve is observed to behave in an asymptotic manner. For example in the case for $d_{gap} = c$, then it implies $x = 1\,309\,366.22$. As this is an approximation, it means that at around this value, then the number of primes will be zero, that is $n_1 = 0$, therefore $n_2 = c = d_{gap}$.

In noting the shape of the curve, generally we may assume that;

$$n_2 - n_1 = kx^m, \text{ where } m < 1$$

And that at approximately;

$$c = kx^m$$

then $n_1 = 0$. In that case, there is always a value x at which we expect to find no prime numbers in the space $[a,b,x]$. This value is always comparatively large as we expect $m < 1$ and k is determined from the definition of the space $[a,b]$. Note the similarity of the curves for the P-index and the difference gap curve.

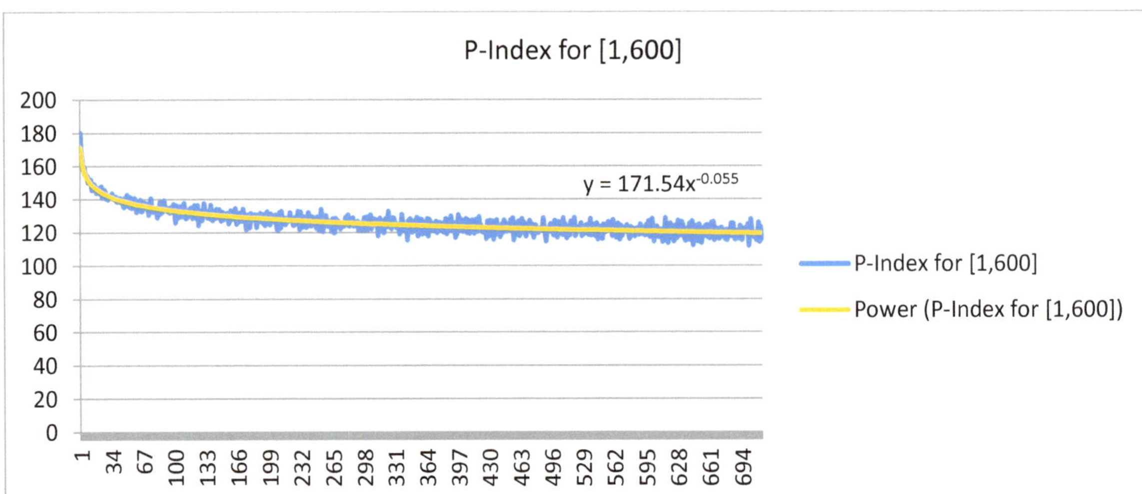

Figure 89

Primality Testing

Primality is the biggest challenge in dealing with primes. Classification provides different ways of addressing that challenge.

OUTCOMES

Developing the Algebraic and Mbulawa primality testing methods.

9 The Algebraic Primality Test

This primality test is a sequential test. It recognizes that there are three types of numbers; the composite, the pseudoprime, and the prime numbers. It differs from the conventional approach by taking all three into consideration. It has five basic steps that must be followed in sequence, where the last step is the one that determines whether we have a prime number or not. By the time it arrives at this step, the probability that the number is prime is one.

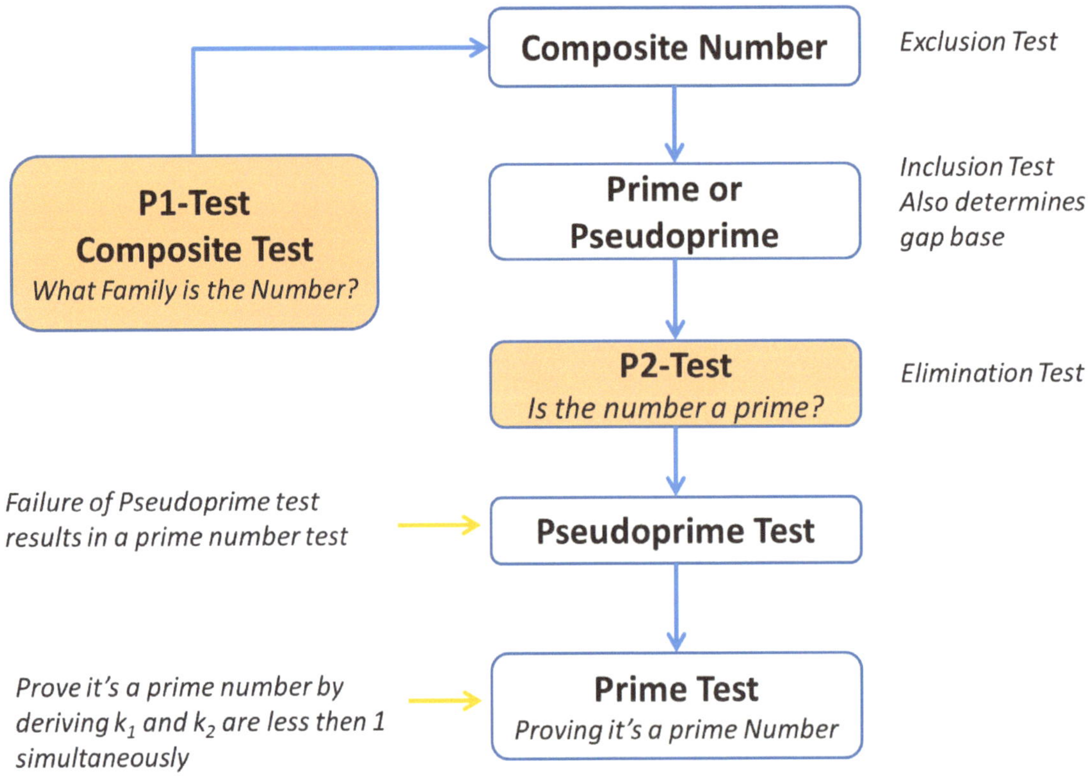

Figure 90

These five steps are divided into two basic tests, the P1-Test and the P2-Test. The first is the P1-Test, and this is also divided into two; the P1-Test (Exclusion) and P1-Test (Inclusion). The P1-Test (Exclusion) is also called the composite test. So basically the P1-Test finds out if the number entered is inside or outside the algebraic sieve. If it is outside, then it is definitely not prime, hence the exclusion. It is the fastest test, no matter how large the number. For the number N to be tested, the generation equations are;

1. $6x - 1$ for GB-4, so we test with $(W + 1)/6$
2. $6x + 1$ for GB-2, so we test with $(W - 1)/6$

An example is given from numbers chosen at random.

	Number Entered	GB-4	GB-2
1	79,887,651,237,731	13,314,608,539,622.0000	13,314,608,539,621.7000
2	91,113,167,458,797	15,185,527,909,799.7000	15,185,527,909,799.3000
3	10,008,809,713	1,668,134,952.3333	1,668,134,952.0000

Table 114

Hence the second number entered is definitely not a prime number because it gives a decimal for both GB2 and GB4, the number is a composite. This is referred to as a composite test because of that. The first number belongs to GB4 and the third number to GB2. These then are possible primes or pseudoprimes, hence the inclusion to now test for primality. Once an inclusion is determined, the gap base of the number whether it belongs to GB-2 or GB-4 is established, then we move onto the second primality test, the P2-Test.

The P2-Test is an elimination test, we want to know whether the number is a prime or pseudoprime. In the sieving process, we noted that x defines the sieve and k defines the sieving process while x is held constant. Hence, for the P2-test, we reverse the process, and we set x as a variable while we look for a corresponding k.

If $6x + 1$ gave an integer, then p is an element of gap base 2. However, if $6x - 1$ is an integer, then p is an element of gap base 4. Both cannot be an integer simultaneously in the two equations. This step determines which equations to use to determine the primality. As shown in the graphic, the power of the test lies in determining things according to the gap base. The values of k_1 and k_2 should be in the same gap base to determine primality is the additional condition.

	GB-2	GB-4	
	a_1	b_1	Result
k_1	Integer	Fraction	Pseudoprime
	Fraction	Integer	Pseudoprime
	Fraction <1	Fraction < 1	Prime
k_2	Integer	Fraction	Pseudoprime
	Fraction	Integer	Pseudoprime
	Fraction < 1	Fraction < 1	Prime

k_1 and k_2 are less than one and number is in gap base 2

k_1 and k_2 are less than one and number is in gap base 4

Figure 91

9.1 Equations for Primality Testing

It was shown for the algebraic sieve that $a_1 = x + (6x + 7)k_1$ and that $b_1 = x + (6x + 11)k_2$ for gap base 2 and 4 respectively. These are x-sieve and y-sieve equations. Generally we can state;

a) $a_1 = x + f_1(x)k_1$, where $f_1(x)$ is for gap base 2
b) $b_1 = x + f_2(x)k_2$, where $f_2(x)$ is for gap base 4.

This implies we can use other equations besides $6x + 7$ and $6x + 11$. For example, we can substitute for $6x - 1$ gap base 4 and $6x + 1$ gap base 2 and achieve the same result.

In each of the given gap base, for k_i an integer and (x_i, y_i) the initial point on the sieve, then a "hit" $W = a_2$ in the sieve occurs according to the gap base 2 row in gap base 2 or 4, and in gap base 4 row in gap base 2 or 4.

	Gap Base 2 -Primality	Gap Base 4 - Primality
Gap Base 2 "Hit"	$a_1 = x_1 + k_1 y_1$, at $W = f(a_1)$, where $-\infty \leq k_1 \leq \infty$	$b_1 = x_2 + k_2 y_1$, at $W = f(b_1)$, where $-\infty \leq k_2 \leq \infty$
Gap Base 4 "Hit"	$b_1 = x_3 + k_3 y_2$, at $W = f(b_1)$, where $-\infty \leq k_3 \leq \infty$	$a_1 = x_4 + k_4 y_2$, at $W = f(a_1)$, where $-\infty \leq k_4 \leq \infty$

Table 115

		x-sieve		y-sieve on x	
		x+(6x+1)k	Hit	-x+(6x+1)k	Hit
x	k	x1	f(x1)	x2	f(x2)
1	0	1	7	-1	-5
1	1	8	49	4	25
1	2	15	91	9	55
1	3	22	133	14	85
1	4	29	175	19	115
1	5	36	217	24	145
1	6	43	259	29	175
1	7	50	301	34	205
1	8	57	343	39	235
1	9	64	385	44	265
1	10	71	427	49	295

Table 116

The above table shows the sieves (1,7) and (-1,-5). This implies $a_1 = 1 + 7k_1$ and $b_1 = -1 - 5k_1$ for W in gap base 2, where the numbers are divisible by 7 and by 5. Now the 5 is from gap base 4, but it looks for factors in gap base 2, hence the reason why we say the y-sieve acting on x.

However, when we enter a number W and we test for primality, it is either gap base 2 or 4, hence our interest is in the equations that are in the gap base 2 column and the column for gap base 4 equations. Hence for gap base 2, $k = (k_1, k_3)$ where k_1 is from the x-sieve and k_3 is from the y-sieve acting on the x-sieve. For gap base 4, then $k = (k_2, k_4)$ where k_2 is from the y-sieve and k_4 is from the x-sieve acting on the y-sieve. Secondly, the primality test algorithm is based on the fact that a prime number implies no hit has occurred, which means no integer value of k_i can be established. Hence if k_i is less than one, this establishes a certainty that the number is prime. Consequently, the algorithm proves that for a given number, k_i is an integer or its is a decimal.

Therefore assuming we know the number $W$and that it is in gap base 2, then define;

a) $a_1 = \frac{W-1}{6}$, and $b_1 = \frac{W+1}{6}$ for W in gap base 2
b) $b_1 = \frac{W+1}{6}$, and $a_1 = \frac{W-1}{6}$, for W in gap base 4

Hence, from sieve equations;

1. For W in gap base 2, then $(a_1, a_2) = (x + (6x+1)k_1, 6\{x + (6x+1)k_1\} + 1)$ for the x-sieve, hence

$$a_1 = x + (6x+1)k_1$$
$$a_2 = 6a_1 + 1$$

Similarly, it implies that;

$$k_1 = \frac{a_1 - x}{6x+1}$$

Now $(b_{3,}b_4) = (\frac{1}{6}(-f(x) - 1) + f(x)k_3, -f(x) - 1 + 6(f(x)k_3) + 1)$ where this is the y-sieve acting on the x-sieve. Note $f(x) = 6x - 1$ as this is a gap base 4 operation, and b_3 is looking for a hit in W.

$$b_3 = \frac{1}{6}(-f(x) - 1) + f(x)k_3 = -x + (6x-1)k_3$$

$$b_4 = -f(x) - 1 + 6(f(x)k) + 1$$

That is;

$$b_3 = \frac{W+1}{6} = b_1$$

$$k_3 = \frac{b_1 + x}{6x-1}$$

2. For the y-sieve we apply a similar approach. That is; $(b_{1,}b_2) = (x + (6x-1)k_2, 6\{x + (6x-1)k_2\} - 1)$ for the y-sieve. In this case W is in gap base 4.

$$b_1 = x + (6x-1)k_2$$

$$b_2 = 6b_1 - 1$$

Therefore;

$$k_2 = \frac{b_1 - x}{6x-1}$$

Hence for the x-sieve on the y-sieve; $(a_{3,}a_4) = (\frac{1}{6}(-f(x) + 1) + f(x)k, -f(x) + 1 + 6(f(x)k) - 1)$ for the x-sieve acting on the y-sieve. Note $f(x) = 6x + 1$ as this is a gap base 2 operation.

$$a_3 = \frac{1}{6}(-f(x) + 1) + f(x)k = -x + f(x)k_4$$

$$a_4 = -f(x) + 1 + 6(f(x)k) - 1$$

Therefore;

$$a_3 = \frac{W-1}{6} = a_1$$

$$k_4 = \frac{a_1 + x}{6x+1}$$

The power of the primality test lies in varying the value of x to look for corresponding values of k. Since $k = (k_1, k_2)$ or $k = (k_3, k_4)$, for the decision criteria we may assume that $k = (k_1, k_2)$ in either gap base 2 or gap base 4. The interpretation code in a general format for the decision on whether it's a prime or pseudoprime is as given in the table. It shows that basically we depend on a_1 and b_1 for the P1-Test and on k_1 and k_2 for the P2-Test.

Test Type	a_1	b_1	**Decision Outcome**
P1-test (exclusion)	Non-integer	Non-integer	Not prime
P1-test (inclusion)	Integer	Non-integer	Prime or Pseudoprime
	Non-integer	Integer	Prime or Pseudoprime
P2-test	k_1	k_2	
	Integer	Non- Integer	Pseudoprime
	Non- Integer	Integer	Pseudoprime
	Decimal	Decimal	Prime

Table 117. Decision criteria.

Hence, a_1 and b_1 gives us the value of x that generates the number we want to test. Now we know that when k is an integer (that is, k_1 or k_2 are integers), this defines a pseudoprime hit. So the first thing the test should do is to look for pseudoprimes, and secondly, a pseudoprime failure implies a prime number success. In the P2-test, failing to find an integer value of k then implies that we have a prime number, and the probability is one if both k_1 and k_2 are less than one, since no positive integer value of k can be defined. This is the basis of the primality test.

As x varies, then possible sieves occur that correspond to a given k. The value of k decreases as x increases. For example for $6x + 7$ and $6x + 11$, then $x = -1$ gives us the greatest values of k_1 and k_2. Also, if $x = -1$, then k_1 and k_2 will always be an integer for gap base 2 and 4 respectively. Therefore, we only consider the k_2 value for gap base 2 and k_1 for gap base 4. This leads to the following condition called the primality condition.

Definition of Primality Condition

Let x vary to from minus one to z, then if at z, $k_1 < 1$ *and* $k_2 < 1$ *is the first occurrence in a given gap base, then p is a prime number.*

It is the first occurrence of k_1 and k_2 that matters in a given gap base that is less than one, not subsequent ones. Uniqueness of the solution is determined on this basis. There will always be a sieve that satisfies this condition. Using the number 133, where $6x + 7$ and $6x + 11$, we demonstrate a test result.

Number = 133		$a_1 = 21$		
X	Sieve	k_1	k_2	Sieve
-1	1	22	4.6	5
0	7	3	1.142857	11
1	13	1.538462	1.470588	17
2	19	1	1.130435	23
3	25	0.72	0.931034	29

Table 118

The number is a gap base 2 number because $t_1 = 21$; hence, we look at k values in gap base 2.The highest values of k_1 and k_2 occur at x = -1, but we ignore $k_1 = 22$ because the 1-sieve is an identity while $k_2 = 4.6$ tells us that we don't have a pseudo-hit. The sieves being used are 1 and 5 respectively. Then $x = 0$ gives us a pseudo-hit with sieve 7, so we have a pseudoprime in 133. Now note that if the progression continued at $x = 3$, we have both k_1 and k_2 being less than one. That is, we test for a pseudoprime first, followed by a failure condition that allows a prime to exist. Hence, $x = 3$ here false as a prime number test is because k_1 or k_2 was satisfied first for the pseudoprime condition. The test is sequential in structure.

Using the prime 41 as an example, we see that the P1-test will give us a value of $t_2 = 5$. This implies that the number is in gap base 4. At this stage, the information is that the number is either a prime or pseudoprime. Therefore, we then pass on to do the P2-test to do primality operation. This gives the following result.

x	k_1	k_2
-1	1.2	7
0	0.454545	1.142857
1	0.235294	0.692308

Table 119. The k values for testing primality.

Note that $k_2 = 7$ is an integer. We ignore this when $x = -1$ for any k_2 as it will always be an integer. But k_1 tells us that it is not a pseudoprime, so we proceed to the next x. The pseudoprime condition is a failure until we get k_1 and k_2 being less than one, giving us a prime at $x = 1$.

Many equations can be used to apply the Primality Condition. The equations $6x - 1$ and $6x + 1$ are not unique as they are not the only equations for generating the sieve. The example below shows using the equations $6x + 7$ and $6x + 11$ in testing for primality of the number 432167.

Primality Test Example Using 6x+7 and 6x+11

Enter number		432167
	x	14405
The sieve is		86437
Gap 2, x-sieve	14405	86437
Gap 4, x-sieve on y-sieve	-14408	-86437
The sieve is	86441	
Gap 4, y-sieve	14405	86441
Gap 2, y-sieve on x-sieve	-14408	-86441

P1-Test	t1	t2
Finding t1, t2	72026.67	72026.000

P2-Test	t1		t2	
k 1=	0.66663196	*by x sieve*	0.666593399	*by y-sieve*
k2 =	0.99992673	*by y-sieve on x*	0.999965	*by x-sieve on y*

x	6x+7	6x+11
-1	1	5
14400	86407	86411
14401	86413	86417
14402	86419	86423
14403	86425	86429
14404	86431	86435
14405	86437	86441
14406	86443	86447
14407	86449	86453
14408	86455	86459

If both t1 and t2 are not integer, then number is not prime
if k1 **OR** k2 is an integer, then number is a pseudoprime.
if k1<1 **AND** k2<1, then number is a prime.

Figure 92

P1-Test

t1	t2
72026.67	72026.000

← Integer result in gap base 4

P2-Test

This takes place in gap base 4

Varying x	14400	14401	14402	14403	14404	14405	14406	14407	14408
k1	0.666883	0.666825	0.666767	0.666709	0.666651	0.666593	0.666536	0.666478	0.66642
k2	1.000255	1.000197	1.000139	1.000081	1.000023	0.999965	0.999907	0.99985	0.999792

↑ This is the unique answer of the P2-Test, the first values less than one simultaneously.

Figure 93

The number 432167 is a prime number. It has the interesting characteristic that it starts with 4 going to 1 and the last two digits are also consecutive starting at 6. However, the number 4321567 is in gap base 2 and it is a pseudoprime.

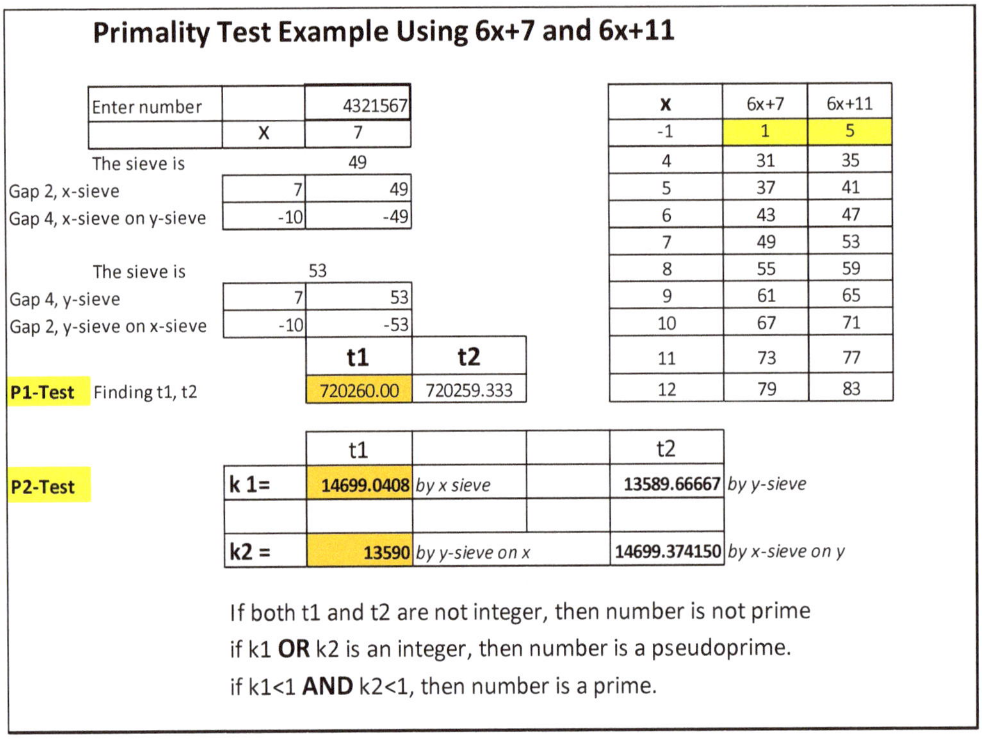

Figure 94

Note that the test here is indeed very fast in determining what type of number we have. The value of $x = 7$ establishes the required result, yet the number 432167 is smaller than 4321567. As it is a divisibility problem, smaller sieves are valid and hence the operation is independent to the size of the number.

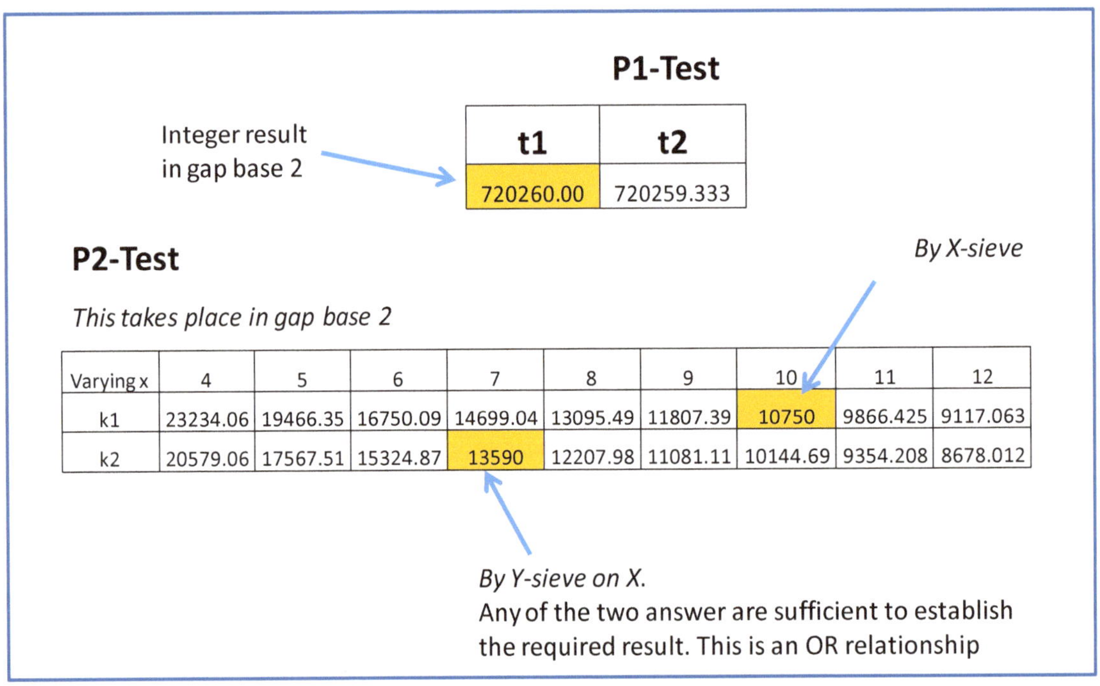

Figure 95

9.2 The Mbulawa Primality Test

Now consider the rule that, if $a, b \in GB2$, $c, d \in GB4$, then;

1. $(a)(b) = ab \in GB2$
2. $(a)(c) = ac \in GB4$
3. $(c)(d) = cd \in GB2$

These are the only possibilities in terms of a product occurring in the algebraic sieve. Now;

1. $a = f(x) = 6x_1 + 1$
2. $b = f(x) = 6x_2 + 1$

Therefore $ab = (6x_1 + 1)(6x_2 + 1)$ where this simplifies to $f(x_3)$ in gap base 2.

$$f(x_3) = 36x_1x_2 + 6(x_1 + x_2) + 1 = 6x_3 + 1$$

Therefore dividing by 6 we have $6x_1x_2 + (x_1 + x_2) = x_3$. In a similar way, we can consider the product $ac = (6x_1 + 1)(6x_2 - 1)$ where this simplifies to $f(x_3)$ in gap base 4.

$$f(x_3) = 36x_1x_2 - 6(x_1 - x_2) - 1 = 6x_3 - 1$$

This yields $6x_1x_2 - (x_1 - x_2) = x_3$. When we consider the last product taking place in GB4, then we have $cd = (6x_1 - 1)(6x_2 - 1)$, that is:

$$f(x_3) = 36x_1x_2 - 6(x_1 + x_2) + 1 = 6x_3 + 1$$

That is $6x_1x_2 - (x_1 + x_2) = x_3$. The above three equations show that given two values x_1 and x_2 we can find the third value x_3 even if we do not use the actual numbers involved. This gives an exact relationship in terms of positions of pseudoprimes in the algebraic sieve that are derived from products.

By doing the operation in reverse, the value of x_3 can be known where the three equations can now be applied to find out what the possible factors of x_3 are. That is, we know the number W;

$$W = f(x_3) = 6x_3 + 1$$

Hence $x_3 = (W - 1)/6$, where x_3 is an integer. If it is not an integer, then the number is not a pseudoprime and a method for composites must be used. We need to get two other integers x_1 and x_2, so we select values of x_2 such that $1 \leq x_2 \leq \infty$, then we derive the integer value $x_1 = k_1$ as we vary x_2. For the $GB4 \times GB4 = GB2$ result, then $x_1 = k_2$. A similar approach is used for gap base 4 product result, where $x_3 = (W + 1)/6$ and $x_1 = k_3$. Hence in each case x_2 will be a factor of W to give use the second factor x_1.

Base	Base	Base Result	Solving
GB2	GB2	GB2	$k_1 = \dfrac{x_3 - x_2}{6x_2 + 1}$
GB4	GB4	GB2	$k_2 = \dfrac{x_2 + x_3}{6x_2 - 1}$
GB2	GB4	GB4	$k_3 = \dfrac{x_3 - x_2}{6x_2 - 1}$

Table 120

Now as x_2 increases, the value of x_1 decreases because of the multiplicative factor of six. The smallest integer value occurs for $x_1 = 1$, where this implies if $x_1 < 1$, then x_3 must be a prime number. Consequently, the test condition is based on the following being true;

a) $x_3 - x_2 < 6x_2 + 1$, gap base 2
b) $x_3 + x_2 < 6x_2 - 1$, gap base 2
c) $x_3 - x_2 < 6x_2 - 1$, gap base 4

This will give us either integers or decimals as we vary x_2. As can be observed from the table, gap base 2 has two equations, that is, a number in gap base two can be a consequence of two different types of products. Now in a primality test condition, the number W that we enter to determine primality, it has no pre-determined product. As a result, we cannot determine which equation in gap base 2 to apply, whereas in gap base 4, there is only one equation, which implies that the condition $x_1 < 1$ is sufficient to define primality. For the gap base 2 scenario, we need to include an "and" condition to define primality.

Definition – Mbulawa Primality Condition

a) For primality in gap base 2, then $k_1 < 1$ and $k_2 < 1$ simultaneously
b) For primality in gap base 4, then $k_3 < 1$

In both cases, this must be a first occurrence is sufficient to satisfy the condition. This is the first difference that we note between the algebraic primality test and the Mbulawa primality test. It can also be demonstrated that the conditions occur earlier in the algebraic test compared to the Mbulawa test, but the difference is very small so as to be insignificant.

Consider for example $W = 91$. This belongs to gap base2, that is, it is a gap base 2 product. Now if you enter this, the integer value occurs for k_2, hence it is a GB2 x GB2 product. Since we have an integer value at $x_2 = 1$, that is $k_2 = 2$. The integer values shows that 91 has factors, consequently it is not a prime number.

Enter Number	W =	91

X3	15.00	Gap Base 2
	15.33333333	Gap Base 4

X2	1

				Product Type
X1	k1	3.2	GB2	GB4xGB4
	k2	2	GB2	GB2xGB2
	k3	2.866666667	GB4	GB2xGB4

Figure 96

Now let us consider the number $W = 131$. This is a gap base 4 number, which immediately means it is from a product of gap base 2 and 4. Starting from x_2 is 1 to 4, there is no such integer occurrence. Therefore the value $k_3 = 0.7782608696$ is the first occurrence at $x_2 = 4$. As a result, it means $k_3 < 1$ defines a prime number 131 in gap base 4.

If we consider the number $W = 103$, we find that it is a gap base 2 number. At $x_2 = 3$, only $k_2 < 1$ is less than 1, hence we are not definite that the number is prime. However, we can suspect that it is prime because one of the conditions for primality has been satisfied. However, at $x_2 = 4$, then both $k_1 < 1$ and $k_2 < 1$ are now less than one, so definitely have a prime number at 103.

Enter Number W =	131

X_3	21.67	Gap Base 2
	22	Gap Base 4

X_2	4

				Product Type
X_1	k_1	1.115942029	GB2	GB4xGB4
	k_2	0.706666667	GB2	GB2xGB2
	k_3	0.782608696	GB4	GB2xGB4

131 is definitely a prime

Figure 97

Enter Number	W =	103

X_3	17.00	Gap Base 2
	17.33333333	Gap Base 4

X_2	3

X_1	k_1	1.176470588
	k_2	0.736842105
	k_3	0.843137255

103 is probably a prime number

Enter Number	W =	103

X_3	17.00	Gap Base 2
	17.33333333	Gap Base 4

X_2	4

X_1	k_1	0.913043478
	k_2	0.52
	k_3	0.579710145

103 is definitely a prime number

AND

Figure 98

Hence the above examples illustrate the basic mechanism of the test, and the differences to the algebraic primality test. This test is related to products whereas the algebraic test is related to sieve positions. There is a difference in emphasis but the modus operandi of both tests is actually similar; it's based on finding whether there are integers or not that form a product in the sieve.

For each of the equations, we can determine the end point of the iteration for x_2. This must occur when x_1 is minimum, that is at $x_1 = 1$ where this will derive for us the maximum value of x_2. We solve for x_2 and this derives the following end points x_2 for a given x_3. For example, for $W = 103, x_3 = 17$, and there will be two possibilities.

a) $x_2 = Int\frac{x_3+1}{5} = Int\frac{17+1}{5} = Int(3.6) = 4$

b) $x_2 = \frac{x_3-1}{7} = Int\frac{17-1}{7} = Int\ 2.2857 = 2$

So once it passes the integer $x_2 = 2$ (since there is a decimal 2.2857), we know that the first condition must be fulfilled, and this happens at x_2 = 3. At x_2 = 4, the second condition is fulfilled.

Base Result	Solving for exact endpoint	Endpoint smaller than W by approximately
GB2	$x_2 = Int\{\frac{x_3+1}{5}\}$	30 times
GB2	$x_2 = Int\{\frac{x_3-1}{7}\}$	42 times
GB4	$x_2 = Int\{\frac{x_3+1}{7}\}$	42 times

Table 121

Similarly consider the number $W = 432167$, then $x_2 = 10289$. Since this is a decimal at 10289.85714, the condition $k_3 < 1$ will occur at exactly $x_2 = 10289$.

We can also find out the relationship of this value x_2 to W in terms of relative size. Since $W = 6x_3 \pm 1$, this approximates as $W = 6x_3$ and if we divide by the endpoint x_2, that is we substitute in $6x_3/x_2$, we get the right column showing relative size. For example when $W = 103$ which is gap base 2, then we count up to approximately (103/42 = 2.45 and $(103/30) = 3.4$, that is for 3 and 4. The exact calculation is that at $x_2 = 4$ we have the correct answer. For $W = 432167$, we divide by 42, and we get 10289.69048.

The Universal Number Distribution Theorem

The section shows how the integer number distribution theorem is derived and shows its usefulness in interpreting integer behavior and distribution in an integrated manner.

10 The Integer Product Law

There is but a small difference in how numbers are regarded in the UNDT framework compared to the conventional approach. However this small difference results in a lot of diversity in how numbers are understood and classified whereas the conventional framework is uniform in approach. The UNDT framework is a structure where all the behavior of integers is understood in multiples of six. Pretty strange!

Conventional Approach	UNDT Approach
Defines the set of integers Z	Defines the set of integers Z
There are negative and positive Integers	There are negative and positive Integers
Zero is between negative and positive numbers in terms of sequence	Zero is between even numbers in terms of classification
There are no negative prime numbers	There are negative prime numbers
There are positive prime numbers inclusive of 2 and 3, and 1 is not prime	The numbers 2 and 3 are not prime numbers, but 1 is a prime
Pseudoprimes have varying definitions depending on the type of prime numbers being studied	Pseudoprimes are a specific set that is distinctly related to spatially to the set of prime numbers. Only one type of pseudoprime exists.
Composites are odd and even. A composite is any number that is not prime, that is, inclusive of pseudoprimes.	Composites are odd and even. Any number that is not prime or pseudoprime is composite.

Table 122

Further interesting observations can be made in regard to product properties involving primes, pseudoprimes and composites. From the fact that the gap base structure was observed as a 2-4-6 sequence, by implication we can assume a sequence 0-2-4-6. That is, the set of composites is defined as gap base zero. Whereas 2-4-6 is repeatable as 8-10-12 and so on, the zero is not repeatable as a base. The implication is that such a set is unique in some way. For the composite number, there are two types as indicated by the definition below.

Definition

a) A pure composite is any number that has 2 or 3 as its product factors.

b) A general composite is any number that has a prime or pseudoprime making a product with a composite.

There is a particular product pattern for all the integers. This observation is summarized by the Integer Product Law.

Integer Product Law

For any two integers p_1 and p_2, then the product p_1p_2 is either a composite in gap base zero, or a pseudoprime in gap base 2 or 4.

Normally, a composite has no gap base, and it is defined as any integer that has factors other than one and itself, or simply as any number that is not prime. But by considering them as a gap base, we can then state particular properties of such a gap base and investigate them with the same vigor as the prime numbers.

The above rule can be elaborated further in the following manner.

1. [composite] x [composite] = [composite], where this is a pure composite or a general composite
2. [composite] x [prime] = [composite], where this defines a general composite
3. [composite] x [pseudoprime] = [composite], where this defines a general composite

4. [pseudoprime] x [pseudoprime] = [pseudoprime]
5. [prime] x [pseudoprime] = [pseudoprime]

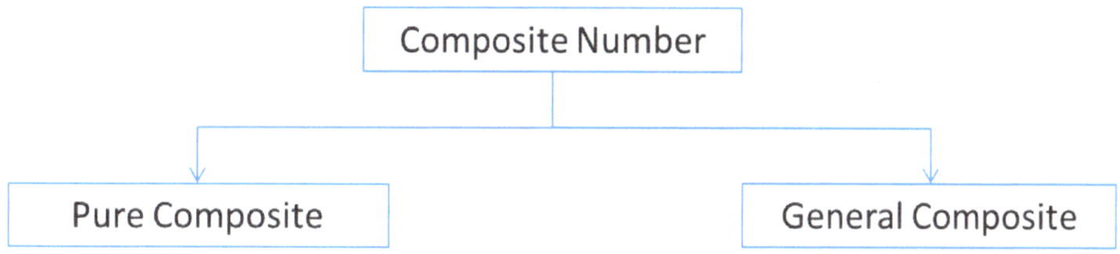

Example:
36 is a pure composite
where $36 = 2^2 \cdot 3^2$

Example:
252 is a general composite
where $252 = 2^2 \cdot 3^2 \cdot 7$

Figure 99

In terms of a product involving a composite, this is a closed set. As long as a composite is involved then the answer will be in the same set, the composite set. Composites define a gap base zero but surprisingly are not considered to have random behavior. Indeed, we can predict the set of all even numbers or odd numbers. However, in the context of the gap base approach, we see such numbers as a distinct set, and form the following hypothesis.

Hypothesis

There exists no single formula to define the nth consecutive composite number.

The composite number can be either pure or general. The hypothesis introduces an interesting scenario that if primes and pseudoprimes have a random behavior, then the same applies to the set of composites. That is, composite numbers are random, and the nth composite number cannot be predicted in sequential formulaic form. Indeed it is true that we can find a formula to predict the next even number or odd number, but there is none to predict the next composite as defined in this concept. For example $2n + 1$ is a single formula that predicts all odd numbers. This does not mean they do not have a structured relationship that has a predictive framework. Surprisingly, we also note that composites have a structured relationship between each other the same way as the primes and pseudoprimes. We can also find the composite gap, and do a similar analysis as the prime numbers.

We further apply the law as follows to the set which consists of primes and pseudoprimes.

1. [prime] x [prime] = [prime], provided $p_1 = 1$ and p_2 is prime
2. [prime] x [prime] = [pseudoprime]
3. [prime] x [pseudoprime] = [pseudoprime]
4. [pseudoprime] x [pseudoprime] = [pseudoprime]

Hence if you multiply any two numbers in gab base 2 and 4, and the result is a pseudoprime, hence there is a possibility that one of them could be a prime number. The law also emphasizes the argument that the composite number should be treated separately from the pseudoprime. Note that 2 and 3 are composites, and as such, no product involving these two numbers will belong to the G-set. The following can also be derived in regard to the G-set.

1. If p_1 and p_2 are elements of gap base 2, then p_1p_2 is an element of gap base 2
2. If p_1 and p_2 are elements of gap base 4, then p_1p_2 is an element of gap base 2
3. If p_1 is an element of gap base 2, and p_2 is an element of gap base 4, then p_1p_2 is an element of gap base 4

This means the rule is also binding on gap base 6, since this gap base is either a subset of gap base 2 or 4.

10.1 Generating Gap Base Zero

This does not imply that there is no gap between composites, rather it denotes the fact that this is a different class of numbers with respect to the primes and pseudoprimes. Now the set of all even numbers can be described by $2x$, while the set of all odd numbers can be described by $2n + 1$. However, some of the odd numbers are also prime or pseudoprime, hence we need to find a way of describing a relation for composites only.

Definition

Any number that is composite is an element of gap base zero.

Therefore the gap base zero consists of two families, family 1 being the odd numbers, and family 2 being the even numbers. We observe that all family 2 composites are generated by $2x$ to give an even number. So the even composite can easily be described. How can we identify the odd composite number? We can identify odd composites by using $6x \pm 1$ since this is the equation that defines whether a number is prime or pseudoprime. When we divide the odd number by this equation, if we get a decimal for both conditions, then the number is composite. For example, for the odd number 7, then we have;

- $7 = 6x - 1$, which means $x = 1.333333$
- $7 = 6x + 1$, which means $x = 1$

Since both results are not decimal, this implies we have a prime or pseudoprime. There is no composite here. Now consider the odd number 9. Similarly;

- $9 = 6x - 1$, which means $x = 1.666667$
- $9 = 6x + 1$, which means $x = 1.333333$

In this case, both results are decimal, so we have a composite. This can be done for all the odd numbers as shown in the table. This shows an interesting pattern, that family 1 composites are also regular in occurrence. The last column shows the composite occurrence. Now if we plot the occurrences against x, we can establish a relationship with respect to n. The values of x are observed from the yellow coloring with decimal result, these being $1, 4, 7, 10, 13, \ldots$ and so on. They are defined by the equation $3x - 2$, therefore in general $n = 3x - 2$. Hence, deriving the odd number composite generator is given by the equation;

$$2n + 1 = 2(3x - 2) + 1 = 6x - 3$$

Therefore we can now describe the family 1 in terms of x to give a table in the same format like gap base 2 and 4. Consequently, we can predict the next odd composite. So we now have two sets, the odd set of composite and the even set of composites, where these are family 1 and family 2 sets respectively. We can compare these two sets to find out if there is a relationship between them. In doing so, we note that there is a growing gap between the family 1 and 2 composites. This gap grows in step of four as shown by the graphic. OE gap is the odd-even gap. This is contrast to the behavior of gap base 2 and 4, which always have a constant difference.

	Even	Odd n = x	Composte Testing		Composite
x	2x	2n+1	6x-1	6x+1	Occurance
1	2	3	0.666667	0.333333	1
2	4	5	1	0.666667	
3	6	7	1.333333	1	
4	8	9	1.666667	1.333333	2
5	10	11	2	1.666667	
6	12	13	2.333333	2	
7	14	15	2.666667	2.333333	3
8	16	17	3	2.666667	
9	18	19	3.333333	3	
10	20	21	3.666667	3.333333	4
11	22	23	4	3.666667	
12	24	25	4.333333	4	
13	26	27	4.666667	4.333333	5
14	28	29	5	4.666667	
15	30	31	5.333333	5	
16	32	33	5.666667	5.333333	6
17	34	35	6	5.666667	
18	36	37	6.333333	6	
19	38	39	6.666667	6.333333	7
20	40	41	7	6.666667	

Table 123

	Class 1	Class 2		
	Even	Odd	OE Gap	Gap
x	2x	6x-3	4x-3	
1	2	3	1	
2	4	9	5	4
3	6	15	9	4
4	8	21	13	4
5	10	27	17	4
6	12	33	21	4
7	14	39	25	4
8	16	45	29	4
9	18	51	33	4
10	20	57	37	4
11	22	63	41	4
12	24	69	45	4

Table 124

10.2 The Integer Distribution Cluster Theorem (IDCT)

Let us now combine the two sets of odd and even composites to form one set denoted as gap base zero. Then we note that there is also a fixed pattern when we combine family 1 and 2 of the composites.

X	Gap Base	Prime or Pseudoprime	Gap	Composites	Composite Gap
15				87	1
				88	1
	GB4	89	1	90	2
	GB2	91	1	92	2
16				93	1
				94	1
	GB4	95	1	96	2
	GB2	97	1	98	2
17				99	1
				100	1
	GB4	101	1	102	2
	GB2	103	1	104	2
18				105	1
				106	1
	GB4	107	1	108	2
	GB2	109	1	110	2

Table 125

a) This gives us a (1,1) and (2,2) pattern for gap difference to define two composite types of families. This means we can classify composites using this methodology.
b) We also observe that primes or pseudoprimes occur at specific points in the distribution, as shown by the left column. The primes and pseudoprimes have a (1,1) gap to the set of composites.
c) The odd numbers that are not prime or pseudoprime occur at specific points as colored orange.
d) The difference between a prime or pseudoprime and the next is always two. This implies that if the difference is more than two, then the consecutive occurrence has been pseudoprimes. Consequently pseudoprimes are the ones responsible for gaps between prime numbers, and prime numbers will always try to occur as twin primes.

Therefore we define a gap 1 composite and a gap 2 composite as a classification technique. There is a specific relationship between primes, pseudoprimes, and composites, where this provides further evidence for validating the classification methodology chosen. The graph of the composites is a stepwise graph, where this interprets a constant regular behavior of the composites in their distribution.

All of these observations lead to the following conjecture.

Axiomatic Assumptions on Composites (Gap Base Zero)

1. *Composite of the same family are exactly predictable algebraically.*
2. *The gap between composite is 2 for even numbers, and six for odd composite numbers.*
3. *Between any two odd composites, there are either:*
    a. *two prime numbers*
    b. *or two pseudoprimes*
    c. *or a prime and a pseudoprime*
4. *There are always three even composites between any two odd composites*

Indeed this is an astonishing regularity, arrangement and relationship in behavior in terms of number distribution. It represents a different way of demonstrating the number distribution theorem in terms of gaps. The pattern derives the integer distribution cluster theorem (IDCT), where the aim of the theorem is to show how the composite gap affects the distribution behavior of numbers.

Consequently we can define a number cluster in terms of x. These are numbers in the distribution that can be derived using one variable. For example $x = 300$ gives us the cluster:

{1797, 1798, 1799, 1800, 1801, 1802}

X		Prime or Pseudoprime	Composites
300			1797
			1798
	GB4	1799	1800
	GB2	1801	1802

Table 126

The above observations may be formulated into a theorem as follows.

Integer Distribution Cluster Theorem (IDCT)

All integer numbers are distributed into a cluster set through a gap pair pattern of (1,1) and (2,2).

The proof of this theorem lies in establishing the universal number distribution framework. The importance about a number cluster is that it has the possibility of containing a twin pair for a given x as well as showing that numbers are a repeatable pattern in the form of a cluster of numbers. Furthermore, the concept emphasizes the fact that prime numbers always try to occur as a twin prime. This can be expressed as an axiom as follows.

Twin Pair Axiom

The minimum gap between any two twin prime pairs is four.

As a result, if one twin prime occurs, we can expect another one to occur within a difference of four as a minimum occurrence.

X		Prime or Pseudoprime	Composites	Composite Gap
15			87	1
			88	1
	GB4	89	90	2
	GB2	91	92	2
16			93	1
			94	1
	GB4	95	96	2
	GB2	97	98	2

Twin prime gap occurrence

Natural prime gap occurrence

Table 127

The composite gap (1,1) is said to define the natural gap occurrence, while the (2,2) composite gap is said to be aligned to the twin prime gap occurrence. Therefore;

a) A twin prime can only occur between a gap base 4 number and a gap base 2 number, and the twin pair will always starts at gap base 4 as already shown by cyclic transform structures.
b) The prime gap of four (called a natural prime gap) will always occur between a gap base 2 number and a gap base 4 number, and that this will always start at gap base 2.

All of this information is carried by the number distribution cluster.

10.3 Composite Groups

From the gap pattern (1,1) and (2,2), there are basically two groups of composites. That is, let group one of composites be those with a gap of one, and group two be those with a gap of two. The gap base 2 and 4 are already defined even though they fall under the pattern (1,1). Therefore this gives us groups that are defined in the same manner as in the case of prime numbers in the delta space.

Definition

Let $(c, 1)$ *and* $(c, 2)$ *form composite elements of set* C_0.

That is, (3,1), (4,1), (9,1) (10,1) are composites in the same group. Similarly, (0,2), (2,2), (6,2), (8,2) and so on are in the same composite group.

Composite	Gap Pattern
0	2
2	2
3	1
4	1
6	2
8	2
9	1
10	1
12	2
14	2
15	1
16	1
18	2
20	2
21	1
22	1
24	2
26	2
27	1
28	1

The (1,1) group of composites

The (2,2) group of composites

Table 128

Furthermore, we note that $(c, 1)$ has two types of elements, and odd number and an even number, whilst $(c, 2)$ has only even elements.

Definition

Let $(c, 1)$ *denote Class 1 composite elements, and* $(c, 2)$ *denote Class 2 composite elements.*

Then this implies this implies that the Class 1 set can be further categorized into two subgroups A and B, where group A is the odd numbers and B the even numbers. The same is true for Class 2, where group A stand for even numbers that multiple of six, and group B the set of even numbers with a difference of six.

Now the question we can ask ourselves is, which is the first composite if we ignore the negative numbers. By looking at the gap pattern, it appears that zero must be the first composite, then we have the pattern as follows.

Composite Sequence	Gap Pattern
0	2
2	2
3	1
4	1

Table 129. The gap pattern for numbers 0,2,3

If we combine the arguments raised in gap base 2 and 4 about the numbers 2 and 3, we see that these numbers fit perfectly in the descriptive pattern of gap base zero. In other words, this is where they belong. The number 2 and 3 belong to gaps base zero. All of these arguments are made in respect of pattern methodology and classification constructs.

10.4 UNDT

For generation of the composites in terms of a classification system, consider the odd composites as a base class which has been denoted as gap base zero. We note that in the gap occurrence between composites, the occurrence of gap 1 is defined by $6x - 3$, which implies that we can use this as a base class. Now the base class plus one gives us the next composite with gap 1, and the base class plus three and five respectively gives us the composites where the gap difference is two. Lets us take $x = 4$ to illustrate the point. Then:

- $6x - 3$ gives 21, this is a composite, a gap one number. We call this **Class 1A**.
- $21 + 1 = 22$, an even composite. We call this **Class 1B**.
- $22 + 2 = 24$, since 24 is a gap 2 number. This is the same as $21 + 3 = 24$; a composite. We call this **Class 2A**.
- $24 + 2 = 26$, since 26 is a gap two number. This is the same as $21 + 5 = 26$; a composite. We call this **Class 2B**.

We can then apply this as a general approach to derive all the numbers that we need using $6x - 3$ as our base class. The graphic below demonstrates using gap theory to produce the rest of the composites for a given range.

	Family-1	Family -2		
x	6x-3	Plus 1	Plus 3	Plus 5
	Class 1A	Class 1B	Class 2A	Class 2B
0	-3	-2	0	2
1	3	4	6	8
2	9	10	12	14
3	15	16	18	20
4	21	22	24	26
5	27	28	30	32
6	33	34	36	38
7	39	40	42	44
8	45	46	48	50
9	51	52	54	56
10	57	58	60	62

Table 130. The different composites classes

All the even composites define a family, called family 2, and the odd composites define family 1. Hence family 2 has three classes; 1B, 2A and 2B as a the condition for family is defined by an even number. However, family 1 has only one class consisting of odd numbers. However, class 1A and 1B defines group one composites because they belong to gap 1 composites. Similarly class 2A and 2B define group two composites as they are of gap 2. When we combine all the generating formulae in terms of x, then this gives us the universal number distribution theorem as illustrated graphically. That is, we assume three families that describe the distribution of numbers, where family-3 consists of numbers that are from gap base 2 and 4.

The basis of the universal distribution theorem is taken from the multi-value function $f(x) = (y_1, y_2, \dots y_n)$ where $(y_1 \cap y_2 \cap \dots \cap y_n) = \emptyset$.

Universal Number Distribution Theorem (UNDT)

Let $-\infty \leq x \leq \infty$ *be an integer, then all integer numbers are universally distributed as composite, prime and pseudoprime by* $f(x) = (y_1, y_2, y_3, y_4, y_5, y_6)$, *where* $y_i = 6x + n$, *for* $n = \{-3, -2, -1, 0, 1, 2\}$.

Since $(y_1 \cap y_2 \cap \dots \cap y_n) = \emptyset$, the sets y_i are unique, but this is not necessarily a ditation set. Similarly, we observe that;

- $n = \pm 1$ defines the set of primes and pseudoprimes
- $n \neq \pm 1$ defines the set of composite numbers or gap base zero.

The implication of the theorem is that all numbers are algebraic, they can be expressed as an algebraic entity in relation to each other. It is now possible to understand where the equations of the form $6x \pm 1$ come from, or how it is relevant to the generation of prime numbers. It also illustrates the reason behind the six factor in prime numbers or the distribution as a whole. For this theorem, the issue is not the counting properties of numbers, rather the focus is on the distributive properties of the numbers including the relationship that defines the distribution.

From the theorem, we derive other equations that define the distribution patterns. That is;

1. $6x - 3\ for\ y = -3$, Class 1A, Family 1
2. $6x - 2\ for\ y = -2$, Class 1B, Family 2
3. $6x - 1\ for\ y = -1$, Gap Base 4, Family 3
4. $6x\ for\ y = 0$, Class 2A, Family 2
5. $6x + 1\ for\ y = 1$, Gap Base 2, Family 3
6. $6x + 2\ for\ y = 2$, Class 2B, Family 2

	Family-1	Family -2			Family-3	
x	6x-3	6x-2	6x	6x+2	6x-1	6x+1
	Class 1A	Class 1B	Class 2A	Class 2B	GB-4	GB-2
-8	-51	-50	-48	-46	-49	-47
-7	-45	-44	-42	-40	-43	-41
-6	-39	-38	-36	-34	-37	-35
-5	-33	-32	-30	-28	-31	-29
-4	-27	-26	-24	-22	-25	-23
-3	-21	-20	-18	-16	-19	-17
-2	-15	-14	-12	-10	-13	-11
-1	-9	-8	-6	-4	-7	-5
0	-3	-2	0	2	-1	1
1	3	4	6	8	5	7
2	9	10	12	14	11	13
3	15	16	18	20	17	19
4	21	22	24	26	23	25
5	27	28	30	32	29	31
6	33	34	36	38	35	37
7	39	40	42	44	41	43
8	45	46	48	50	47	49
9	51	52	54	56	53	55
10	57	58	60	62	59	61

Table 131. The general distribution of numbers with "6" factor

Going down increasing the value of x, the factor of the difference of six is universal. The difference between all numbers along the column is six. Note that the composites behave like the prime numbers when the negative values are involved. That is, the positives of class 1B become the negatives of class 2B, a similar behavior of gap base 2 and 4. However class 1A and class 2A positives are merely negated.

There is an obvious relationship between the integer cluster theorem and UNDT as shown by the diagram below.

X	Gap Base	Prime or Pseudoprime	Equation	Composites	Composite Gap	Class	Equation
1				3	1	1A	6x-3
				4	1	1B	6x-2
	GB4	5	6x-1	6	2	2A	6x
	GB2	7	6x+1	8	2	2B	6x+2
2				9	1	1A	6x-3
				10	1	1B	6x-2
	GB4	11	6x-1	12	2	2A	6x
	GB2	13	6x+1	14	2	2B	6x+2
3				15	1	1A	6x-3
				16	1	1B	6x-2
	GB4	17	6x-1	18	2	2A	6x
	GB2	19	6x+1	20	2	2B	6x+2

Table 132

The UNDT is based on a difference of 6 between the x values whereas the ICDT is based on showing the repeat pattern and the composite gap formation.

10.5 Arithmetic Model For Prime Number Generation

A simple model may be suggested as a fundamental structure for the set of primes and pseudoprimes from the arithmetic context. Firstly the model suggests a gap base structure, and secondly a generative structure. That is, we can develop a 4-2 approach or a 2-4 combination as a progression difference to create two sets of numbers. Both systems will give the same gap base system, and the maximum difference between elements of the same set will always be 6. For example, if $p_1 = 5$ then we have a 2-4 format, that is 5-7-11. If $p_1 = 1$ then we have a 4-2 pattern, for example 1-5-7.

From the model it is then possible to generate the set of prime and pseudoprimes without resorting to algebraic equations, instead arithmetic is applied to a given structure. Hence we have a given generation;

1-5-7

7-11-13

13-17-19

19-23-25

25-29-31, ...

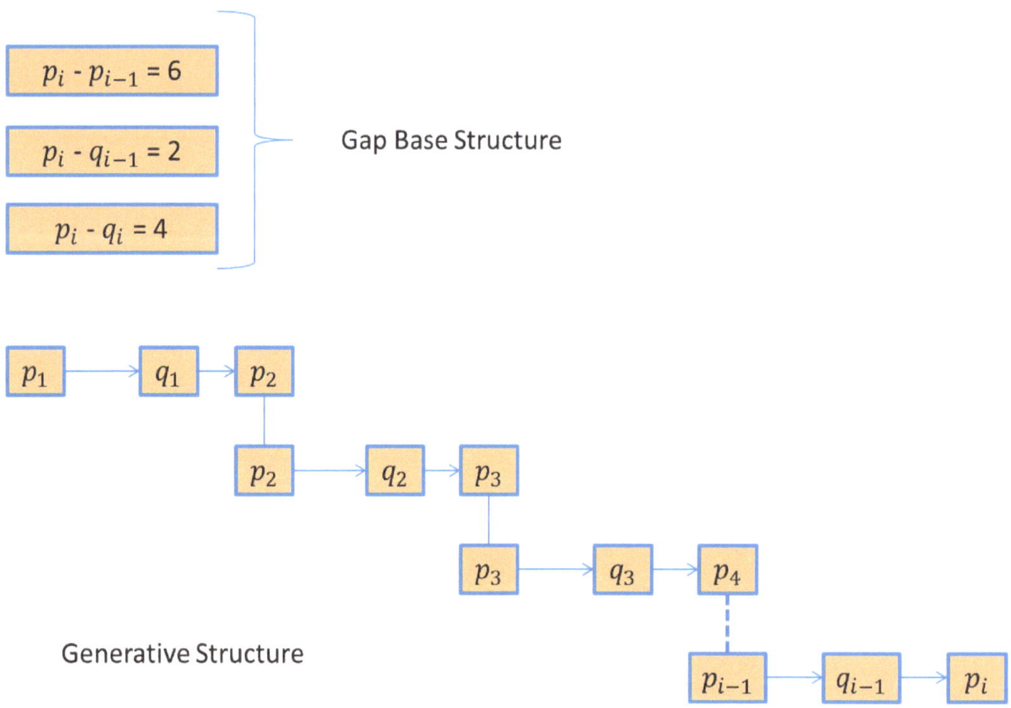

Figure 100. Fundamental structure of prime and pseudoprime generation

To start generating this sequence all you need is the first term (1,5). If it is 2-4, then we have;

5-7-11

11-13-17

17-19-23

23-25-29, ...

This is all done by addition of 2 and 4, or 4 and 2. Whilst this has been through of in specific terms, it can be developed further for a general case through a definition.

The definition of the arithmetic model for integer number distribution is as follows.

Definition – Arithmetic Distribution Model

Let $p_i \in A$ *and* $q_i \in B$ *where;*

a. $p_i - q_{i-1} = 2$
b. $p_i - q_i = 4$
c. $p_i - p_{i-1} = 6$

and $(p_1, q_1) = (a, b), 1 \leq i \leq \infty.$

The model is expressed in general terms through the use of the term (a, b). Hence;

a) For $(a, b) = (1,5)$, the model will generate a set consisting of primes and pseudoprimes.
b) For $(a, b) = (0,2)$, the model will generate Class 1B and Class 2A
c) For $(a, b) = (2,4)$, the model will generate Class 2A and Class 2B
d) Similalrly $(A \cup B)' = C$, where this defines the set Class 1A.

Again the model shows that there are three types of odd numbers, and three types of even numbers in integer distribution.

The model gives us the structure GB2-GB4-GB2 . The other structure would then be GB4-GB2-GB4. The substructure given by GB4-GB2 then defines the possibility of a twin prime occurrence for both GB4 and GB2 being prime. That is, in the structure GB2-GB4, a twin prime occurrence will never happen.

10.6 Consequence of the UNDT

The major consequence of the theory is to offer a general distribution context for all numbers, and in particular in the definition and relationship of prime, pseudoprime and composite. It has the same role as a unifying theory as to how numbers are classified and distributed, both negative and positive numbers. It is a unifying theory because it is not only concerned about the distribution of prime numbers, it concerns the entire number spectrum of the integers. It acts like the periodic table used in chemistry that classifies elements. Furthermore, it consolidates the concept of defining a prime, a pseudoprime, and a composite in an algebraic context. The framework of UNDT explains the regularity of prime numbers and the integers as a whole. Though random, the distribution is highly organized!

The distribution can be arranged in terms of the number family, or in an ascending order. The ascending order shows the power of the theory as it puts all odd numbers together, the prime and pseudoprime numbers in one place, as well as the even numbers according to their type. In particular note the position of 0, 1, 2 and 3 in the distribution pattern.

Position of 2 and 3 in the distribution

X	Class 1A	Class 1B	GB-4	Class 2A	GB-2	Class 2B
0	-3	-2	-1	0	1	2
1	3	4	5	6	7	8
2	9	10	11	12	13	14
3	15	16	17	18	19	20
4	21	22	23	24	25	26
5	27	28	29	30	31	32
6	33	34	35	36	37	38
7	39	40	41	42	43	44
8	45	46	47	48	49	50
9	51	52	53	54	55	56
10	57	58	59	60	61	62
11	63	64	65	66	67	68
12	69	70	71	72	73	74
13	75	76	77	78	79	80

Odd Numbers

Prime or Pseudoprime

Table 133

The sets that are generated by UNDT also have interesting properties. For example if you divide all the numbers in Class 1A by 3, then we get the set of primes or pseudoprimes. Not only that, the division process derives the same set, which is Class 1A.

Class 1A	Divide by 3
-3	-1
3	1
9	3
15	5
21	7
27	9
33	11
39	13
45	15
51	17
57	19
63	21
69	23
75	25

The set of Class 1A is derived again

Set of odd numbers not prime or pseudoprime

Set of primes or pseudoprimes both gap base 2 &4

Table 134

The property of producing the sets gap base 2 and 4 is unique to Class 1A. The other sets can also be divided by 2 to derive some interesting observations. The process demonstrates that the sets are related just like in the same way that odd and even numbers are related. Class 2A has the same property as 1A, when divided by two it derives itself. In addition to that it derives Class 1A.

Class 1B and 2B have the common characteristic that they each derive gap base 4 and 2 respectively when divided by 2. The two sets interchange, that is, Class 1B also derives 2B and 2B derives 1B.

Dividing by 2 generates Class 1A and Class 2A

Dividing by 2 generates Class 1B and the set gap base 2

X	Class 2A	Divide by 2		Class 1B	Divide by 2		Class 2B	Divide by 2
0	0	0		-2	-1		2	1
1	6	3		4	2		8	4
2	12	6		10	5		14	7
3	18	9		16	8		20	10
4	24	12		22	11		26	13
5	30	15		28	14		32	16
6	36	18		34	17		38	19
7	42	21		40	20		44	22
8	48	24		46	23		50	25
9	54	27		52	26		56	28
10	60	30		58	29		62	31
11	66	33		64	32		68	34
12	72	36		70	35		74	37
13	78	39		76	38		80	40
14	84	42		82	41		86	43
15	90	45		88	44		92	46
16	96	48		94	47		98	49

Dividing by 2 generates Class 2B and the set gap base 4

Table 135

Class 2A is the only set that can be divided by both 2 and 3. As a result it yields three different types of sets. It gives;

- Class 1A when divided by 2, and also generates itself
- The set of even numbers when divided by 3
- The set of integers when divided by 3 followed by 2, or by 6 as one operation.

X	Class 2A	Div by 2	Div by 3	Div by 2
0	0	0	0	0
1	6	3	2	1
2	12	6	4	2
3	18	9	6	3
4	24	12	8	4
5	30	15	10	5
6	36	18	12	6
7	42	21	14	7

Table 136

The third set can be said to consist of primes, pseudoprimes and composites. This can be defined as a unique property of set 2A. It is not surprising as it is the one that contains zero and all numbers are multiples of six.

The last observation that we may make is to note whether composites are random or not. From the behavior pattern of prime numbers and pseudoprimes, it has been noted that structure in which these numbers exist is consistent and predictable, but the consistent structure does not imply predictable values within the structure. Hence the conclusion that prime numbers are random, the same being true for pseudoprimes. A similar argument holds for the composites. They exist in a consistent structure because there are precise equations to derive such composites, and the gap patterns are also consistent. However, all that this consistency does is to define a position in the given structure, but not the nth value of that position if we were to arrange them in a sequence. Secondly, we work with the subsets in order to find values, but the universal set of composites has no equation to describe the value of a composite. This leads to a conjecture.

Conjecture

It is not possible to establish the sequential value of the nth composite by a direct equation.

If we have all the integer numbers in sequence, we cannot derive an expression that describes the position of the next composite. That is, regularity is positional and is a consequence of structure, but the actual value of the composite is independent of the position. The implication is that composites are also random, though they exist in a consistent structure. This is actually an inverse situation; if primes are random in distribution, the inverse applies to composites.

10.7 Structure of UNDT

A theory has three things that define it; the structure, then the system of the structure, and the model which is a consequence of the structure and system. The general structure of the UNDT framework can be made in terms of odd and even numbers. The structure gives us six types of numbers where;

a) Three types of odd numbers are derived
b) Three types of even numbers are derived

The difference to the next number in the same group or type is always six. It is a bit of a mystery why six, because the system will not work with 3, 4, 5, 7, of 8. It has a perfect way of relating all the integers in the context of six.

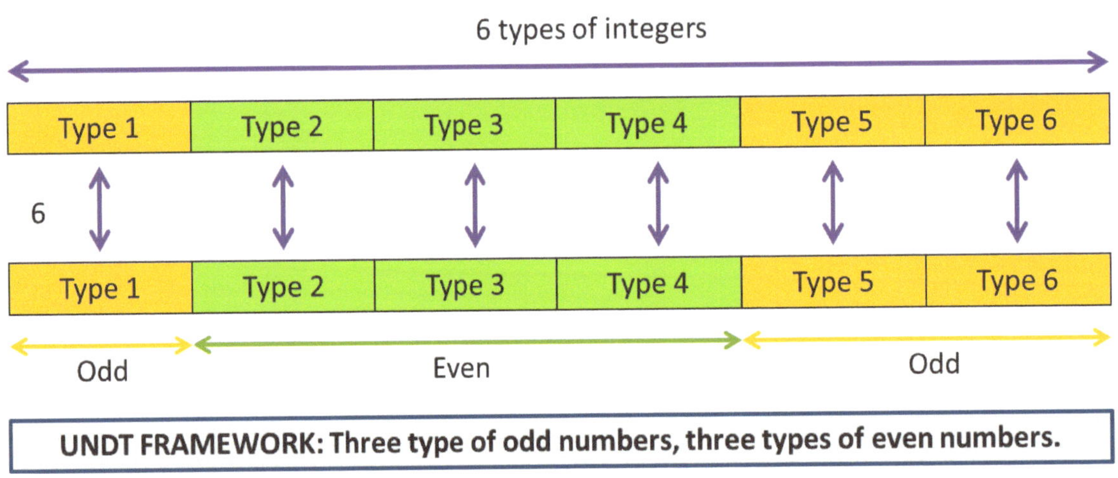

Figure 101

We can go deeper to consider other properties of the structure. From the various observations and definitions made, the UNDT has an interesting structure that consists of two linear ditation sets, for the odd numbers and the even numbers. The two sets are actually connected.

There are three types of number under UNDT: **Prime** Numbers in gap base 2 and 4; *Odd Numbers* **Pseudoprimes** in gap base 2 and 4; *Odd Numbers* **Composites** in gap base zero; *Odd& Even Numbers*					
Non-Ditation set		**Ditation Set D_t**		**Ditation Set D_e**	
Family 1	**Family 2**	**Family 3**		**Family 2**	
Gap Base Zero				**Gap Base Zero**	
Gap 1	**Gap 2**			**Gap 1**	**Gap 2**
Class 1A	**Class 2A**	**Gap Base 4**	**Gap Base 2**	**Class 1B**	**Class 2B**
6x-3	**6x**	**6x-1**	**6x+1**	**6x-2**	**6x+2**

Table 137

The structure gives seven ways of categorizing odd numbers, and five ways of categorizing the even number to form the UNDT structure. This is quite a difference from an odd number being prime or composite in the older context. Note that gap base zero is equivalent to the concept of a composite under the older context.

Odd Numbers	**Even Numbers**	
Gap Base Zero	Gap Base Zero	
Gap 1	Gap 1	Gap 2
Class 1A	Class 1B	Class 2A
Gap Base 2		Class 2B
Gap Base 4		
Prime		
Pseudoprime		

Table 138

It has already been noted that $f(x) = \{6x - 1, 6x + 1)$ forms a ditation set called D_t. Furthermore a ditation set has sets that are at a particular angle with respect to each other. A similar behavior is noticed for the sets $6x - 2$ and $6x + 2$. That is, we can define a ditation set $f(x) = (6x - 2, 6x + 2)$, and we realize that the condition $f_2(x) = -f_1(x)$ is maintained. This is defined as a ditation set D_e because it consists of even numbers.

Questions of interest would be whether this ditation set has random numbers like the other set D_t. A close observation shows that the set has no random numbers but is consistent in the way its elements are distributed.

	Family 1	**Family 2**	**Family 3**	**Family 3**	**Family 2**	**Family 2**
	Gap 1	**Gap 2**			**Gap 1**	**Gap 2**
	Class 1A	**Class 2A**	**Gap Base 4**	**Gap Base 2**	**Class 1B**	**Class 2B**
x	**6x-3**	**6x**	**6x-1**	**6x+1**	**6x-2**	**6x+2**
-11	-69	-66	-67	-65	-68	-64
-10	-63	-60	-61	-59	-62	-58
-9	-57	-54	-55	-53	-56	-52
-8	-51	-48	-49	-47	-50	-46
-7	-45	-42	-43	-41	-44	-40
-6	-39	-36	-37	-35	-38	-34

-5	-33	-30	-31	-29	-32	-28
-4	-27	-24	-25	-23	-26	-22
-3	-21	-18	-19	-17	-20	-16
-2	-15	-12	-13	-11	-14	-10
-1	-9	-6	-7	-5	-8	-4
0	-3	0	-1	1	-2	2
1	3	6	5	7	4	8
2	9	12	11	13	10	14
3	15	18	17	19	16	20
4	21	24	23	25	22	26
5	27	30	29	31	28	32
6	33	36	35	37	34	38
7	39	42	41	43	40	44
8	45	48	47	49	46	50
9	51	54	53	55	52	56
10	57	60	59	61	58	62
11	63	66	65	67	64	68

Table 139

It is not surprising that this is a ditation set because we note that it has even numbers that are twice of the set D_t. For example the numbers 11 and 13 when doubled give 22 and 26 in the set D_e. The next numbers are 17 and 19 in D_t doubled to 34 and 38 in D_e. Now in between these two numbers are 28 and 32, where these are double 14 and 16. All of these are elements of D_e. This is a consistent pattern as shown in the table.

GB4	GB2	Gap 1	Gap 2	Gap 1	Gap 2
6x-1	6x+1	6x-2	6x+2	6x-2	6x+2
5	7	4	8	4	8
11	13	10	14	10	14
17	19	16	20	16	20
23	25	22	26	22	26
29	31	28	32	28	32
35	37	34	38	34	38
41	43	40	44	40	44
47	49	46	50	46	50
53	55	52	56	52	56
59	61	58	62	58	62
65	67	64	68	64	68
		70	74	70	74
		76	80	76	80
		82	86	82	86
		88	92	88	92
		94	98	94	98
		100	104	100	104
		106	110	106	110
		112	116	112	116
		118	122	118	122
		124	128	124	128

130	134	130	134
136	140	136	140

Table 140

If we trace the distribution of the even numbers from their occurrence, for example 7, 14, 28, 56, 112 and so on, we see that this gap is becoming bigger, and that this is a similar behavior to the formation of pseudoprimes. The numbers from the set D_t produce even numbers at a consistent position in the grid, then the other numbers from the set D_e "fill" in the gaps in between so to speak. The only difference is that no prime numbers are produced. In the D_t set, pseudoprimes take specific positions and the primes are the numbers that fill in the empty spaces.

The Class 1A and Class 2A are not ditation sets. These two sets do not have the property $f_2(x) = -f_1(x)$. If we double any number in Class 1A, then the number will be in the set Class 2A.

10.8 Arithmetic Structure - Multiplication and Addition Between Sets

The integer product law finds the relationship in products between the three type of numbers; the primes, the pseudoprimes and the composites. This gives us a framework for additive and multiplicative operations for the linear ditation set, that gap base 2 and 4. We can discriminate further the type of answer we get in a product by considering the different types of sets. With the exception of Class 2B, all other products within a set form a closed operation. The distribution of the product class is probably a result of the algebraic structure of numbers since a distinct pattern of resulting products can be observed.

	Class	Class	Resulting Product Class
Multiplying elements in the same set	Class 1A	Class 1A	Class 1A
	Class 1B	Class 1B	Class 1B
	Class 2A	Class 2A	Class 2A
The Even Number Ditation set	Class 2B	Class 2B	Class 1B
	Class 1B	Class 2B	Class 2B
Different sets but the result is in Class 2A	Class 1A	Class 1B	Class 2A
	Class 1A	Class 2A	Class 2A
	Class 1A	Class 2B	Class 2A
	Class 1B	Class 2A	Class 2A
	Class 2A	Class 2B	Class 2A
Composite set multiplying a gap base set (Odd Ditation Sets). Note that the result is an even ditation set with GB2 and GB4.	Class 1A	GB-2	Class 1A
	Class 2A	GB-2	Class 2A
	Class 1B	GB-2	Class 1B
	Class 2B	GB-2	Class 2B
	Class 1A	GB-4	Class 1A
	Class 2A	GB-4	Class 2A
	Class 2B	GB-4	Class 1B
	Class 1B	GB-4	Class 2B
Product between gap base sets; the ditation sets	GB-2	GB-2	GB-2
	GB-4	GB-4	GB-2
	GB-2	GB-4	GB-4

Table 141. Product distribution

The product can further be classified as being gap 1 or gap 2 if the result is a composite, or in terms of family if we are considering the framework of the UNDT. We realize that there are 21 product categories, and consequently we may observe the distribution and probability of a given product.

	1A	1B	2A	2B	GB2	GB4	
Occurrence	3	4	8	3	2	1	21
Probability	14%	19%	38%	14%	10%	5%	

Table 142

In terms of addition, the UNDT framework is a modulo six operation for the type of class. For example, starting with Class 1A, if we add 1 to any element here we get an element of Class 1B. We can continue until we come to class 2B.

Addition is also considered between the sets, but a different approach is used. The addition is defined to take place in terms of the generating functions of the sets. For example: $60542 + 1911 = 62453$, where are are adding the two numbers from Class 2B and Class 1A respectively to get a third number. When we do the addition in terms of the generating functions, then we use the value of x, and the second number we denote it with the value of y. Now the corresponding value of x and y for the two numbers in the example are $x = 10090$, and $y = 319$. Similarly the result 62453 corresponds to the value $z = 10409$, and we note that $10409 = 10090 + 319$. Now the result is in gap base 4. This operation is then denoted as follows:

$$f(x \in 2B) + f(y \in 1A) = f(z \in GB4)$$

That is;

$$f(10090) + f(319) = f(10409)$$

Therefore we generally assume that $x + y = z$.

Taking f(x)	Adding f(y)	The result is:
Adding within sets		
1. $f(x \in 1A)$	$f(y \in 1A)$	$f(z+1 \in 2A)$
2. $f(x \in 1B)$	$f(y \in 1B)$	$f(z-1 \in 2B)$
3. $f(x \in 2A)$	$f(y \in 2A)$	$f(z \in 2A)$
4. $f(x \in 2B)$	$f(y \in 2B)$	$f(z+1 \in 1B)$
5. $f(x \in GB2)$	$f(y \in GB2)$	$f(z \in 2B)$
6. $f(x \in GB4)$	$f(y \in GB4)$	$f(z \in 1B)$
Adding between sets		
7. $f(x \in 1A)$	$f(y \in 1B)$	$f(z-1 \in GB2)$
8. $f(x \in 1A)$	$f(y \in 2A)$	$f(z \in 1A)$
9. $f(x \in 1A)$	$f(y \in 2B)$	$f(z \in GB4)$
10. $f(x \in 1A)$	$f(y \in GB4)$	$f(z-1 \in 2B)$
11. $f(x \in 1A)$	$f(y \in GB2)$	$f(z \in 1B)$
12. $f(x \in 1B)$	$f(y \in 2A)$	$f(z \in 1B)$
13. $f(x \in 1B)$	$f(y \in 2B)$	$f(z \in 2A)$
14. $f(x \in 1B)$	$f(y \in GB4)$	$f(z \in 1A)$
15. $f(x \in 1B)$	$f(y \in GB2)$	$f(z \in GB4)$
16. $f(x \in 2A)$	$f(y \in 2B)$	$f(z \in 2B)$
17. $f(x \in 2A)$	$f(y \in GB4)$	$f(z \in GB4)$
18. $f(x \in 2A)$	$f(y \in GB2)$	$f(z \in GB2)$
19. $f(x \in 2B)$	$f(y \in GB4)$	$f(z+1 \in 1B)$
20. $f(x \in 2B)$	$f(y \in GB2)$	$f(z+1 \in 1A)$
21. $f(x \in GB4)$	$f(y \in GB2)$	$f(z \in 2A)$

Table 143

10.9 UNDT Prime Generation

One of the interesting relationships that can be immediately picked up is the following two.

1. $f(x \in 1A) + f(y \in 1B) = f(x \in GB2)$
2. $f(x \in 1A) + f(y \in 2B) = f(x \in GB4)$

For example;

a) 29 = 14+15
b) 67 = 33+34
c) 49 = 27+22
d) 95 = 50+45

Originating sets

X	Class 1A	Class 1B	Class 2B
1	3	4	8
2	9	10	14
3	15	16	20
4	21	22	26
5	27	28	32
6	33	34	38
7	39	40	44
8	45	46	50
9	51	52	56

Addition

1A+1B	1A+2B	1A+1B	1A+2B
		7	11
13	17	19	23
25	29	31	35
37	41	43	47
49	53	55	59
61	65	67	71
73	77	79	83
85	89	91	95
GB2	GB4	GB2	GB4

Table 144

That is, the sum is as unique as the generating functions we have used before in producing a prime or pseudoprime using the function of x and y approach. This can be simplified to a function by substitution and assuming $x = y$.

$$f(x \in 1A) = 6x - 3$$

$$f(y \in 1B) = 6x - 2$$

$$Therefore\ f(x \in GB2) = 12x - 5$$

And similarly;

$$f(x \in 1A) = 6x - 3$$

$$f(y \in 2B) = 6x + 2$$

$$Therefore\ f(x \in GB4) = 12x - 1$$

Surprisingly we can switch the roles around and still get a similar result by adding 1. That is;

$$f(x + 1 \in 1A) = 6(x + 1) - 3$$

$$f(y \in 1B) = 6x - 2$$

$$Therefore\ f(x \in GB4) = 12x + 1$$

And adding to set 2B where the assumption has been $x = y$, similarly;

$$f(x \in 1A) = 6x - 3$$

$$f(y + 1 \in 2B) = 6(x + 1) + 2$$

$$Therefore\ f(x \in GB2) = 12x + 5$$

	1A+1B	1A+2B	1A+1B	1A+2B
	12x-5	12x-1	12x+1	12x+5
X	GB2	GB4	GB2	GB4
0	-5	-1	1	5
1	7	11	13	17
2	19	23	25	29
3	31	35	37	41
4	43	47	49	53
5	55	59	61	65
6	67	71	73	77
7	79	83	85	89
8	91	95	97	101
9	103	107	109	113

6X-1	6X+1
GB4	GB2
-1	1
5	7
11	13
17	19
23	25
29	31
35	37
41	43
47	49
53	55

Table 145

This actually gives us alternative equations to apply in deriving the primes and pseudoprimes in gap base 2 and 4. Finding the sum can also be done where we derive sets in terms of $12x \pm y$, for $y = \pm 1, y = \pm 5$. The result is also used as a factorization engine instead of $6x \pm 1$.

10.10 Natural Primality Testing

The gap base system provides us with a simple but effective way of being certain that some numbers are definitely not prime, where this is called natural primality testing. Using the gap base system, there are three types of odd numbers, since this is where the prime number occurrence is observed. These are;

a) Gap base 4 odd numbers defined by $6x - 1$, where this is called O_1. Here there are two sub-categories of odd numbers, a prime or a pseudoprime.
b) Gap base 2 numbers defined by $6x + 1$, where this is called O_2. Here there are two sub-categories of odd numbers, a prime or a pseudoprime.
c) Class 1A numbers that are also called composites defined by the fact none of them satisfy $6x \pm 1$, where this is called O_3.

Prime number occurrence takes place under $6x \pm 1$ for O_1 or O_2 but O_3 will never be a prime number. Therefore the UNDT framework is said to define natural primality testing from O_3.

Hence instead of looking for prime numbers, we can also focus on looking for odd numbers that are certainly not prime numbers. This is particularly useful for large numbers. Consider the following number and knowing its classification.

W = 789900541289753188975009123215467980768572971397525750310229845601221

This number has 69 digits altogether, and the question of interest is whether it has the possibility of being prime, or whether we can find a number after it that we are hundred percent sure it is not a prime number. With classification this becomes a simple task to perform.

The first step is to convert the number to a string so that we can break it into manageable parts in order to perform division by 6.

i. If (6x-1) is integer, then the number is gap base 2
ii. If (6x+1) is integer, then the number is pgap base 4
iii. If (6x+3) is integer, then the number is Class 1A.

Secondly, we note that this is a large number and adding or subtracting a number less than 10 will not influence the rest of the number. So we choose a four digit number that we call the classification tail from the right, and we need only perform the arithmetic operations on this number to determine our final result.

Classification Tail = 1221

The following table then shows the process of division to determine the outcome. For n=0, that is our starting point, so the remainder is 0. We then truncate the number into equal parts of 9 digits since the computer takes up to ten digits.

	Truncated Number Strings	Put Remainder in front of number	Divide the number	Integer Answer without decimal	Change answer to string to form answer	This is the remainder as number	Change Remainder to string
n=0						0	0
n=1	789900541	0789900541	131650090.2	131650090	131650090	1	1
n=2	289753188	1289753188	214958864.7	214958864	214958864	4	4
n=3	975009123	4975009123	829168187.2	829168187	829168187	1	1
n=4	215467980	1215467980	202577996.7	202577996	202577996	4	4
n=5	768572971	4768572971	794762161.8	794762161	794762161	5	5
n=6	397525750	5397525750	899587625	899587625	899587625	0	0
n=7	310229845	0310229845	51704974.17	51704974	51704974	1	1
n=8	60	160	26.66666667	26	26	4	4

Table 146

When we have our final division remainder, we use this on the classification tail to determine the classification of the number as follows.

		TAIL				
Start Tail Addition	n=9	1221.00	Number	Put Rem in front	Divide by 6	Remainder
Class 1A = 6x-3	Add 3	1224.00	1224.00	41224	6870.666667	4
Gap Base 4= 6x-1	Add 1	1222.00	1222.00	41222	6870.333333	2
Gap Base 2 = 6x+1	Subtract 1	1220.00	1220.00	41220	6870	0

Table 147

When we have a remainder that is 0, this indicates an integer. Therefore the number is a gap base 2 number, and it has the possibility of being a prime number. This we are definite about. Now we can use this information to determine the gap base 4 number as well as the Class 1A number. Using the UNDT framework, the difference between these is 2 and 2 from START point.

GB4	789900541289753188975009123215467980768572971397525750310229845601219	Minus 2
GB2	789900541289753188975009123215467980768572971397525750310229845601221	Start
Class 1A	789900541289753188975009123215467980768572971397525750310229845601223	Plus 2

Table 148

We can further develop our question and demand five numbers after W are definitely not prime. Outside the UNDT framework, this would be a major challenge, but under UNDT, it only takes a minute to determine these numbers. We can generate a general approach in the following manner. Everything in this framework is systematic.

n	GB-4	GB2	Class 1A
1	1219	1221	1223
2	1225	1227	1229
3	1231	1233	1235
4	1237	1239	1241
5	1243	1245	1247
6	1249	1251	1253

Table 149

All the numbers that are definitely not prime will be found in Class 1A, so the tail addition is made on Class 1A. However, the table also shows for gap base 2 and 4 in case we have an additional interest. Therefore the numbers are as follows.

789900541289753188975009123215467980768572971397525750310229845601223
789900541289753188975009123215467980768572971397525750310229845601229
789900541289753188975009123215467980768572971397525750310229845601235
789900541289753188975009123215467980768572971397525750310229845601241
789900541289753188975009123215467980768572971397525750310229845601247
789900541289753188975009123215467980768572971397525750310229845601253

Table 150

The above numbers are not prime, and this is a natural primality test as a consequence of the classification system. In looking at the odd numbers, the pattern is obviously following a specific pattern. The starting odd number is gap base 4, followed by gap base 2, then class 1A. The difference between each of them is 2 in a cyclic manner.

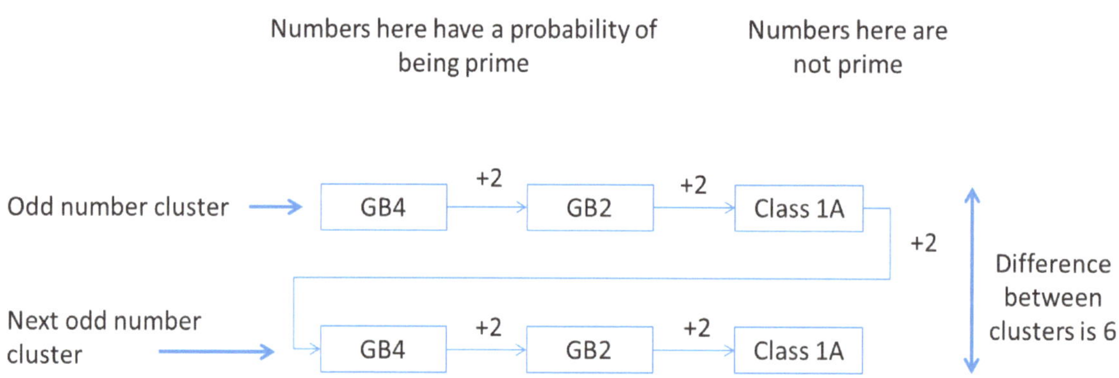

Figure 102

Each cycle with a difference of 2 is said to form an odd number cluster, and the difference between each cluster in the same column is six. As a result of this type of odd number structure, we can then accurately predict the set of all numbers that are odd and definitely not prime, which we refer to as natural primality testing.

This operation is simplified greatly through the concept of merge operations.

Structured Factorization

The major application of the UNDT is the development of structured or systematic factorization. This is a new approach that eventually leads to the concept of factor pairs.

11 Factorization under UNDT

The fact that 2 and 3 are not defined to be prime numbers, and 1 is a prime number does not conform to the traditional understanding of the prime number for reasons already outlined. The numbers 2 and 3 fit the definition of being natural numbers and being divisible by one and themselves only. The concept of a prime number has been solely based on this premise. Its major consequence has been the fundamental theorem of arithmetic which states that all positive numbers can be written as factors of primes. This develops the notion of prime numbers as being building blocks for the natural numbers. Similarly the number 1 is not considered prime because a prime number divides by one and itself, whereas here "itself" is one and therefore creates a repetition. It may be noted that from the conventional perspective, a specific classification theory has never been used as a methodology to define the prime number. This has a consequence then as to how they are also regarded in terms of the factorization process, where the process even includes negative numbers.

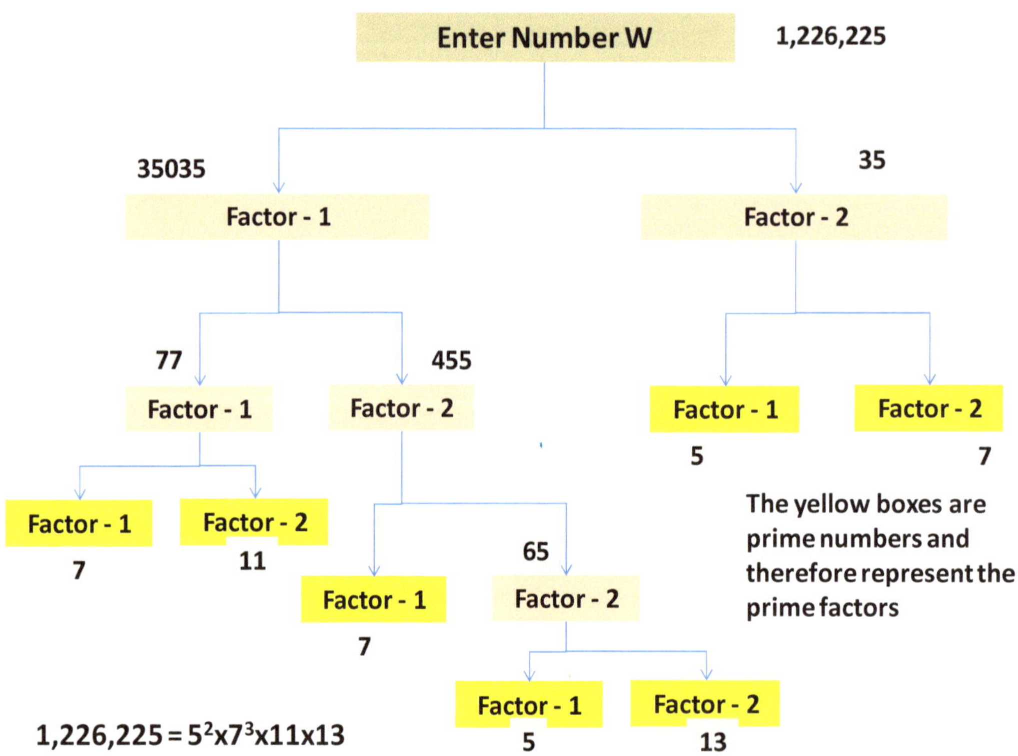

Figure 103

Conventionally two methods are used to factorize numbers. The first is the trail division method defined through dividing consistently by the smallest possible prime number until the process stops. The second method involves using a tree to look for the highest products that define the number, and doing the same process to the branch until all the prime factors have been derived. Of the two methods, the tree method is probably faster.

The interest is to find out the effect of factorization in the context of the universal number distribution theorem (UNDT). This theorem has five main consequences;

- the distribution of the prime numbers within the set of integers,

- the distribution of pseudoprimes
- the distribution composites
- the relationship between the three types of numbers described in terms of defining families.
- The fifth consequence would be how it affects the understanding of factorization of integers.

The following patterns were then derived by factorizing numbers in each of the UNDT sets. This represents the first effort in trying to identify a particular factorization pattern.

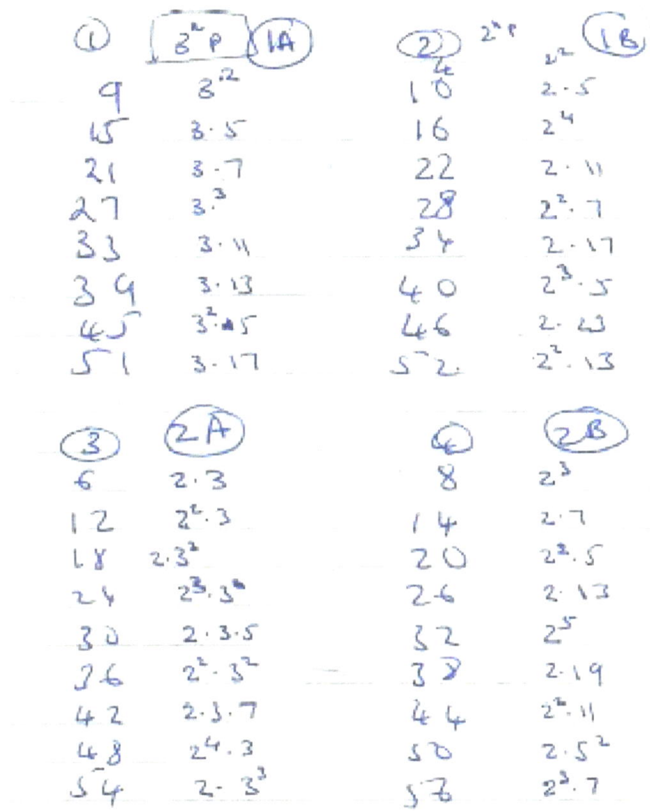

Figure 104

A clear pattern emerges when numbers from each class are factored as shown in the graphic. For example class 1A results in factors of 3 as a pattern, class 2A has 2 and 3 as the basic pattern whilst 1B and 2B have a 2-pattern. It seems that the 5-pattern is also a common feature in the factorization. In terms of observation, the following patterns are therefore noted.

a) 2^n
b) 2^n3^m
c) 2^n5^m
d) 3^n
e) 3^n5^m
f) 5^n

One may argue that the pattern makes a case for 2 and 3 to be regarded as prime numbers in the context of the Fundamental Theorem of Arithmetic. In terms of pattern methodology, this is where 2 and 3 can claim some credibility. Indeed if we do not classify numbers, and only assume the rule that a prime must be divisible by one and itself only, the argument is valid.

It is also clear that elements of gap base 2 and 4 consist of prime numbers and pseudoprimes only, hence another structure called $[P]$ is defined.

Definition

Let $[P] = p_1^{x_1} p_2^{x_2} \dots p_n^{x_n}$ *where* p_i *is a prime number sequence in its natural order and that* $0 \leq x \leq \infty$, *and* $0 \leq n \leq \infty$.

The set of primes that define $[P]$ are called the prime number factorization set. This is a pseudoprime set. Therefore if *n* = 4, then we have as an example $[P]$ = $7^2 x 13^4 x 17^3 x 19$, implying the rest of x_i are zero. The table that follows is derived in terms of the universal number distribution theorem to show the factorization pattern of any integer that is not a prime number.

The factorization set then is based on three types of sets, the set C_0 consisting of composites, and the set $[P]$ consisting of pseudoprimes, and elements of the set $C_0 \times [P]$. The following is noted.

a) The set C_0 derives composite factors of an integer
b) The set [P] derives prime factors of an integer

This is a new dimension of interpretation for the factorization theory since 2 and 3 are not prime numbers.

Format	Class	Number Type	Generator
a) $[P]$	Gap Base 2, or Gap Base 4	Pseudoprime Odd	$6x + 1$ *or* $6x - 1$
b) $2^n[P]$	Class 2B	Composite, Even	$6x + 2$
c) $2^n[P]$	Class 1B	Composite, Even	$6x - 2$
d) $2^n 3^m [P]$	Class 2A	Composite, Even	$6x$
e) $3^n[P]$	Class 1A	Composite, Odd	$6x - 3$

Table 151

Clearly 2 and 3 are still part of factorization in the UNDT, but both of them are not involved in factoring numbers that are pseudoprimes, that is the set $[P]$. They are actually used to produce the set of even numbers. The table shows that there is a specific factorization pattern. That is;

1. the pseudoprimes are defined only by $[P]$
2. factorization occurs by class in C_0
3. in each class, the set of pseudoprimes forms a factoring base for the rest of the other numbers in $C_0 \times [P]$.

Hence $[P]$ is referred to as a factorization base. Such a base is defined by either a prime number or a pseudoprime number. Note that even though $2^n[P]$ defines a particular pattern, it appears in two different classes.

Furthermore, the implication is that you can have numbers produced without the base by using;

a. 2^n
b. 3^n
c. $2^n 3^m$

The three covers the set of all composites C_0, and consequently define a framework for developing a factorization algorithm. These numbers are referred to as the factorization keys. For example the series;

$$2, 4, 8, 16, 32, 64, 128, \cdots$$

is a factorization key based on 2^n. When we look for prime factors in the key, this is referred to as primary factorization, and when we look for primes in the base this is referred to as secondary factorization.

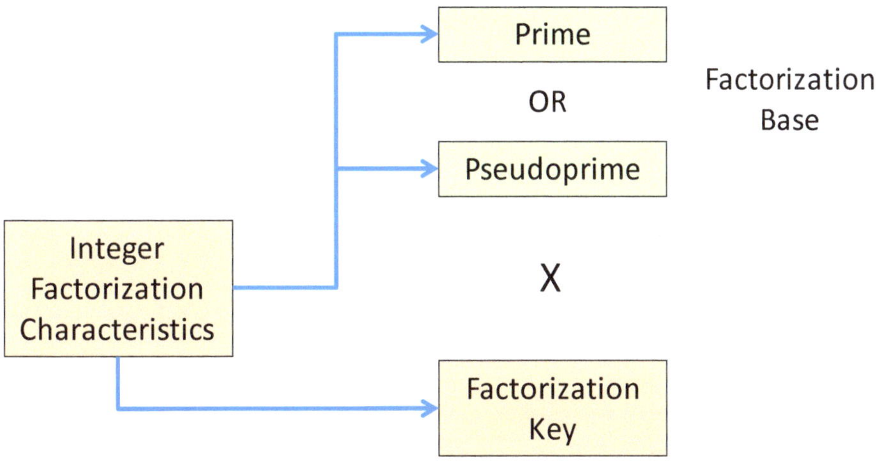

Figure 105

Consequently we have prime based or pseudoprime based integers, and composite based integers. For example 14 is a prime based integer; $14 = 2 \times 7$. Similarly 70 is a pseudoprime based integer; $70 = 2 \times 35$, and 12 = 2×6 Is a composite based integer. The series $13 \cdot 2^n$ for $1 \leq n \leq \infty$ is prime based through 13. We can factor a number into two components under UNDT; the factorization key and the factorization base.

a) A factorisation key consists of composite factors of a number.
b) A factorisation base consists of prime factors of a number.

Consider the number 56435659063056, the factorization key which is 432 and the factorization base which is 130638099683. Hence we can specify a factorization base as$[P]$ = 7^2x13^4x17^3x19 and then define the key to the base as $2^n 3^m$ for $0 \leq n \leq 6$ and $0 \leq m \leq 3$. Then $2^n 3^m$ gives the following numbers that multiply $[P]$;

1 2 3 4 6 8 9 12 16 18 24 27 32 36 48
54 72 96 108 144 216 288 432 864

For $n = m = 0$ gives an identity, hence the result is only $[P]$. Note for example that $432 = 2^4 3^3$, the required factorization key.

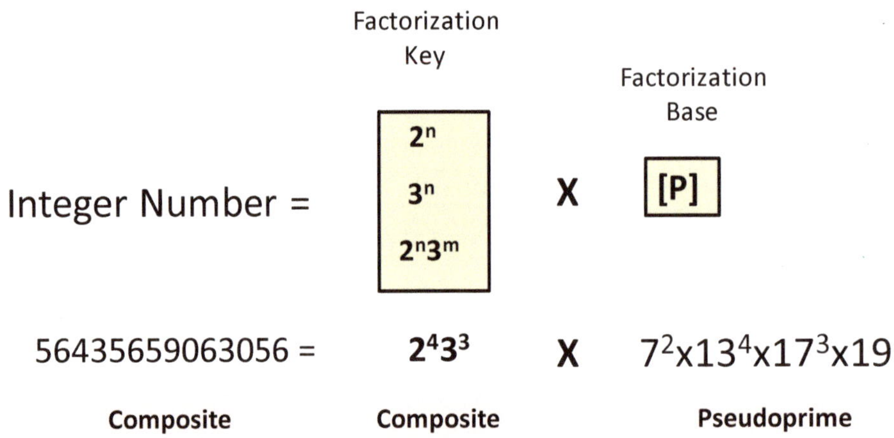

Figure 106

The general factorization pattern is also shown by the examples below.

Generator	Class	Form	Examples
$6x \pm 1$	GB2, GB4	$[P]$	1. $2737 = [7 \cdot 17 \cdot 23]$ 2. $12805 = [5 \cdot 13 \cdot 197]$ 3. $28025 = [5^2 \cdot 19 \cdot 59]$
$6x + 2$	2B	$2^n[P]$	4. $3404 = 2 \cdot 2 \cdot [23 \cdot 37]$ 5. $272 = 2 \cdot 2 \cdot 2 \cdot 2 \cdot [17]$
$6x$	2A	$2^n 3^m [P]$	6. $3402 = 2 \cdot 3 \cdot 3 \cdot 3 \cdot 3 \cdot 3 \cdot [7]$ 7. $464070 = 2 \cdot 3 \cdot [5 \cdot 31 \cdot 499]$
$6x - 2$	1B	$2^n[P]$	8. $464068 = 2 \cdot 2 \cdot [11 \cdot 53 \cdot 199]$ 9. $268 = 2 \cdot 2 \cdot [7]$
$6x - 3$	1A	$3^n[P]$	10. $14067 = 3 \cdot 3 \cdot [521]$ 11. $3399 = 3 \cdot [11 \cdot 103]$

Table 152

In the context of the UNDT, the numbers 1, 2, 3 and 5 are defined as fundamental UNDT integers since they result in a specific factorization pattern of the integer numbers.

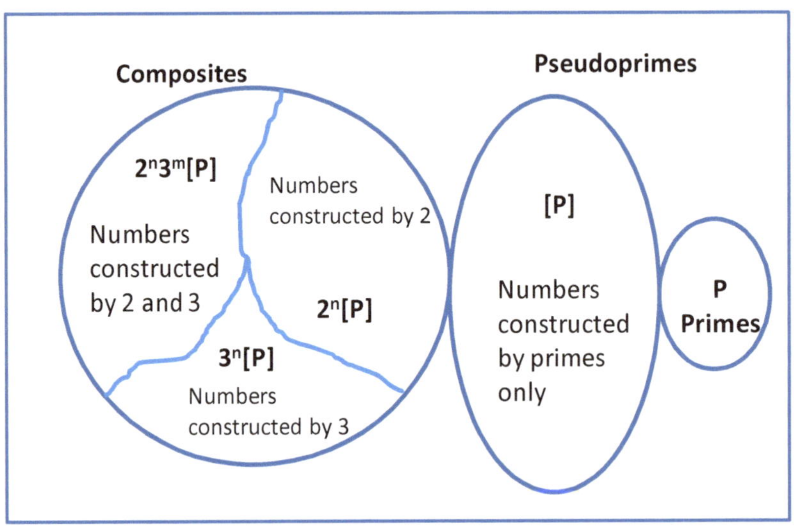

Figure 107

They are used to get the factorization key of any integer that is composite or pseudoprime as shown in the set diagram. The significance of the diagram is that it demonstrates that there is a specific pattern to factorization. If we develop a factorization structure, then we can do it along this pattern to maximize efficiency of the process.

We note that the form follows the pattern of primes, pseudoprimes and composites in the way factorization is defined structurally.

One of the amazing things about the UNDT is the type of regularity that it produces in the behavior patterns of numbers. When we divide any set with the smallest possible factor 2 or 3, then we get the same sets repeating themselves. That is, if you divide set 1A by three, you get elements of GB2, GB4 and 1A again, and this happens consistently for the second and third time and so on. This is illustrated in the table.

Set	1st Format	2nd Format	3rd Format	4th Format	
1A	GB4:GB2:1A	GB4:GB2:1A	GB4:GB2:1A	...	
2A	1A:2A	1A:2A	1A:2A	1A:2A	...
1B	GB4:2B	GB2:1B	GB4:2B	GB2:1B	...
2B	GB2:1B	GB4:2B	GB2:1B	GB4:2B	...

Table 153

This leads to the transfer principle in factorization.

11.1 The Structured Factoring Method (SFM)

Factorization is one of the arithmetic challenges in mathematics that has been met with the mighty force of high speed super computers. It can take years for a computer to find prime factors of a given large number, and this has resulted in using such numbers for cryptography. Hence the interesting question is whether classification will offer a quicker and more methodical approach to factorization.

As has been noted, the distribution of prime factors is also an organized behavior though the behavior is basically random. They occur according certain classes. Consequently it is possible for one to determine what numbers will occur and their class.

The SFM is based on the fact that we have three distinct number sets; the primes, pseudoprimes and the composites. Now a prime number does not have factors by definition, which implies that the method should only consider the following possibilities.

a) Factoring a pseudoprime. The method looks for powers of the prime factors in a pseudoprime.
b) Factoring a composite. The method looks for powers of the composites 2 and 3.
c) Factoring a composite that is a product of pseudoprime, composite and prime. The method looks for both prime factors and 2 and 3.

Basically if the method is developed for the pseudoprime, and then for the composite, the combination of products will mean the two systems coming together to derive the required factors. This then derives what is called the structured factorization method (SFM). The method is structured because each procedure defines a factorization path, where factors are derived algebraically.

11.2 SFM for [P]

The set of pseudoprimes belongs to the algebraic sieve, therefore we can use the properties of the sieve in order to develop a methodology for factorization.

As an example, consider the number 4661. Using $6x \pm 1$, 777 implies that it is in gap base 4. We then note that there is a direct relationship between the number 777 and the values of x that define the multiplicands. That is;

$$f(x_1) \cdot f(x_2) = f(x_3)$$

$$f(10) \times f(13) = f(777)$$

This translates to the same as $59 \times 79 = 4661$, therefore x_1 = 13, x_2= 10 and x_3 = 777. Hence in the factorization process, we make use of this fact, as well as using equations to establish integer values that indicate the gap base.

INPUT	GB2	GB4
3481	580	580.3333
3599	599.6667	600
3721	620	620.3333
3953	658.6667	659
4087	681	681.3333
4189	698	698.3333
4453	742	742.3333
4661	776.6667	777

x2	x1	x3		
10	13	777	GB-4	4661
59	79	776.6667	GB-2	4661

	x2	x3
	10	777
		x1
Testing input in GB2		13.33898
Testing input in GB2		12.57377
Testing input in GB4		13

Table 154

Secondly, there is the pair theorem stated thus;

For W a pseudoprime, then $f(x_a)$ and $f(x_b)$ are ditation factors of W if and only if $W = f(x_a) \cdot f(x_b)$.

This gives us a framework for looking for factor pairs on the basis of Factor-1 and Factor-2. As a result we begin by assuming a simple factorization taking recognition of the UNDT environment. If there is a number W, then we have to know the gap base of this number, and the gap base of the factors from the product rule as shown in the table.

Factor-1	Factor-2	Result
GB2	GB2	GB2
GB2	GB4	GB4
GB4	GB4	GB2

Table 155

The table shows that a product is defined in terms of gap base, not just in terms of two numbers multiplying each other. Even though the table shows only two factors multiplying each other, we can develop the model deeper by considering other product options.

1st Order

Factor-1	Factor-2	Result
GB2	GB2	GB2
GB2	GB4	GB4
GB4	GB4	GB2

2nd Order

Factor-1	Factor-2	Result
GB2 GB2	GB2	GB2
GB2 GB2	GB4	GB4
GB4 GB4	GB2	GB2
GB4 GB4	GB4	GB4
GB2 GB4	GB4	GB2

3rd order

Factor-1	Factor-2	Result
GB2 GB2	GB2 GB2	GB2
GB2 GB2	GB2 GB4	GB4
GB2 GB2	GB4 GB4	GB2
GB2 GB4	GB4 GB4	GB4
GB4 GB4	GB4 GB4	GB2

Table 156

The first order product is between any two numbers, and the second order product is between any three numbers, while the third order product is between any two pairs of numbers. The assumption is that this pattern will define factorization pattern even if more than two pairs are involved as further simplification can always take place within this framework. This is illustrated by taking a product of eight numbers.

		GB2			Result
		GB2	GB2		2 numbers
	GB4	GB4	GB2		3 numbers
GB2	GB4	GB4	GB2	GB2	5 numbers
GB4 GB4	GB4	GB2 GB4	GB4 GB4	GB2	8 numbers

Table 157

Consequently we can confine our methodology to the first three orders of products as our basic framework of SFM. Then it implies that the first order is the solution that we are looking for, when no further factorization can take place. The two possible structures for further factorization are the second and third order. Secondly, we always assume that we are starting with W being in the third order as that is the most complex structure for consideration, and then work our way down to the first order relationship.

To operationalize the structure, we use the primality test as the basic since the aim anyway is to look for prime factors. The structure only concerns itself with the set $[P]$, which consists of pseudoprimes. The type of primality tests are:

1. The algebraic sieve, which looks for factors in terms of the punching out the non-prime numbers. This will not be effective in terms of speed and efficiency.
2. The algebraic primality test. The advantage of this test is that it can find the primality of a number each time it finds a factor. But it does not search for factors, it only tests for primality. It is a reasonable strategy to pursue.
3. The Mbulawa primality test. In terms of the framework of the SFM proposed above, this is probably the best approach because we can test for primality and at the same time we can do a factor search. The operation can iterate between these two processes until it finds all the prime factors. The prime factors found will define the base for the factorization process, since $[P]$ defines the base.

The steps of the process are as follows.

1. Find the gap base of the number W
2. Find the factors of W
3. Determine the product format of W
4. Establish that all factors are in the first order structure through testing for primality.

A comment may be made here in passing on the fruitfulness of doctrine. In developing the SFM approach, the power of the gab base framework is evident all the way. Factorization takes place in terms of the structure of gap base theory as already illustrated. This stems from the fact that the product structure itself occurs according to the framework. Therefore the SFM is an illustration of fruitfulness of the doctrine of gap base theory.

We already have a basic format from the Mbulawa primality test. We vary x_2 and look for the integer value x_1, knowing that these two define a set of factors. We make use of the following equations based on $f(x_1) \cdot f(x_2) = f(x_3)$.

a) $x_3 = 6x_1x_2 + (x_1 + x_2)$, defining a GB2xGB2 product
b) $x_3 = 6x_1x_2 - (x_1 - x_2)$, defining a GB2xGB4 product
c) $x_3 = 6x_1x_2 - (x_1 + x_2)$, defining a GB4xGB4 product

The above equations can be written in the following manner in the context of the model for factorization.

a) $x_3 = 6x_1x_2 + (k_1 + x_2)$
b) $x_3 = 6x_1x_2 - (k_2 - x_2)$
c) $x_3 = 6x_1x_2 - (k_3 + x_2)$

However, the last result is obtained from $(6x_1 + 1)(6x_2 - 1)$ that is GB2xGB4. It will be different when we consider it the other way round, the result is not commutative. Hence;

$$6x_1 - 1)(6x_2 + 1) = 36x_1x_2 + 6x_1 - 6x_2 - 1$$

This simplifies to $x_3 = 6x_1x_2 + x_1 - x_2$, that is; $x_3 = 6x_1x_2 + (k_3 - x_2)$.

11.2.1 Primitive Factorization Space

Our starting point is $f(x_3) = W$, the number we want to factorize. Since we know W, then we can determine x_3, and then vary x_2 to get x_1. Sincex_3 is a product of x_1 and x_2, then it implies;

$$x_1 < x_3 \text{ and } x_2 < x_3$$

So in the count function of x_2, we are certain that we will not count more than x_3. This condition defines the primitive factorization space, and this is similar to the trial division method of looking for prime numbers. Since our main concern is counting through x_2, we then define the verification level as $x_2 < x_3$. By verification level we mean we have counted such that we arrive at a value of x_2 where we are certain the entire prime factors for the given number can be found. The smaller this level is, the better. Since $W = 6x_3 \pm 1$, and assuming the bigger value $W = 6x_3 + 1$, then in general we can set $x_3 = (W - 1)/6$ where this defines the verification level for a given W.

When we know what x_1 is, we then substitute in $f(x_1) = 6x_1 \pm 1$ according to gap base to get $Factor1$, knowing that $\frac{W}{f(x_1)} = Factor2$ to give us the other factor. We distinguish the type of product and gap base by solving in terms of $k = x_1$ from the following table.

GB2xGB2	GB4xGB4	GB2xGB4	GB4XGB2
$k_1 = \frac{x_3 - x_2}{6x_2 + 1}$	$k_2 = \frac{x_2 + x_3}{6x_2 - 1}$	$k_3 = \frac{x_3 - x_2}{6x_2 - 1}$	$k_3 = \frac{x_3 + x_2}{6x_2 + 1}$

Table 158

Then $Factor1 = 6k_1 + 1\ or\ 6k_2 + 1$ for gap base 2, and $Factor2 = W/Factor1$. Similarly, for gap base 4, then $Factor1 = 6k_3 - 1$ and $Factor2 = W/Factor1$. The factor pairs (Factor-1, Factor-2) are called trial pairs, and the process produces as many as possible. For example in $W = 595$, then $x_3 = 99$ and when $x_2 = 1$, this gives us $x_1 = 20$ or $x_1 = 14$. We note that W is a gap base 2 number, which means it is a result from the product GB4xGB4 or GB2xGB2. Now;

$$f(20) = 6(20) - 1 = 119, \text{and } \frac{W}{119} = 5$$

$$f(14) = 6(14) + 1 = 85, \text{and } \frac{W}{85} = 7$$

When we enter x_2 = 1, we get two factor pairs that satisfy the condition, these being (119,5) and (85,7) both in GB-2. The third factor we ignore because this is not a GB4 and GB2 product and $k_3 = 19.666667$, a decimal. Infact line one and line three for the factors will always be the same, and we determine the correct result by considering the gap base of the product. The result is a second order factorization, hence we now iterate on 85 and 119.

Enter Number W =	595

x3		
	99.00	Gap Base 2
	99.333333	Gap Base 4

x2	1

				Factors Found	
			Product Type	GB2	GB2 or GB4
x1	20	GB2	GB4XGB4	119	5
	14	GB2	GB2XGB2	85	7
	19.666667	GB4	GB2XGB4	119	5

Figure 108

Enter Number W =	85

x3		
	14.00	Gap Base 2
	14.3333333	Gap Base 4

x2	1

				Factors Found	
			Product Type	GB2	GB2 or GB4
x1	3	GB2	GB4XGB4	17	5
	1.85714286	GB2	GB2XGB2	12.14286	7
	2.66666667	GB4	GB2XGB4	17	5

Enter Number	W =	119

x3		
	19.67	Gap Base 2
	20	Gap Base 4

x2	3

				Factors Found	
			Product Type	GB2	GB2 or GB4
x1	1.33333333	GB2	GB4XGB4	7	17
	0.87719298	GB2	GB2XGB2	6.263158	19
	1	GB4	GB2XGB4	7	17

Figure 109

Therefore let $W = 85$ and $W = 119$. Again $x_2 = 1$ gives us factors 17 and 5 for 85, and $x_2 = 3$ gives us factors 7 and 17 for 119. These are GB4xGB4 and GB2xGB4 products respectively. Hence the model is able to also establish immediately the type of product trace we have. Since these are the prime factors we are looking for, hence 595 = 5x7x17. Rather than iterating this way, we can do one process since we know that looking for integer values of x_2 has an initial point and an end point.

11.2.2 The Standard Factorization Space

We now try to establish the necessary verification levels x_2. For the primitive space, we know that $x_1 < x_3$ and $x_2 < x_3$. From $W = 6x_3 \pm 1$, and assuming the bigger value $W = 6x_3 + 1$, then $x_3 = (W - 1)/6$, which implies x_3 is smaller than W by a factor of 6. For example if $W = 8569$, then we need only count up to 1428 to establish all the prime factors of W. This is about 17% of the value of W, and derives the standard factor pairs for W.

When $x_2 = 1$ this gives us the maximum value of x_1 that can be found, therefore we can define what is referred to as a bounded factor search using this technique. That is, for a given number W, factors of W can be established by looking for integer values of x_1 as shown in the table. Solving for the endpoint is done by assuming $x_1 = 1$. Interestingly, for GB2(1), the value of x_1 when $x_2 = 1$ is also the maximum point. That is;

$$x_1 = (x_3 + x_2)/(6x_2 - 1)$$

Hence if $x_1 = 1$, then $x_2 = (x_3 + 1)/5$. If $x_2 = 1$, then $x_1 = (x_3 + 1)/5$. This is a useful relationship because we need only make $x_2 = 1$ to know how far to reach the maximum value.

Base Result	Solving for endpoint for factors of W; $x_1 = 1$	Endpoint smaller than W by about
GB2(1)	$x_2 = (x_3 + 1)/5$	30 times
GB2(2)	$x_2 = (x_3 - 1)/7$	42 times
GB4	$x_2 = (x_3 + 1)/7$	42 times

Table 159

Using that approach, we derive the following table for $W = 8569$. If we divide by the smallest number to find the count that satisfies us that we have all the prime factors, then $\frac{8569}{30} = 286$. That means we do not have to go beyond 286 in looking for factors of 8569 when we vary x_2. Indeed, when $x_2 = 1$ in the structure, then $x_1 = 130$, which means that will be the maximum value of x_2 before x_1 becomes less than one. Remember $x_1 < 1$ defines primality, so for factors of W it implies $x_1 \geq 1$.

W = 8569

x_2	x_1	Factor-1	Factor-2	Type
2	130	779	11	GB4 GB4
3	75	451	19	GB2 GB2
7	35	209	41	GB2 GB2
35	7	41	209	GB2 GB2
75	3	19	451	GB2 GB2
130	2	11	779	GB4 GB4

Table 160

This is a very useful observation indeed. For example if we have a number such as 130638099683 which is a GB4 number, then we automatically know that we can vary x_2 up to about 3110430945 in order to establish its factors. The number 3110430945 represents a verification level.

In other words, if we do not substitute through an iteration process, we can vary x_2 from the initial point to the endpoint to derive the factors that we are looking for. This is an ideal situation for computer programming as one just makes a loop to look for factors. Note the factors, where they appear in the table. The pattern is such that it always starts with factor-2 being the smallest, and then switches for factor-1 to start from a larger factor to a smaller factor.

We now have a framework for the structured factoring method (SFM), and we can explore its strengths and particularities. All the observations establish the type of tool we have, with the aim of making it computer friendly for programming.

11.2.3 Factorization Observations

Observation 1

The numbers in yellow are always a product of two numbers in orange box. Looking at the example, then:

a) 119 = 7x17
b) 85 = 5x17
c) 35 = 5x7

As this will always be there case, we use such numbers to establish the factors, that is 595 = 5x7x17. We multiply from the top of factor-2.

x_2	x_1	Factor-1	Factor-2
1	20	119	5
1	14	85	7
3	6	35	17
6	3	17	35
14	1	7	85
20	1	5	119

Table 161

We also note that x_1 and x_2 have a symmetric behavior, they vary in their values in a "mirror-effect" manner. The blue colored values of x_1 and x_2 is called the factor switch, because factors on the right go left and those on the left go right. That is initially $Factor1 > Factor2$, and a factor switch occurs then $Factor1 < Factor2$. The occurrence of the factor switch tells us that all unique factor pairs have been identified. Therefore in computer programming, when a factor switch occurs, the computer would know that all unique factor pairs have been found.

Observation 2

It is particularly interesting when the factor switch occurs immediately alone at the start of the operation as shown in the following example. The factoring of 319 immediately goes into a factor switch. This means this is the only pair that is defined for the number.

W= 319

x_2	x_1	Factor-1	Factor-2
5	2	11	29
2	5	29	11

Table 162

In such a factor switch, the factors are always prime or pseudoprime. A general pattern can be established from the following table.

W	x_2	x_1	**Factor-1**	**Factor-2**	Type
779	7	3	19	41	GB2GB4
817	3	7	43	19	GB2GB2
	7	3	19	43	
1411	3	14	83	17	GB4GB4
	14	3	17	83	

Table 163

A factor switch is a standard occurrence. Prime numbers that are from the same gap base result in a factor switch, but if two primes are from different gap bases, then we will have only one factor pair in our result as shown with the number 779.

The main observation is that for both cases, this tells us that W is only a product of two pairs of numbers, we should not expect any more. It is also a verification test for the number of factor pairs available. The result also tells us that there is no factor switch for W if it is of the product form GB2×GB4. However, GB4×GB4 may also result in the absence of a factor switch.

Observation 3

The obvious question is; are the prime factor pairs always yielding results that have prime factors only? The answer is no, and we can see from the example of $W = 14399$. The first result gives us 31213x17, which we also factor and its first values are 4459 and 7.

W = 14399

x_2	x_1	**Factor-1**	**Factor-2**
3	5202	31213	17
1	743	4459	7
2	400	2401	13
8	106	637	49
15	57	343	91
57	15	91	343
106	8	49	637
400	2	13	2401
743	1	7	4459

Table 164

It is clear from this example that *31213 = 7×13×49×91*, and 49 and 91 can be factored further. Therefore the power of the SFM lies in the fact that it can provide us the factor pairs in a short time, but a check is necessary to ensure that the values are all prime.

Observation 4

When $x_1 < 1$, then there are no more factors pairs to be found in the bounded factor search. The condition $x_1 < 1$ means $x_1(1) < 1$ and $x_1(2) < 1$ for gap base two, and $x_1(3) < 1$ for gap base four. However, this does not mean that further factors cannot be found in the pairs themselves. But we are hundred percent certain that no more factor pairs exist. This makes SFM a powerful mathematical tool for factorization.

For example consider $W = 17423$, a gap base 4 number.

W = 17423

x_2	x_1	**Factor-1**	**Factor-2**	**Type**
22	22	133	131	GB2 GB4

Table 165

In this case 133 has factors. The factors of 133 are 7 and 19.

Observation 5

Assume $W = xy$, and $x \in GB2$, $y \in GB4$, then we have the relationship that;

a) If x and y are prime or pseudoprime, then x_2 will generate only one factor pair.

779

x_2	x_1	**Factor-1**	**Factor-2**
7	3	19	41

Table 166

That is, there is no factor switch if the product results from elements of a different gap base.

b) If either x or y is prime, then several pairs will be generated according to $1 \leq x_2 \leq (x_3 + 1)/7$. The difference between the operation in GB4 and GB2 number is that GB4 is not symmetric in the value of x_1 and x_2 in its factoring pattern. An example is given with $W = 1001$ = 7x11x13. However, the prime factors found follow a similar pattern.

W=1001

x_2	x_1	**Factor-1**	**Factor-2**	Type
2	15	91	11	GB2XGB4
13	2	13	77	GB2XGB4
24	1	7	143	GB2XGB4

Table 167

There is no factor switch.

Let us take another example, $W = 7007$. Now GB-4 has the limitation that it does not have a factor switch as demonstrated by the example above. If we run this number, we observe the same thing as it is a GB-4 number. But what is challenging about GB-4 numbers is the ability to pick up the factors as shown in this example. The power of SFM seems lost.

W=7007

x_2	x_1	**Factor-1**	**Factor-2**	**Type**
2	106	637	11	GB2GB4
13	15	91	77	GB2GB4
24	8	49	143	GB2GB4
90	2	13	539	GB2GB4
167	1	7	1001	GB2GB4

Table 168

It seems clear then that the best method for GB-4 is the use of the fast method, as this delivers the results in the format we want. Consider the numbers $W = 1001$ and $W = 7007$.

	x_2	x_1	**Factor-1**	**Factor-2**	**Type**
W=1001	2	15	91	11	GB2GB4
91	1	2	13	7	GB2GB2
	x_2	x_1	**Factor-1**	**Factor-2**	**Type**
W=7007	2	106	637	11	GB2GB4
637	1	15	91	7	GB2GB2
91	1	2	13	7	GB2GB2

Table 169

Observation 6

As stated before, factor pair occurrence when W is in GB2 has a symmetric effect, the values of x_1 and x_2 transpose position such that the point of transposition indicates the number of prime factors involved. An example is given from $W = 8569$. We note that:

a) 779 = 19x41
b) 451 = 11x41
c) 209 = 11x19

	x_2	x_1	Factor-1	Factor-2	Type
	2	130	779	11	GB4 GB4
	3	75	451	19	GB2 GB2
W = 8569	7	35	209	41	GB2 GB2
	35	7	41	209	GB2 GB2
	75	3	19	451	GB2 GB2
	130	2	11	779	GB4 GB4

Table 170

This gives us a way of knowing where the factor switch occurs. We multiply the numbers from the bottom if we use factor-1 or from the top if we use factor-2. For example *11x19 = 209*, so we know that the factor switch occurs after 209.

Another example from $W = 805$, where we have *5x7 = 35*, so the factor switch occurs after this number.

	x_1	x_2	Factor-1	Factor-2	
	1	19	115	7	GB2 GB2
	1	27	161	5	GB4 GB4
	4	6	35	23	GB4 GB4
W = 805	6	4	23	35	GB4 GB4
	19	1	7	115	GB2 GB2
	27	1	5	161	GB4 GB4

Table 171

Both 8569 and 805 are examples of a perfect factor switch because in the factorization path there is no GB2×GB4 product.

Observation 7

We can have a deviation from the above where the factors have to be confirmed. For $W = 7315$, we still have the factor switch, but 35, 55, 77 does not follow the expected pattern, and as such they are referred to as a factor creep. In a factor creep, the product of factor-2 numbers results in a number in factor-2 instead of factor-1. In the standard approach of SFM, we cannot ignore a repeat value of x_2, therefore we include (1,174) as a valid pair, but in the iterative approach, it can be ignored.

W = 7315

x_2	x_1	Factor-1	Factor-2
1	244	1463	5
1	174	1045	7
2	111	665	11
3	64	385	19
6	35	209	35
9	22	133	55
13	16	95	77
16	13	77	95
22	9	55	133
35	6	35	209
64	3	19	385
111	2	11	665
174	1	7	1045
244	1	5	1463

Table 172

We still get our result that 7315 = 5×7×11×19, but the creep has to be taken into account when doing the SFM. It is easy to detect because one of the numbers will be in factor-2. That is, 5×7 =35, 7×11=77, and 5×11=55. This does not mean factor pairs from a creep are not valid, they are actually correct and in place because 77×95 = 55×133 = 35×209 to give us 7315.

Observation 8

We can develop a standard method of factorization and a faster method of factorization through iteration. The standard method would use the above technique of looking for factor pairs by continuously increasing the value of x_2 according to the bounded search parameters.

The fast method breaks this cycle and instead uses the lowest possible values of x_2 in its search for factor pairs. An example is made from $W = 888511171$. Using the standard approach, we would get 14 factor pairs. The first value is 52265363, and we iterate on this.

W=888511171

Without iteration

X2	X1	Factor-1	Factor-2
3	8710894	52265363	17
8	3150749	18904493	47
23	1080914	6485483	137
133	185338	1112029	799
388	63583	381499	2329
1073	22998	137989	6439
1353	18244	109463	8117

With iteration

X2	X3	Factor-1	Factor-2
3	8710894	52265363	17
8	185338	1112029	47
23	63583	381499	137
1353	1073	6439	8117

18244	1353	8117	109463
22998	1073	6439	137989
63583	388	2329	381499
185338	133	799	1112029
1080914	23	137	6485483
3150749	8	47	18904493
8710894	3	17	52265363

Table 173

Looking at the values of x_2, we are likely to count x_2 to 8710894 the bounded search parameter. Using iteration method, we get 3 more pairs and have arrived by x_2 = 1353. This time we have a factor switch at (1353,18244). Note that the maximum value of x_2 is 1353, yet we are dealing with a very large number.

Observation 9

The fast method also avoids the factor creep and the factor switch. Whereas it took 14 steps to do 7315, it takes only 3 steps to do it with the fast method. In the fast method, we enter 7315, then 1463, then 133.

W = 7315

x_2	x_1	Factor-1	Factor-2
1	244	1463	5
2	22	133	11
1	3	19	7

Table 174

The search ends at (19,7) because 19 is prime. The fast method always ends with a prime number in factor-1.

Observation 10

The SFM approach works like a coordinate system in identifying elements of the set $[P]$. It is actually a form of sieve methodology, and is the "secret" behind the technique. For example we note the following, where base 1 is gap base 4, and base 2 is gap base 2.

Base	x2	x1	Factor-1	Factor-2
1	0	0	1	-1
2	0	0	1	1
1	1	0	1	5
2	1	0	1	7
1	2	0	1	11
2	2	0	1	13
1	3	0	1	17
2	3	0	1	19
1	4	0	1	23
2	4	0	1	25

Base	x2	x1	Factor-1	Factor-2
1	0	0	1	-1
2	0	0	1	1
1	1	1	5	5
2	1	1	7	7
1	2	2	11	11
2	2	2	13	13
1	3	3	17	17
2	3	3	19	19
1	4	4	23	23
2	4	4	25	25

Table 175

This brings us back to the concept of a multi-valued function when we have elements such as (1,0) = 5 and (1,0) = 7. We realize that it has meaning if two elements may have the same coordinate but differ in the base, that is (1,1,0) = 5 and (2,1,0) = 7 where 1 and 2 define a base system. This results in uniqueness within the multi-value mapping as it assigns an attribute that causes differentiation in terms of space.

Note also that it lays the claim that our first prime number is actually -1, defined by (1,0,0) = -1, where (2,0,0) = 1, and the gap between them is two. This agrees with the UNDT approach where $x = 0$ derives the first elements, where -1 is the first element in GB4.

Composites | Composites By Gap | Universal Number Distribution Theorem | Prime Mapping | Pseudop

By Family | Order

X	Family-1	Family-2			Family-3	
	Class1A	Class1B	Class2A	Class2B	GB-4	GB-2
0	-3	-2	0	2	-1	1
1	3	4	6	8	5	7
2	9	10	12	14	11	13

Figure 110

The importance of this system is that it explains the behavior of the SFM construct. For example, it gives us understanding why we can have repeat values for x_2 as in the example shown. For $W = 7315$, then 7315 = 1463×5=10457, hence we have a (1,1) occurrence for x_2.

x_2	x_1	Factor-1	Factor-2
1	244	1463	5
1	174	1045	7

Table 176

The values of $x_2 = 1$ occur at the same time because (1,0) = 5 and (1,0) = 7 are defined by the same variable. Therefore every value of x_2 corresponds to the SFM coordinate of the factor we seek. In this example:

a) (244,0) = 1463
b) (174,0) = 1045

In the fast method, we can ignore the second occurring value and take only the first one, but the same may not hold for the standard method.

Observation 11- Differentiating Theorem

We also have a condition where $x_1 = x_2$ defines a product between two elements of [P] that are the same, that is a square product. Therefore:

a) $W \in [P]$
b) $W = y^2$, where $y = 6x \pm 1$
c) $x_1 = x_2$

What this says is that the square root of W which is a pseudoprime will never be a composite number, and therefore the square root of a composite number will never be a prime number.

Differentiating Theorem

The square root of a composite will always be a composite.

For example the square root of 3364 is 58, and this is a composite in Class B. It is another way of saying one can never multiply two composites to get a pseudoprime, which is the integer product law.

11.3 The Factor Pair Theory

All of the above observations can now be used to develop a formal theoretical structure for the factorization process. In summary, we establish the SFM construct to possess properties as follows. The properties are used to conceptualize a systematic approach in doing factorization and primality testing called the factor pair theory.

a) It derives all the elements of $[P]$, that is in gap base 2 and 4. This becomes an alternative presentation of the primes and pseudoprimes.
b) It derives a unique set of factor pairs. This is its most important function.
c) In each prime factor pair, the SFM approach looks for the largest Factor-1 and the smallest Factor-2. That is $Factor1 > Factor2$.
d) The first run is to derive the factor pairs, and then to do a primality test on each factor. The method gives us exact numbers to test, reducing the factorization process time taken.
e) It conducts a bounded search for the pairs starting from 1 to $x_1 = (x_3 + 1)/5$ or $x_1 = (x_3 - 1)/7$ in gap base 2 and $x_1 = (x_3 + 1)/7$ in gap base 4. The bounded search gives us precision in the prime factor search, hence reducing time taken to identify factors.
f) The bounded search can have a factor creep in which factor-2 numbers multiply each other. Numbers above the factor creep in factor-2 form the expected factors or those below in factor-1.
g) When $x_1 < 1$, then there are no more factors pairs to be found in the bounded factor search. The condition $x_1 < 1$ means $x_1(1) < 1$ and $x_1(2) < 1$ for gap base two, and $x_1(3) < 1$ for gap base four. This is the certainty condition of the prime factor search.
h) When $Factor1 < Factor2$, this is called a factor switch.
i) Gap base two products have a factor switch, where factor-2 numbers become factor-1 numbers.
j) The product between two numbers in factor-1 or factor-2 will give the number after which a factor switch occurs.
k) When a factor switch is a single occurrence, it defines a selective primality test, that is, the two factors will be prime, or at least one will be prime.
l) The product between factor-2 numbers results in an answer that is a factor-1 number, the converse being true after a factor switch unless there is a factor creep.
m) We can produce a standard version of SFM and a fast version through iteration.
n) The square root of a composite will never be a prime number, hence giving us the differentiating theorem.
o) In the context of the SFM framework, -1 is the first prime number if we apply pattern methodology. This provides a basis for defining the negative prime numbers.

Consequently the above observations lead to the development of a factor pair theory through the formation of a systematic model.

11.3.1 The SFM Coordinate System

The basis of the SFM coordinate system is the universal number distribution theorem (UNDT). Every number in this system has a specific position in the grid such that it can be described. For example $x = 4$ gives us two numbers in Family-3, and that is 23 in GB4 and 25 in GB2. That is, the coordinate may be described in terms of;

- The value of x
- The family
- The class in the family

Hence this provides a unique relationship between numbers as a system. Consequently it results in a powerful factorization theory because there is a precise relationship between numbers.

X	Family-1	Family-2			Family-3	
	Class1A	Class1B	Class2A	Class2B	GB-4	GB-2
0	-3	-2	0	2	-1	1
1	3	4	6	8	5	7
2	9	10	12	14	11	13
3	15	16	18	20	17	19
4	21	22	24	26	23	25
5	27	28	30	32	29	31
6	33	34	36	38	35	37
7	39	40	42	44	41	43

Table 177

The challenge then is the development of a factorization engine by applying the UNDT framework. As noted before, we make use of the following equations based on $f(x_1) \cdot f(x_2) = f(x_3)$.

a) $x_3 = 6x_1x_2 - (x_1 - x_2)$, defining a GB2xGB4 product
b) $x_3 = 6x_1x_2 - (x_1 + x_2)$, defining a GB4xGB4 product
c) $x_3 = 6x_1x_2 + (x_1 + x_2)$, defining a GB2xGB2 product

The engine has the following structure as shown, using $W = 7$ as an example.

Enter Number	W =	7

x3	1.00	Gap Base 2
	1.3333333	Gap Base 4

x2	1

				Factor-1	Factor-2
			Product Type	GB2	GB2 or GB4
x1	0.4	GB2	GB4XGB4	1.4	5
	0	GB2	GB2XGB2	1	7.00
	0.0666667	GB4	GB2XGB4	1.4	5

Table 178

What the engine does is to create a coordinate system by which to identify every pseudoprime number, and then uses this to connect in the product definition. First of all it tests the gap base of the number, then runs values of x_2 starting at 1, and looks for integer values of x_1. Every integer value of x_1 has a corresponding factor labeled 1 and 2. If one enters W as a prime number or as a pseudoprime, then a coordinate structure is derived as shown. For example, entering $W = 7$ gives a coordinate when $x_2 = 1$, then $x_1 = 0$, $factor1 = 1$ and $factor2 = 7$. The same result is obtained for $W = 5$, what differs is the gap base.

x2	1

			Factor-1	Factor-2
		Product Type	GB2	GB2 or GB4
x1	0.3333333	GB2	GB4XGB4	1	5
	-0.047619	GB2	GB2XGB2	0.714286	7.00
	0	GB4	GB2XGB4	1	5

Table 179

If we keep on entering we get a coordinate table to infinity. Note that $x = 3$ from the UNDT framework gives us 17 and 19 in the engine, where $x_2 = 3$ and $x_1 = 0$.

x2	x1	Factor-1	Factor-2
0	0	1	-1
0	0	1	1
1	0	1	5
1	0	1	7
2	0	1	11
2	0	1	13
3	0	1	17
3	0	1	19
4	0	1	23
4	0	1	25

Table 180

Hence applying this in product form, it implies $17 \cdot 34 = (17{,}3) \cdot (23{,}4) = (391{,}65)$. In carrying out factorization, we are then reversing this process knowing 391 and finding (17,3) and (23,4).

In order to make this into a system, the equations are re-arranged to reflect x_1 differently as follows, so as to be able to identify which x_1 we are dealing with. This develops three variables instead of two.

a) $x_3 = 6x_1x_2 - (k_1 - x_2)$
b) $x_3 = 6x_1x_2 - (k_2 + x_2)$
c) $x_3 = 6x_1x_2 + (k_3 + x_2)$

Let us consider the example $W = 3595163 = 139 \cdot 151 \cdot 167$, that is $(139{,}23) \cdot (151{,}25) \cdot (167{,}28)$. From the UNDT framework, we have;

X	GB2	GB4
23	137	139
24	143	145
25	149	151
26	155	157
27	161	163
28	167	169

Table 181

When we run the engine, then we get the result below.

X2	K3	Factor-1	Factor-2

28	3498	20989	167
3869	25	151	23213
4203	23	139	25217

Table 182

Furthermore in terms of UNDT, it implies that other numbers relate in the following manner. These can be found from $6x - 1$ as they are GB4 numbers.

x	Pseudoprime
3498	20989
3869	23213
4203	25217

Table 183

Hence there exist a factorization coordinate system that allows the factor pair theory structure to be defined.

Coordinate Factorization Theorem

Every $W \in D_t$ has a unique factorization coordinate (x_2, k_1) or (x_2, k_2) in gap base 2 and (x_2, k_3) in gap base 4.

The application of this theorem is the derivation of factor pairs as a consequence of the coordinate system defining the structure.

11.3.2 Transient Localized Databases

In trial division, factors of a number can be found by dividing with all possible prime numbers less than the square root of that number. Hence for example, for W we then divide by prime numbers less than or equal $\sqrt{W}$. The disadvantage is that one builds up a database of prime numbers that continues to grow as W becomes larger.

Definition

A factor pair system is defined as the creation of a localized transient database from which factorization is performed or primality of a number is determined.

Therefore, instead of a database of prime numbers, one can build a database consisting of factor pairs from which to factorize and look for prime numbers. It is a transient database because it only exists to perform the operation, after the operation it is non-existent. It is localized because the database is specifically built in relation to one number rather than a set of numbers.

The construct then becomes an effective tool for factorizing elements of $[P]$. The computer program tests the gap base of the number, and then uses the appropriate equations to look for factor pairs. Note for example that if $W = 65\,111\,009$, then the product of the pair will always yields W.

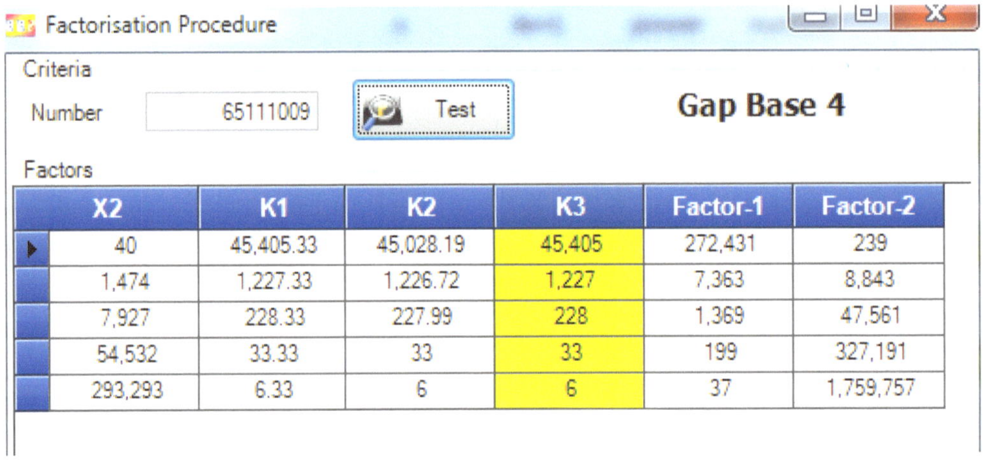

X2	K1	K2	K3	Factor-1	Factor-2
40	45,405.33	45,028.19	45,405	272,431	239
1,474	1,227.33	1,226.72	1,227	7,363	8,843
7,927	228.33	227.99	228	1,369	47,561
54,532	33.33	33	33	199	327,191
293,293	6.33	6	6	37	1,759,757

Figure 111

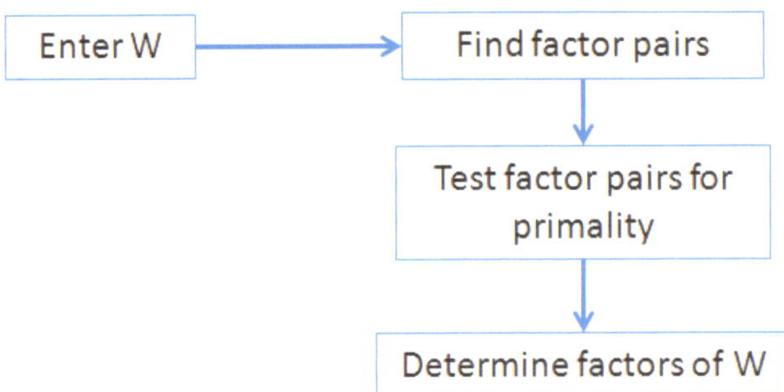

Figure 112

The standard factor pair arrangement is always in ascending or descending order such that it is possible to identify the smallest elements within the set of factor pairs. For example in this case, then we have 37 and 199 for Factor-1 and 239 and 8843 for Factor-2 being the starting points. The derivation of a factor pair database is premised on the following structure.

This clearly shows that UNDT influence, since the factors are derived in terms of the gap base of the number. It is not an arbitrary process, rather it is highly structured. Some numbers produce a set of unique factor pairs as shown by this value of W, while it has been observed that in some cases the factor pairs can duplicate by change of position.

W is GB2	**Factor-1**	**Factor-2**
	Gap base 2	Gap Base 2
Result	Pseudoprime or Prime	Prime
	Prime	Pseudoprime or Prime
W is GB2	**Factor-1**	**Factor-2**
	Gap Base 4	Gap Base 4
Result	Pseudoprime or Prime	Prime
	Prime	Pseudoprime or Prime
W is GB4	**Factor-1**	**Factor-2**
	Gap Base 2	Gap Base 4
Result	Pseudoprime or Prime	Prime
	Prime	Pseudoprime or Prime

Table 184

The derivation of the database is then defined on the basis of two types of sets, defined as set A and set B, where set A gives factor-1 and set B gives factor-2.

Definition

Let there be a set $A = \{a_1, a_2, \cdots a_n\}$ *and a set* $B = \{b_1, b_2, \cdots b_n\}$ *then* $W = a_i \cdot b_i$ *,where* $W \in D_t$ *, and* (a_i, b_i) *is called a factor pair of* W *for* $1 \le i \le n$.

From the definition, it is clear that;

$$n(A) = n(B)$$

Now if (a_i, b_i) is a factor pair, and $W \in D_t$,then a_i defines a specific set of numbers, say set A. The same being true of b_i, which we may say defines set B. Hence the following conditions are observed in forming the factor pair.

a) $a_i > 1$ *and* $b_i > 1$
b) $n(A) = n(B)$, where n is the total number of elements.
c) Let a_1 be the smallest factor of W in set A, then b_n will be the biggest factor of W in set B.
d) In general; $a_i < a_{i+1}$, and $b_{i+1} < b_i$.
e) Consequently b_1 is the smallest factor and a_n the greatest factor in set B and A respectively.

The sets A and B are said to form the factor sets. From these conditions, the following is further observed in regard to distribution of factor pairs. Another definition is given on symmetric and non-symmetric pairs, the focus of this one is with respect to the factor pairs themselves.

Definition

For $W = a_i \cdot b_i$*;*

a) A symmetric relationship is defined if $a_1 = b_n$ *and* $b_1 = a_n$.
b) A non-symmetric relationship is defined if $a_1 \neq b_n$ *and* $b_1 \neq a_n$.

In particular we note for the symmetric relation that if $n - 2$, then there will be a factor switch defined. A factor switch can also be defined for $n > 2$. If $n = 1$ for the same type of relationship, then the factor pair defines a square root made of two prime numbers or pseudoprimes. However, for the non-symmetric relationship, $n = 1$ defines W as having only two prime factors.

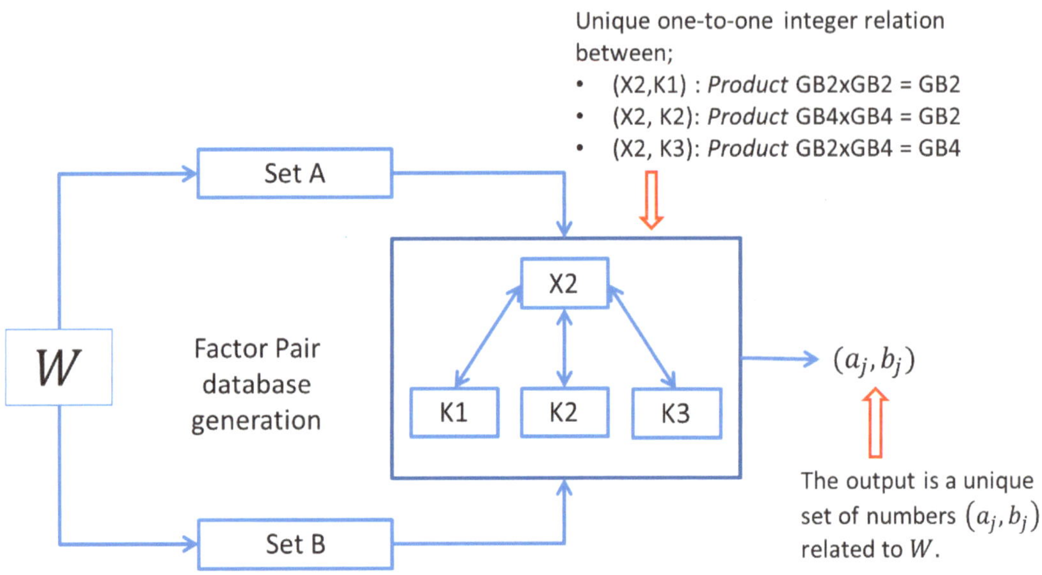

The assumption is that W is a pseudoprime for the factor pair generation process. The elements in set A and set B are chosen through x_2, where x_2 has a one-to-one relationship with either k_1, k_2 or k_3 where these must be integer values. Consequently, this results in a unique set of values in relation to W defined by the relationship;

$$(x_2, k_i, a_j, b_j), \qquad 1 \leq i \leq 3 \; and \; 1 \leq j \leq n$$

where n is the number of factor pairs and $W = a_j \cdot b_j$. These are called database parameters. For example for the number $W = 65\,111\,009$, then (1474, 1227, 7363, 8843) is a unique set of values in relation to W.

Fundamental Theorem of Factor Pairs

For any $W \in D_t$, *then* (x_2, k_i, a_j, b_j) *for* $1 \leq i \leq 3$ *and* $1 \leq j \leq n$ *derives a uniquely defined factor pair* (a_j, b_j) *for* W.

The theroem is similar to the fundamental theorem of arithmetic, but applies to pseudoprimes only. For example, the following factor pairs are unique to 26161213 and 26161235 respectively.

X2	K2	Factor-1	Factor-2
2	335400	2012401	13
335400	2	13	2012401

Table 185

That is $26161213 = 2012401 \cdot 13$ where this is the only pair derived while for example $26161235 = 5232277 \cdot 5$ is one of the pairs.

X2	K3	Factor-1	Factor-2
1	872046	5232277	5
3	256484	1538905	17
16	45897	275383	95
28	26109	156655	167
54	13499	80995	323
81	8990	53941	485
275	2644	15865	1649
529	1374	8245	3173
1536	473	2839	9215

2366	307	1843	14195
2700	269	1615	16199
5222	139	835	31331
44951	16	97	269705
51297	14	85	307781
229486	3	19	1376915

Table 186

It must be noted that this uniqueness is defined by the four variables, whilst factor pairs on their own may not be unique as already noted that they can be symmetric and non-symmetric. This property of uniqueness to a given number W implies that factor pairs can also be used for cryptography. For example from 26161235, then;

5291374-82453173

is a unique number, and all such numbers will be odd. The description of being unique is an assumption that awaits to be proved. However, the uniqueness of this number is made from the basic fact that the factor pairs are uniquely tied to the number W such that 5291374-82453173 is called a fingerprint of W. This number would probably be more difficult to break then looking for factors of 26161235. That is, to get back to 26161235 from 529137482453173 is indeed a great challenge!

11.3.3 Symmetry and Non-Symmetry of Factor Pairs

The question is what is the source of the symmetric and non-symmetric behavior of factor pairs? The answer lies in the product structure of numbers in the set D_t. The product structure governs the way the factorization engine works, and the type of patterns that are derived for factor pairs. Since the engine runs the factorization using a coordinate system, this makes the product format to play a key role in the structural concepts of the factor pair theory.

Factor-1	Factor-2	Product	Structure
GB-2	GB-2	GB-2	Symmetric
GB-4	GB-4	GB-2	Symmetric
GB-2	**GB-4**	**GB-4**	**Non-Symmetric**

Table 187

For any GB4 product, then Factor-1 will always be a GB2 number and Factor-2 a GB4 number. However, for a GB2 product, Factor-1 and Factor-2 are commutative with respect to the product structure, hence the symmetric output of the factor pairs. Therefore the symmetric nature of the distribution is a consequence of the commutative property in the product of the two numbers.

Lack of the commutative property arises because Factor-1 and factor-2 are defined structures of the factorization engine. That is, factor-1 has specific properties, and so does factor-2. As already noted, factor-1 will always be a GB2 number in a non-symmetric distribution.

This illustrates an interesting problem with regard to structure, that within a given structure number multiplication is not altogether commutative. Therefore a GB4 product is not commutative because all of factor-1 must be a GB2 number and all of factor-2 must be a GB4 number - the structure is rigid in this respect.

Generally;

- $GB2 \cdot GB2 \cdots GB2 = GB2$, that is, a product of all GB2 numbers results in a GB2 number.

- $GB4 \cdot GB4 \cdots GB4 = GB4$, that is if the product consists of an odd number of GB4 numbers, then the result will be a GB4 number
- $GB4 \cdot GB4 \cdots GB4 = GB2$, implies that the product consists of an even number of GB4 numbers, and the result will be a GB2 number.

It is this characteristic that makes the product structure to be significant in factor pairs, and the type of factorization variable that will describe W, and whether it is symmetric or non-symmetric.

Definition

A factor pair distribution is;

a) symmetric if for any x_2 there exist an $x_2 = k_1$ or $x_2 = k_2$ provided $k_1 \neq k_2$ for any two factor pairs.
b) non-symmetric if all factor pairs are unique provided for any x_2 then $x_2 \neq k_3$.

This definition is more precise and general than the one previously given because it includes negating conditions that define an asymmetric factor pair distribution. Perfect symmetry of factor pairs occurs whenever W is either defined by;

1. k_1 only, where W is a product of only two gap base 4 numbers that are prime, and this derives only 2 factor pairs.
2. k_2 only, where W is a product of two or more gap base 2 numbers that are prime.
3. k_1 and k_2 provided $k_1 \neq k_2$ for some factor pair, and W is derived from a product consisting of the following basic product structure. The number of factor pairs in general will be a function of the number of prime factors of W.
    i. $(GB2) \cdot (GB4 \cdot GB4)$, where this always derives 6 factor pairs
    ii. $(GB2 \cdot GB2) \cdot (GB4 \cdot GB4)$, where this always derives 14 factor pairs
    iii. $(GB4 \cdot GB4) \cdot (GB4 \cdot GB4)$, where this always derives 14 factor pairs

The product (GB4·GB4) provided both GB4 numbers are pseudoprimes results in a k_1 and k_2 distribution though it may not be symmetric because of the pseudoprime condition. Similarly for a product like $(GB2 \cdot GB2) \cdot (GB4)$ this results in numbers defined by k_3 in GB4. Provided factor-1 and factor-2 are prime, the following table indicates the factorization tables involved. Odd means the number of factors is odd, and the same applies for even.

Factor-1	Factor-2	Product	Factorization Variable
GB-2	GB-2	GB-2	k_2
GB-4	GB-4	GB-2	k_1
GB-2	GB-4	GB-4	k_3
$GB2 \cdot GB2 \cdots GB2$	$GB4 \cdot GB4 \cdots GB4$ (odd)	GB-4	k_3
$GB2 \cdot GB2 \cdots GB2$	$GB4 \cdot GB4 \cdots GB4$ (even)	GB-2	k_1 and k_2

Table 188

For example for $W = 77970283$, the factors are 71 and 83 in GB2, and 101 and 131 in GB4. And there are no two factor pairs such that $k_1 = k_2$. The factor pair symmetry is perfect.

X2	K1	K2	Factor-1	Factor-2
12	183029		1098173	71
14	156567		939401	83
17	128664		771983	101

22	99199		595193	131
982		2205	13231	5893
1195		1812	10873	7171
1397		1550	9301	8383
1550		1397	8383	9301
1812		1195	7171	10873
2205		982	5893	13231
99199	22		131	595193
128664	17		101	771983
156567	14		83	939401
183029	12		71	1098173

Table 189

For example the number $W = 41629$. This number is defined by k_2 only. The prime factors are 7, 19, 313 and these are all GB2 numbers.

X2	K2	Factor-1	Factor-2
1	991	5947	7
3	365	2191	19
22	52	313	133
52	22	133	313
365	3	19	2191
991	1	7	5947

Table 190

The prime factors of a number may repeat, hence it is of interest whether this affect the symmetry of the number in gap base 2, and the asymmetric property of the number in gap base 4. It appears that the structure of the factorization still remains either symmetric or asymmetric, what changes is the expected number of factor pairs that we get given the prime factors of the number.

For example consider the number W=5988947 with prime factors 17, 17, 23 and 53 . The distribution is asymmetric, the number of factor pairs is 7 instead of the expected 15 from the limit factor pair theorem.

X2	K3	Factor-1	Factor-2
3	58715	352291	17
4	43398	260389	23
9	18833	112999	53
819	203	1219	4913
1108	150	901	6647
2553	65	391	15317
3454	48	289	20723

Table 191

Hence the symmetric and asymmetric properties in distribution are controlled mostly by the behavior of k_1, k_2 and k_3. Now these variables are linked to the coordinate system, which explains why difference is such properties are reflected if the coordinates are equal.

11.3.4 Asymmetry of Factor Pairs

Factor pair distribution can occur in three ways;

- It can be symmetric, as in the case of GB2
- It can be non-symmetric, as in the case of GB4

- It can be asymmetric as in the case of GB2 and GB4

The asymmetric nature is easily detected in the gap base 2 factor pairs, but not so easily in gap base 4. Asymmetric distribution occurs when the same coordinate is used to describe related numbers in D_t. For example consider the number $W = 678727975$.

X2	K1	K2	Factor-1	Factor-2
1	22624266		135745595	5
3	6654196	5953754	39925175	17
5953754		3	19	35722525
6654196	3		17	39925175
22624266	1		5	135745595

Table 192

The factor 19 is responsible for the asymmetric property, and we note that this is the same coordinate as the factor 17, that is coordinate 3. Therefore both $k_1$and k_2 must be integers for asymmetry to occur, hence there will be values such that $k_1 = k_2 = x_2$. In this example, this value is the coordinate 3. Alternatively you can have the condition as $k_1 = x_2$ and $k_2 = x_2$, where this would be true for $k_1 = x_2 = 6654196$ and $k_2 = x_2 = 5953754$. The difference is the "and" condition.

X	Family-3	
	GB-4	GB-2
0	-1	1
1	5	7
2	11	13
3	17	19
4	23	25
5	29	31
6	35	37
7	41	43

Table 193

A variation of an asymmetric distribution occurs in gap base 2 when we have numbers that do not have the same coordinate, but their products do. For example $W = 5 \cdot 5 \cdot 23 \cdot 29 = 25 \cdot 23 \cdot 29$, where 23 and 25 have the same coordinate, that is 4. Hence this will give us an asymmetric distribution in gap base 2.

X2	K1	K2	Factor-1	Factor-2
1	556		3335	5
4	121	111	725	23
5	96		575	29
19		24	145	115
24		19	115	145
96	5		29	575
111		4	25	667
121	4		23	725
556	1		5	3335

Table 194

For gap base 4, the same reasoning applies though the result is different because GB4 has unique factor pairs that are non-symmetric. As a result, the asymmetry is not observable from the factor pair distribution. For example consider the number $W = 23579$.

X2	K3	Factor-1	Factor-2
3	231	1387	17
54	12	73	323
207	3	19	1241

Table 195

The factor pairs show no change in distribution pattern or their uniqueness nature. However, we have the situation where $k_3 = x_2$. Hence this is similar to gap base 2, and so forms a basis for a common definition.

Definition

A factor pair distribution is asymmetric if at least at one point;

a) $x_2 = k_1 = k_2$
b) $x_2 = k_3$

Therefore a gap base 4 asymmetric distribution is noticed from $x_2 = k_3$. It may be noted that asymmetric occurrences are true if the numbers involved are pseudoprime, that is not all numbers are necessarily prime.

X2	K3	Factor-1	Factor-2
1	1974	11845	5
4	429	2575	23
86	19	115	515
96	17	103	575
395	4	25	2369

Table 196

The asymmetry is observed from (23,4) and (25,4) for $W = 59225$. Therefore asymmetry is defined for $x_2 = k_1 = k_2$ and $x_2 = k_3$ irrespective of whether the product is between prime numbers or pseudoprime numbers only. However some asymmetric behavior from pseudoprimes is traced back to the prime numbers. For example in the product $133 \cdot 145 = 19285$, then the primes 5 and 7 define the asymmetry of the number.

X2	K3	Factor-1	Factor-2
1	1387	8323	5
5	239	1435	29
6	198	1189	35
7	169	1015	41
34	34	205	203
48	24	145	287
991	1	7	5945

Table 197

In multiplying $203 \cdot 205 = 41615$ both pseudoprimes, we have (203,34) and (205,34) to define the asymmetry, as well as a trace back of (5,1) and (7,1). Hence asymmetry can also be described from product pattern. That is, if for example $(GB4 \cdot GB4)$ and provided both GB4 numbers are pseudoprimes, then there is a likelihood that the result will be asymmetric.

11.3.5 Factor Pair Characteristics and Factor-2 Theorem

Consider the number $W = a_i \cdot b_i$, then it has been defined that $a_1 = b_n$ *and* $b_1 = a_n$ for a symmetric relationship and $a_1 \neq b_n$ *and* $b_1 \neq a_n$ for a non-symmetric relationship. The following observations can be made. Secondly, not all numbers have the same factorization characteristic. Gap base 2 numbers differ in property to the gap base 4 numbers, and from the computer program, it appears that gap base 4 numbers even take longer to factorize than gap base 2 numbers of the same range. What is more interesting is to observe that type of patterns for the gap base 2 and 4 numbers in terms of factorization, and consequently derive techniques for process improvement.

Gap Base 4

a) Factor-2 is b_1, where this is always a prime number in gap base 4.

b) For gap base 4, we can have pesudoprimes from a_1 to a_n and $a_n > b_1$. Also a_1 will be the largest pseudoprime, and a_n will be the smallest prime or pseudoprime provided $largest\ x_2 < largest\ k_3$.

X2	K3	Factor-1	Factor-2
1	79277	475663	5
4	17234	103405	23
3447	19	115	20681

Table 198

c) For gap base 4, $a_1 > b_1$ or $a_1 < b_1$, and $a_n < b_1$ or $a_n > b_1$. This is due to the non-commutativity of the factor pair operation in gap base 4. But $a_1 < b_1$ depending on product of a_3 and a_2 being less than b_1.

X2	K3	Factor-1	Factor-2
1605	41	247	9629
20863	3	19	125177
30492	2	13	182951

Table 199

d) Gap base 4 has unique values of both x_2 and k_3. That is, the system is uniquely (x_2, k_3) implying no commutatvity between x_2 and k_3. The distribution is not asymmetric.

X2	K3
1605	41
20863	3
30492	2

Table 200

The gap base 4 class is only defined by the $K3$ variable, unlike gap base 2 that uses $K1$ and $K2$. When there exists at least a value where $X2 = K3$, then the distribution is asymmetric as shown for the case $X2 = 1$ and $K3 = 1$.

X2	K3	Factor-1	Factor-2
1	872040	5232241	5
6	124577	747463	35
622886	1	7	3737315

Table 201

e) For gap base 4, if $b_1 = 5$, then a_n is pseudoprime that is a multiple of 5 or prime. For example 139 is prime.

X2	K3	Factor-1	Factor-2
1	2570411	15422467	5
116	18492	110953	695
151	14201	85207	905
511	4193	25159	3065
20966	102	613	125795

71006	30	181	426035
92461	23	139	554765

Table 202

f) For gap base 4, if $b_1 \neq 5$, then b_1 is prime, and a_n is prime or pseudoprime.

X2	K3	Factor-1	Factor-2
7	9668	58009	41
48	1381	8287	287
56628	1	7	339767

Table 203

g) For gap base 4, if $n = 1$ then a_1 is prime and b_1 is prime. This is a self-testing procedure.

Factor-1	Factor-2
8089	9533

Table 204

Definition

Numbers factorized by $K3$ are $K3$ numbers.

It is evident since $K3$ is the only variable involved, therefore it implies all $K3$ number are gap base 4 numbers. Hence if we say W is $K3$, it implies it is a gap base 4 number. Gap base 4 has a unique condition that only occurs in it. We note that $13 < 311$ and $169 < 311$, similarly also $28 < 674$.

X2	K3	Factor-1	Factor-2
52	28	169	311
674	2	13	4043

Table 205

GB4 Reflex Conditions

For W a $K3$ number, then if;

a) $a_1 < b_1$ *provided* $a_2 \cdot a_2 < b_1$ *or* $a_2 \cdot a_3 < b_1$ *for* $n \leq 3$.
b) $a_1 > b_1$ *for* $n > 3$.
c) $a_n < b_1$ then $largest\ x_2 > largest\ k_3$.
d) $a_n > b_1$ then $largest\ x_2 < largest\ k_3$.

This are referred to as the reflex condition of gap base 4, and they do not occur in gap base 2. The first two conditions are independent of x_2 and k_3. This is a distinguishing characteristic between W in GB2 or GB4.

Gap Base 2

h) For gap base 2, $a_i > b_i$ and $a_n = b_1$ for any W. This is a perfect symmetry of factor pairs, and occurs whenever we have the factor pairs uniquely defined by (x_2, k_1) or (x_2, k_2).

X2	K2	Factor-1	Factor-2
1	56638	339829	7
7	9220	55321	43
8	8091	48547	49
50	1317	7903	301

188	351	2107	1129
351	188	1129	2107
1317	50	301	7903
8091	8	49	48547
9220	7	43	55321
56638	1	7	339829

Table 206

i) For gap base 2, factor pairs can be generated by (x_2, k_1) and (x_2, k_2) and be perfectly symmetrical.

X2	K1	K2	Factor-1	Factor-2
1	7368.6	5263	31579	7
4	1602	1473.52	9611	23
27	229	225.8588957	1373	161
229	27	26.62763636	161	1373
1602	4	3.665869136	23	9611
5263	1.333407227	1	7	31579

Table 207

However, it is possible that it is not perfectly symmetrical, that is, asymmetrical. This is observed for the factor pair (43, 55309). This occurs whenever $k_1 = k_2$. There are 13 factor pairs for $W = 2378287$, and $x_2 = k_1 = k_2 = 7$.

X2	K1	K2	Factor-1	Factor-2
3	23316.70588	20862	125173	19
7	9668	9218	58007	41
12	5583	5429.712329	33497	71
130	509	507.3636364	3053	779
136	486.5239264	485	2911	817
225	294	293.2316802	1763	1349
294	225	224.411898	1349	1763
485	136.4269508	136	817	2911
509	130	129.5816694	779	3053
5583	12	11.66596018	71	33497
9218	7.333592493	7	43	55309
9668	7	6.666431071	41	58007
20862	3.333383931	3	19	125173

Table 208

j) Every value of x_2 corresponds to a value of k_1 or k_2 in gap base 2. This is a structural property, and the previous example is used to illustrate this. It is a structural property because it actually gives us a coordinate system with regard to numbers that have prime factors, where these numbers are referred to as factor pairs. For gap base 2, this is (x_2, k_1) or (x_2, k_2). The operation is commutative in terms of coordinates and hence not a unique relationship. That is (x_2, k_1) commutes to a value (k_1, x_2), hence the symmetric behavior in gap base 2.

Note that for $x_2 = 7$, then both of k_1 and k_2 as integer as 9668 and 9218 respectively where this gives us the situation $(x_2, k_1, k_2) = (7,9668,9218)$. Since x_2 is a unique figure, this implies a distinct value of x_2, hence later down the process we now have expected behavior of (x_2, k_1) or (x_2, k_2).

X2	K1	K2
3		20862

7	9668	9218
12	5583	
130	509	
136		485
225	294	
294	225	
485		136
509	130	
5583	12	
9218		7
9668	7	
20862		3

Table 209

k) For gap base 2, if $n = 2$, then a_1 and b_1 are prime. This is called a self-testing primality test.

X2	K2	Factor-1	Factor-2
2	335400	2012401	13
335400	2	13	2012401

Table 210

The gap base 2 has a particular characteristic that for numbers that are factored by $K1$ only or $K2$ only, the factor pairs are symmetrical, and that the self-testing primality test occurs only when the variable is either $K1$ or $K2$ but not both. This implies a number defined by $K1$ and $K2$ cannot have less than three factor pairs. This gives categories within gap base 2.

Definition

Gap base 2 has three types of numbers;

a. *$K1$ numbers, factored through $K1$ only.*
b. *$K2$ numbers, factored through $K2$ only.*
c. *$K1K2$ numbers, factored by both $K1$ and $K2$.*

Both $K1$ and $K2$ can result in self-testing primality, while $K1K2$ numbers do not. From the above, the general observation is that the first factor-2 is always prime for gap base 2 or 4. This leads to the following structural theorem on factor pair theory.

Factor-2 Theorem

Let $W \in D_t$ with factor pair (a_1, b_1), hence;

i. $a_1 \in GB2$ *if* $W \in GB4$.
ii. *b_1 is the smallest prime number of W, if $W \in GB2$*
iii. *For n factor pairs, then b_1 is a prime number and a_1 contains all the remaining prime factors of W.*

This theorem gives (a_1, b_1) a special position in the factor pair theory. For gap base 4, there are situations where $a_1 < b_1$, hence a_1 may not necessarily be the biggest pseudoprime. However both GB2 and GB4 have b_1 as a prime number, and b_1 is always GB4 if W is GB4 because of the product format $GB2 \cdot GB4$. The GB2 number may be a product of $GB2 \cdot GB2$ and $GB4 \cdot GB4$ hence b_1 may be a GB2 or GB4 number.

11.3.6 Factor Pair Theorem

What the factor pair theory does is to build a quick database of numbers in set A and set B such that one is able to look for prime factors. The database is narrow and restricted because it is based on a bounded search. Consequently one does not have to store a lot of prime numbers to create efficiencies in looking for prime factors.

The factor pair theorem is stated as follows.

Factor Pair Theorem

For a given integer $W \in D_t$, and n is the number of factor pairs, then if;

a) $n = 0$, W is a prime number
b) $1 \leq n \leq 3$, then W is a pseudoprime with two or three prime factors
c) $n > 3$, then W is a pseudoprime with four or more prime factors.

Therefore we can use this theorem as a primality test from $n = 0$. We do not necessarily know how many factor pairs there are according to this theorem. But the proof is trivial for the first part.

If we find no factor pairs for W, then W must be a prime number. Conversely, if we do find factor pairs, then W is definitely not prime, it must be a pseudoprime. This is actually taken as part of the Mbulawa primality test because both use more or less the same information. In the Mbulawa primality test, k_1 and k_2 must be less than 1 or $k_3 < 1$ if the number is gap base 4. The same condition can be used for the factor pair theorem, if this condition is satisfied, then no more factor pairs can be derived. However, whilst this condition is a must to prove a number is prime for the Mbulawa test, for the factor pair theorem it is optional.

Consider for example the number 151 as shown.

x2	k1	k2	k3	Factors	Factors
1	5.2	3.428571	4.866667		
2	2.454545	1.769231	2.121212		
3	1.647059	1.157895	1.313725		
4	1.26087	0.84	0.927536		
5	1.034483	0.513514	0.701149		
6	**0.885714**	**0.756757**	0.552381	No factor	pairs

Table 211

No factor pairs are found for 151, which is the same as saying k_1 and k_2 are less than 1.

Besides primality, factor pairs are also useful for factorization as indicated by the theorem. This theory is a consequence to the fact that any number that can be factorized has a group of numbers that define it through a product. The factor pairs approach makes this more efficient in the UNDT environment. For example strictly speaking 2397 = 47 x 51 is a factor pair, or 2397 = 3 x 17 x 47; but it is not in the UNDT environment. Unlike in the UNDT environment, there is no relationship between 47 and 51, or between 3, 17 and 47. Nothing links them together, and this is what makes factorization to be a formidable task. As a result the factor pair theory would not work in such an environment. Secondly, in the non-UNDT environment, it is not possible to use factor pairs as a method for primality testing.

11.3.7 Factor Pair Factorization Theorem

In terms of structure, we can then summarize the factor pair theory in the following manner.

Condition	Result	Comment
$n = 0$	W is a prime number	Factor Pair Primality test
$n = 1$	a_1 and b_1 are prime, $W \in GB4$	Factorization: Self-testing primality
$n = 2$	a_1 and b_1 are prime, $W \in GB2$	Factorization: Self-testing primality
In (a_1, b_1)	$b_1$is always a prime number, and $a_1$is a product of remaining prime factors	Primality test on b_1 factor only. This is a *de facto* prime factor of W.
$n = 3$	W is pseudoprime with 3 prime factors, $W \in GB4$	Factorization: Self-testing primality
$n > 3$	W is a pseudoprime with four or more prime factors	Factorization of W with primality testing on identified factor pairs.

Table 212

Therefore the fifth condition for $n > 2$ raises the need to develop a further method to derive factors of W. The purpose of the factor pair theory is to express a relationship between the factor pairs in order to facilitate the process of deriving prime factors of W. It has been noted that pseudoprimes for a given W are related to each other.

X2	K3	Factor-1	Factor-2	Prime Factor Relationship
1605	41	$z = 247$	9629	13, 19
20863	3	$y = 19$	125177	13, 9629
30492	2	$x = 13$	182951	19, 9629

Table 213

One of the key relationships is that $W = Factor1 \cdot Factor2$, for example $W = 247 \cdot 9629$, or $W = 19 \cdot 125177$. The next key relationship is that pseudoprimes are related to each other through products between factor-1 and factor-2, for example $125177 = 13 \cdot 9629$. Now if $W = 19 \cdot 125177$, and $125177 = 13 \cdot 9629$, then it implies $W = 13 \cdot 19 \cdot 9629$.

What remains then is to find a method of connecting these relationships together in order to define a factorization path for W. It is noted that in the factor pairs, the smallest value will represent a prime number of W. Infact the smallest two numbers will be prime for gap base 2, whilst this may be true in some cases for gap base 4. Therefore the idea than is to build a connection through the first prime number. This leads to the following theorem as a tool for pseudoprime factorization.

Factor Pair Factorization Theorem

If $W = x \cdot y \cdot z$, such that $x \le y \le z$, and $W \in D_t$, then there exists a factor pair $(a_i \cdot b_i) = (x \cdot y, z)$ or $(a_i \cdot b_i) = (z, x \cdot y)$.

In other words, there will be two numbers that we multiply to give us a third number in one of the factor pairs. Using the same example, hence $13 \cdot 19 = 247$. This defines our $x \cdot y$, such that it is part of a factor pair. Hence $z = 9629$, or we have the factor pair $(247,9629)$ to define prime factors $W = 13 \cdot 19 \cdot 9629$. Consequently this theorem defines the factorization path for W, where W must be a pseudoprime since it defines factors for the number. Defining x as the smallest element in set A or B gives the initial point of the factorization path. The example does not imply that z will always be a prime number, it may also be pseudoprime, but this simply leads to re-entering a new value of W until all the prime factors are established.

W = 22184449

Factor-1	Factor-2
3169207	7
764981	29
215383	103
109283	203
30769	721
20909	1061
7427	2987
2987	7427
1061	20909
721	30769
203	109283
103	215383
29	764981
7	3169207

W = 109283

Factor-1	Factor-2
1061	103
103	1061

Table 214

In the above example $W = 22184449$ as the first entry, and the second entry is $W = 109283$. Now $7 \cdot 23 = 203$, hence this gives the factor pair $(109283{,}203)$. The next step is to factor 109283 which are 103 and 1061. The process describes the factorization path for finding factors of W, clearly showing the link and relationships. This will give us all the factors for W which are 7, 29, 103 and 1061.

In general, W has the following factor pair format.

a) $(x \cdot y, z)$ or $(z, x \cdot y)$
b) $(x \cdot z, y)$ or $(y, x \cdot z)$
c) $(x, y \cdot z)$ or $(y \cdot z, x)$

From the condition $x \leq y \leq z$, then;

a) $x \cdot y \leq z$
b) $y \leq x \cdot z$
c) $x \leq y \cdot z$

Since $W = a_i \cdot b_i$ and $W = x \cdot y \cdot z$, then $(x \cdot y, z) = (a_i , b_i)$. Therefore $a_i = x \cdot y$ and $z = b_i$. Let $x = a_n$, then it will be the smallest factor of W from set A. Secondly, we can look for the next smallest value of y in set A or B as indicated by the prime pair formation to derive a_i and consequently calculate z.

Therefore we produce factor pairs in order to get the values of x and y. Their product then leads us to the factor pair (a_k , b_k), and z will either be a_k or b_k. The next step is to test x, y and z for primality by applying the factor theorem again or using a primality test procedure. Let us look at the theorem in terms of the above example.

a) W is pseudoprime
b) The factor pairs are in ascending and descending order
c) For any factor pair, then $W = a_i \times b_i$, for example $W = 272431 \times 239$
d) The smallest values of x and y are given by $x = 37$, and $y = 199$. This is the condition $x = a_i$, and $y = b_j$.
   I. Now $W = 37 \cdot 199 \cdot z$
   II. $x \cdot y = 7\,363 = a_k$. This means we have a_k, we can look for z.
   III. If $x \cdot y = a_k$, then $z = b_k$, hence $z = 8\,843$. That is $W = 37 \cdot 199 \cdot 8843$

This is the thrust of the theorem, thereafter it can repeat itself after testing for primality. For example 37 and 199 are prime, but 8 843 is a pseudoprime. So the cycle repeats by setting $W = 8\,843$. The factor pair theorem creates a factorization path because it determines which numbers to follow by a systematic manner until all prime factors are derived.

11.3.8 Factor Pair Limit Theorem

This theorem is critical for factorization because it makes it practically a fast and predictable operation. It implies that the operation does not extend to infinity and so we can be deterministic in developing the factorization engine. The application of the factor pair factorization theorem can be extended in the following manner by observing that W can have several factors. For example $W = 31 \cdot 41 \cdot 47 \cdot 61$ and this can be further broken down into a factor pair $W = (31 \cdot 41)(47 \cdot 61)$ as one example. Our pair then is $\{(31 \cdot 41), (47 \cdot 61)\}$. Hence a table is constructed along the same lines. Let there be a prime or pseudoprime w, x, y and z (though the pattern applies even if 2 and 3 are included), then the possible permutations will be as follows.

1	xy	
G-1	(x,y)	(y,x)

2	xyz	
G-1	(x,yz)	(yz,x)
	(y,xz)	(xz,y)
	(z,xy)	(xy,z)

3	wxyz	
G-1	(w,xyz)	(xyz,w)
	(x,wyz)	(wyz,x)
	(y,wxz)	(wxz,y)
	(z,wxy)	(wxy,z)
G-2	(wx,yz)	(yz,wx)
	(wy,xz)	(xz,wy)
	(wz,xy)	(xy,wz)

4	vwxyz	
G-1	(v,wxyz)	(wxyz,v)
	(w,vxyz)	(vxyz,w)
	(x,vwyz)	(vwyz,x)
	(y,vwxz)	(vwxz,y)
	(z,vwxy)	(vwxy,z)
G-2	(vw,xyz)	(xyz,vw)
	(vx,wyz)	(wyz,vx)
	(vy,wxz)	(wxz,vy)
	(vz,wxy)	(wxy,vz)
	(wx,vyz)	(vyz,wx)
	(wy,vxz)	(vxz,wy)
	(wz,vxy)	(vxy,wz)
	(yz,vwx)	(vwx,yz)
	(xz,vwy)	(vwy,xz)
	(xy,vwz)	(vwz,xy)

Table 215

Each permutation can be group as G-1 or G-2 by considering the first pair on the left. The above distribution is said to be symmetric because the second column is a duplicate of the first because of commutability. This pattern is common for gap base 2. When there is no commutative property in the factor pairs, then it is said to be non-symmetric and this gives only one column on the left. This is half the total factor pairs. This is common for gap base 4 elements. One column is said to define the set of unique factor pairs, that is the non-symmetric case.

A count gives the following factor pair distribution.

Product	Number of Prime Factors	Number of Factor Pairs-GB2	Number of Factor Pairs – GB4	Difference in GB2	Difference in GB4
xy	2	2	1		
xyz	3	6	3	4	
wxyz	4	14	7	8	4
vwxyz	5	30	15	16	8

Table 216

It is noted that the difference is in powers of 2, where this leads to the following observation for gap base 2.

$$2^1 = 2$$

$$2^1 + 2^2 = 6$$

$$2^1 + 2^2 + 2^3 = 14$$

$$2^1 + 2^2 + 2^3 + 2^4 = 30$$

Hence if there are x prime factors, the power on 2 will be $(x - 1)$ and we take a sum of the series to give us the total number of possible factor pairs of W. The values 2, 6, 14, 30 and so on are called standard values of n for a symmetric distribution. The standard values occur for $K2$, $K1K2$ numbers, that is gap base 2 numbers whose distribution is determined by $K2$ *or* $K1K2$. The $K1$ number occurs as an exception, and will be explained later. Consequently for example $n = 10$ would be a non-standard value in the symmetric distribution. The formula for gap base 2 is derived as follows, where m is the number of prime factors, n the number of factor pairs;

$$n = 2^m - 2, \qquad where\ 1 \leq m \leq \infty$$

Now $m = 1$ gives us an interesting value, that is $n = 0$, implying that there are no factor pairs. No factor pairs means the number is a prime, and this corresponds to the fact that $m = 1$ means there is only one factor, the number itself if we assume 1 to be a trivial solution.

For gap base 4, then the pattern for maximum number of factor pairs is as follows. Essentially, this is half the factor pairs of a gap base 2 number.

$$2^0 = 1$$

$$2^0 + 2^1 = 3$$

$$2^0 + 2^1 + 2^2 = 7$$

$$2^0 + 2^1 + 2^2 + 2^3 = 15$$

$$2^0 + 2^1 + 2^2 + 2^3 + 2^4 = 31$$

Hence if there are x prime factors, the maximum power on 2 will be $(x - 2)$ and we take a sum of the series to give us the total number of possible factor pairs of W. The values 1, 3, 7, 15, 31 and so on are called standard values of n for a non-symmetric distribution. The standard values occur for $K3$ numbers, that is gap base 4 numbers whose distribution is determined by $K3$. The formula for gap base 4 is derived as follows, where m is the number of prime factors, n the number of factor pairs;

$$n = 2^{m-1} - 1, \qquad where\ 1 \leq m \leq \infty$$

In a similar manner $m = 2$ gives us one factor pair in gap base 4, while $m = 1$ implies the number is prime since $n = 0$.

Because of the different types of numbers and expressions, two distinct theorems are constructed. The theorems make it possible for us to define a factorization process because they define a bounded search for factor pairs for both cases. The gap base 2 theorem describes a symmetric distribution. For all the cases, the theorems assume that W has factors (a_1, b_1), $(a_2, b_2), \cdots (a_n, b_n)$. The range $[1, Int\left\{\frac{W+7}{30}\right\}]$ is a condition where we are certain that all the necessary factor pairs will be derived.

GB2 Factor Pair Limit Theorem

If W is a $K2$ or $K1K2$ number with m prime factors, then it has $n = 2^m - 2$ factor pairs in the range $[1, Int\left\{\frac{W+7}{30}\right\}]$ and $2 \leq m \leq \infty$.

Since $m = 1$ defines a prime number, it is not include as W is said to have factors. The gap base 4 theorem defines a non-symmetric distribution.

GB4 Factor Pair Limit Theorem

If W is a $K3$ number with m prime factors, then it has $n = 2^{m-1} - 1$ factor pairs in the range $[1, Int\left\{\frac{W+7}{30}\right\}]$ and $2 \leq m \leq \infty$.

The two theorems describe a symmetric or non-symmetric distribution, while the following defines an asymmetric distribution for gap base 2. However in gap base 4, the asymmetry has no bearing on the number of factor pairs, but the number is influenced by whether $largest\ k_3 < largest\ x_2$, the reflex condition of gap base 4. Hence the "A" denotes asymmetric and the "R" denotes reflex condition. For the asymmetric distribution, we cannot eliminate the duplicate factor pairs of gap base 2. In the AR theorem, the value of n is less than or equal to a given maximum. For example if we take a number in gap base 4, and it has 3 prime factors, then we expect 7 factor pairs. But if the reflex condition applies, the number may have 5 factor pairs. A similar argument applies for the gap base 2 number.

Consider the $K2$ number 1 794 031, the prime factor 109 is repeated, that is $W = 109 \cdot 109 \cdot 151$. This has 4 factor pairs for 3 prime numbers instead of the expected 6.

X2	K2	Factor-1	Factor-2
18	2743	16459	109
25	1980	11881	151
1980	25	151	11881
2743	18	109	16459

Table 217

This is an asymmetric distribution as X2 = K2 = 18. Repeating prime factors also affect the distribution pattern of factor pairs, and the theorem is defined to reflect this.

AR Factor Pair Limit Theorem

Let $W \in D_t$ with m prime factors in the range $[1, Int\left\{\frac{W+7}{30}\right\}]$, then the number of factor pairs is;

a) $n \leq 2^m - 2$ where $2 \leq m \leq \infty$, if
  - *i. W is a asymmetric*
  - *ii. W is symmetric but one or more prime factors repeat*

b) $n \leq 2^{m-1} - 1$, where $2 \leq m \leq \infty$, if
  - *i. $largest\ k_3 < largest\ x_2$*
  - *ii. W is non-symmetric but one or more prime factors repeat*

When optimization techniques are applied, then the duplicate pairs in gap base 2 are ignored such that the number of pairs in gap base 2 and 4 are the same. The optimizing expression is given by $Int(\sqrt{W} + 1)/6)$. This then gives us a combined theorem expressed as follows.

General Factor Pair Limit Theorem

Let W have m prime factors, then in a range $[1, Int(\sqrt{W}+1)/6)]$ the number of unique factor pairs will be $n \leq 2^{m-1} - 1$, where $1 \leq m \leq \infty$.

If $m = 1$, then $n = 0$, and the number is a prime number, it is its own prime factor. Consider the number 26 161 213 which is gap base 2 with prime factors 13 and 2 012 401. There are only two factor pairs, and it is a $K2$ number.

X2	K2	Factor-1	Factor-2
2	335400	2012401	13
335400	2	13	2012401

Table 218

Then according to the theorem we should have $2^1 = 2$ factor pairs. Since gap base 2 is symmetric, this is the total number of expected pairs. All these pairs must be found in the interval up to $Int\left\{\frac{W+7}{30}\right\} = 872\,041$. However the program finds them at $x_2 = 335\,400$. The number 26 161 135 has five prime factors (5, 7, 11, 13, 5227), which implies we expect $2^1 + 2^2 + 2^3 + 2^4 = 30$ factor pairs. When running the program we get 28 factor pairs because the number is asymmetric, and gets them after counting to $x_2 = 872\,038$ where this is the same as $Int\left\{\frac{W+7}{30}\right\} = 872\,038$. It is observed that the asymmetry arises from $x_2 = 1\ and\ 2$.

X2	K1	K2	Factor-1	Factor-2
1	872038	622884	5232227	5
2	396381	335399	2378285	11
6	124577		747461	35
9		79276	475657	55
...	...	...	...	...
79276		9	55	475657
124577	6		35	747461
335399		2	13	2012395
396381	2		11	2378285
622884		1	7	3737305
872038	1		5	5232227

Table 219

As an example of a $K3$ number, 26 161 385 has 5 prime factors, these being 5, 17, 19, 97 and 167. The factor pairs are non-symmetric as the number is gap base 4 as follows.

	X2	K3	Factor-1	Factor-2
1	1	872046	5232277	5
2	3	256484	1538905	17
3	16	45897	275383	95
4	28	26109	156655	167
5	54	13499	80995	323
6	81	8990	53941	485
7	275	2644	15865	1649
8	529	1374	8245	3173
9	1536	473	2839	9215
10	2366	307	1843	14195
11	2700	269	1615	16199
12	5222	139	835	31331
13	44951	16	97	269705

14	51297	14	85	307781
15	229486	3	19	1376915

Table 220

Therefore the number of factor pairs is 15. That is;

$$(2^1 + 2^2 + 2^3 + 2^4)/2 = \frac{30}{2} = 15$$

Note that x_2 counts to 229 486 and it has found all the factor pairs, whereas k_3 "counts" to 872 046 before it does the same. Now $Int\left\{\frac{W+7}{30}\right\} = 872\ 046$ as indicated in the theorem. If you consider another example from the number 26161289, then the number of factor pairs is 11, and the prime factors are 7, 11, 11, 67 and 461. The number of prime factors is 5, but this is less than 15 factor pairs as demanded by the theorem.

11.3.9 Factors Pairs Occurrence and Primes

The following table shows numbers that are gap base 2 and whose root is class 1A. The aim is to show the pattern of the number of factor pairs against the number of prime factors from a particular group of numbers.

Number	Pairs	No. of Factors	Number	Pairs	No. of Factors
52184503	6	3	52184803	6	3
52184509	6	3	52184809	PRIME	
52184515	6	3	52184815	14	4
52184521	2	2	52184821	2	2
52184527	6	3	52184827	PRIME	
52184533	2	2	52184833	2	2
52184539	6	3	52184839	6	3
52184545	28	5	52184845	6	3
52184551	2	2	52184851	2	2
52184557	PRIME		52184857	2	2
52184563	2	2	52184863	2	2
52184569	2	2	52184869	6	3
52184575			52184875	6	3
52184581	2	2	52184881	2	2
52184587	6	3	52184887	6	3
52184593	2	2	52184893	2	2
52184599	2	2	52184899	14	4

Table 221

There is no particular observed relationship in the way the factor pairs and number of prime factor occurrence, except to state that the theorem is shown to predict the number of prime factors for given pairs. The same approach is used for gap base 4 numbers with a gap base 4 and gap base 2 root.

Number	Pairs	No. of Factors	Number	Pairs	No. of Factors
77112305	7	4	77112401	1	2
77112311	1	2	77112407	1	2
77112317	1	2	77112413	1	2
77112323	1	2	77112419	1	2
77112329	5	4	77112425	17	6

77112335	7	4
77112341	1	2
77112347	3	3
77112353	1	2
77112359	1	2
77112365	7	4
77112371	3	3
77112377	3	3
77112383	1	2
77112389	3	3
77112395	3	3
77112431	7	4
77112437	1	2
77112443	1	2
77112449	PRIME	
77112455	7	4
77112461	1	2
77112467	1	2
77112473	PRIME	
77112479	3	3
77112485	1	2
77112491	3	3
77112497	7	4

Table 222

This gives us the following distributive behavior in terms of frequent factors and factor pairs for the two sets of data values from gap base 2 and 4. The "F" stands for factors, hence "F-6" means a number having 6 factors.

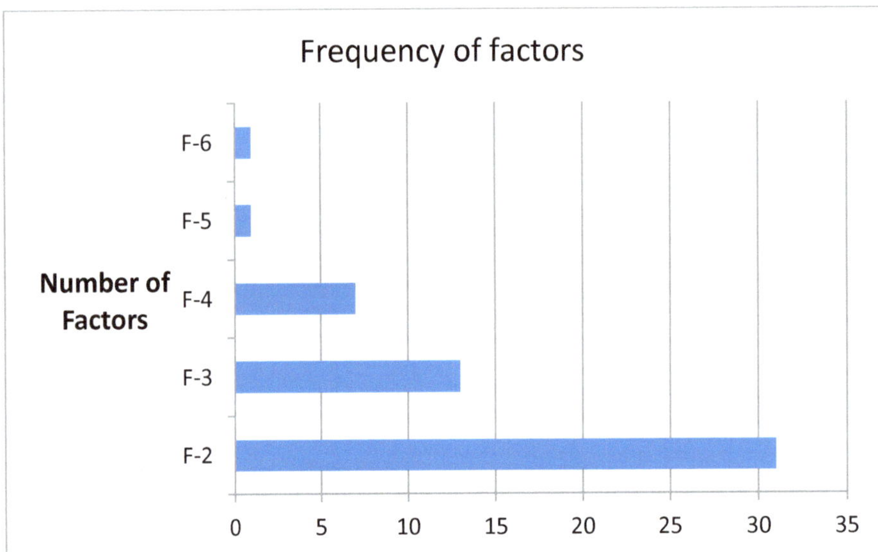

Figure 113

The most frequent is two prime factors for W in the set D_t, followed by three factors. The number of prime factors is more or less evenly distributed between gap base 2 and 4. For example gap base 2 has 15 numbers with two prime factors and gap base 4 has 16 numbers. A similar result is observed for 3 prime factors.

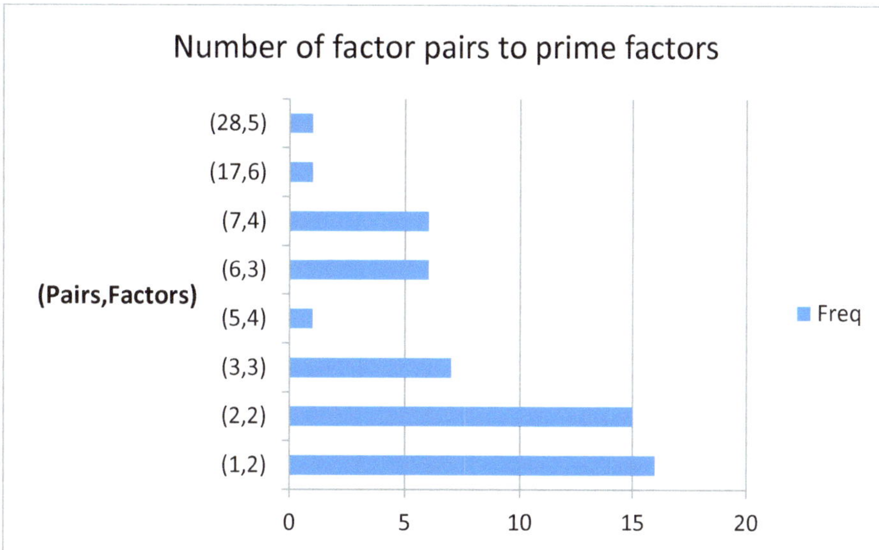

Figure 114

In the notation (3,3) for example this means *(number of factor pairs, number of prime factors)*, hence (3,3) and (7,3) are all looking at 3 prime factors. Gap base 4 factor pairs are the most frequent followed by gap base 2 factor pairs for D_t. This implies 2 prime factors from 1 factor pair are the most common followed by 2 prime factors from 2 factor pairs; that is, gap base 4 has a slightly higher occurrence than gap base 2. Note that (3,3), (6,3) and (7,4) have almost the same frequency of occurrence.

For example consider the base 2 number $W = 2378845$, the number of factor pairs is 5. Asymmetry always arise whenever $k_1 = k_2$ for some x_2. For example $k_1 = k_2 = 1$ for $x_2 = 56639$ and $x_2 = 79295$.

X2	K1	K2	Factor-1	Factor-2
1	79295	56639	475769	5
6	11328	10715.351	67967	35
11328	6	5.666495	35	67967
56639	1.3333402	1	7	339835
79295	1	0.6666632	5	475769

Table 223

11.3.10 Optimization of Factor Pairs

Consider W with two prime factors. Then we note that as the number increases, the time it takes to find factors also increases.

W	Milliseconds
5963	0
352307	6
1019363	17
17354717	136

Table 224

Therefore there is need to optimize the process. When k_1, k_2, k_3 are equal to 1, then the search for factor pairs will end. Hence we can use this condition to define when to stop searching for more factor pairs, and it is called the certainty factor pair condition. The values of k_i can be quiet large.

$k_1 = \frac{x_3 - x_2}{6x_2 + 1}$	$k_2 = \frac{x_2 + x_3}{6x_2 - 1}$	$k_3 = \frac{x_3 - x_2}{6x_2 - 1}$

Table 225

Now the k_1 and k_2 values are related to x_2, hence instead of using k values, we can use x_2 to establish the limiting value. Infact for gap base 2, every value of x_2 must appear as a value of either k_1 or k_2. For example $W = 678727919$ gives the following result.

X2	K1	K2
1	22624264	
16		1166199
81	233240	
233240	81	
1166199		16
22624264	1	

Table 226

The exception is with gap base 4, where every value of k_3 differs from x_2 , they are unique with respect to each other. For example consider $W = 77112329$. Note that k_3 has a smaller counting range compared to x_2, because this becomes a source of processing efficiency. However, $k_3 = 1$ gives us the largest value of x_2.

X2	K3
7089	302
49622	43
262287	8
347353	6
1836008	1

Table 227

Since $0 < W$, then k_1, k_2, k_3 are greater than zero because they define a positive product. Consequently this implies $0 < x_3 - x_2$ or $x_2 < x_3$. Therefore the condition $1 \leq x_2 \leq \infty$ does not apply, we do not have to count x_2 to infinity to determine factorization. This is the first limiting value for x_2. Therefore;

$$1 \leq x_2 \leq x_3$$

Hence as x_2 approaches x_3, then k_1, k_2, k_3 approaches zero. However, the smallest integer value of $k_i = 1$, hence this derives the following for x_2 an integer.

$x_2 = Int\{\frac{x_3 + 1}{5}\}$	$x_2 = Int\{\frac{x_3 - 1}{7}\}$	$x_2 = Int\{\frac{x_3 + 1}{7}\}$

Table 228

Generally, it is safe to consider the smaller divisor to maximize the probability of covering a factor either from gap base 2 or 4. This means we accept the divisor 5. Then between $x_3 = (W - 1)/6$ and $x_3 = (W + 1)/6$, we select the slightly larger value. Therefore we can calculate the exact largest possible value for x_2 as;

$$x_2 = Int\left\{\frac{W + 7}{30}\right\}$$

This is the limiting value that defines a prime pair search, in contrast to the values k_1, k_2, k_3 for a Mbulawa primality condition. From the approximation $W/30$, it is evident that;

$$Int\left\{\frac{W+7}{30}\right\} < x_3$$

Therefore the pairs that are derived using $x_2 = Int\left\{\frac{W+7}{30}\right\}$ as the maximum count are called standard pairs.

Definition

All factor pairs derived for $1 \leq x_2 \leq Int\left\{\frac{W+7}{30}\right\}$ *are standard factor pairs.*

The factor pair limit theorem tells us the maximum number of factor pairs we expect to generate for given W. It is noted that for gap base 2 this includes factors that have been repeated, while gap base 4 produces unique factor pairs only. It is noted that from the integer product law, then for any two integers $p_1$and p_2, then the product of any two numbers p_1p_2 is either a composite in gap base zero, or a pseudoprime in gap base 2 or 4.This has an impact in the way we conceptualize finding the factors of a number W in terms of the square root principle. From this rule, we have the following possibilities in terms of products between two factors of W.

1. [composite] x [composite] = [composite]
2. [composite] x [prime] = [composite]
3. [composite] x [pseudoprime] = [composite]
4. [pseudoprime] x [pseudoprime] = [pseudoprime]
5. [prime] x [pseudoprime] = [pseudoprime]

Normally, if we seek the prime factors of a number W, then we know that at least one of such prime factor will not be more than $\sqrt{W}$. For W not prime, then there exist a and b , where $1 < a, b < W$ such that $W = ab$. Since we cannot have both a and b greater than $\sqrt{W}$ as this would imply $W = ab > W$ a contradiction. So either $a \leq \sqrt{W}$ or $b \leq \sqrt{W}$. Hence W has a prime factor less than or equal to $\sqrt{W}$.

The table shows that pattern for gap base 2 and 4 numbers.

	W	$\sqrt{W}$	Prime Factors
GB2	678727915	26052.40709	5,97 ,1399439
	678727921	26052.4072	2971, 228451
	678727927	26052.40732	727, 933601
	678727933	26052.40743	13, 433, 120577
	678727939	26052.40755	73, 9297643
GB4	678727907	26052.40693	11,13,17,23,61,199
	678727913	26052.40705	167, 359, 11321
	678727919	26052.40716	29, 31, 754981
	678727931	26052.40739	7, 601, 161333
	678727937	26052.40751	19,151, 236573

Table 229

The problem may be considered in terms of the biggest factor-1 and factor-2, and looking at them as (p_1, p_2) representing prime factors of W. That is *(greatest prime factor in Factor-1, greatest prime factor in Factor-2)*. For example for 678727919 a gap base 4 number, we have;

X2	K3	Factor-1	Factor-2
5	3900735	23404411	29
150	125830	754981	899
3649075	5	31	21894449

Table 230

Then for factor-1 we have 754981 and for factor-2 we have 29, that is $(p_1, p_2) = (754981, 29)$. Hence $p_1 > \sqrt{W}$ and $p_2 < \sqrt{W}$. The same applies for the gap base 2 number 678727915, $p_1 > \sqrt{W}$ and $p_2 < \sqrt{W}$, where $(p_1, p_2) = (1399439,97)$. Since the other values are a repeat, the notation does not include them as valid (p_1, p_2).

X2	K1	K2	Factor-1	Factor-2
1	22624264		135745583	5
16		1166199	6997195	97
81	233240		1399439	485
233240	81		485	1399439
1166199		16	97	6997195
22624264	1		5	135745583

Table 231

It is possible that factor-2 in gap base 2 may be the only one with prime factors of W, that is, we have $(0, p_2)$ being the only biggest prime. This implies all the numbers before the factor switch are pseudoprime in factor-1. This is unlikely in gap base 4 because it is not symmetric.

A relationship is then defined showing $\sqrt{W}$ with respect to the biggest prime factor of W in terms of factor-1 and factor-2. Condition 1 and 2 is for the pure composite number, where the factors are only 2 and 3. Condition 2 covers a number W that is a product of a composite and a prime or pseudoprime. In this case, p_1 can take a value that is composite, hence p_2 must be prime or pseudoprime, where this will derive a product that is a general composite. Hence condition 3 is true for gap base 2 and 4, while condition 4 is true for gap base 4 only. This is expressed as a condition. It is interesting to note that there is no case where $p_1 > \sqrt{W}$ and $p_2 > \sqrt{W}$, though we do have the case $p_1 < \sqrt{W}$ and $p_2 < \sqrt{W}$

The condition is useful in understanding how $\sqrt{W}$ relates to finding prime numbers in the context of factor-1 and factor-2.

Square Root Factor Condition

Let p_1 be the greatest in factor-1 and p_2 the greatest in factor-2 where (p_1, p_2) are factors of W, then:

	Factor-1	Factor-2	Type	Factor Type
Condition 1	$p_1 < \sqrt{W}$	$p_2 > \sqrt{W}, \quad W = 6$	Pure Composite	$p_1 = 2, p_2 = 3$. Composite
	$p_1 < \sqrt{W}$	$p_2 < \sqrt{W}, \quad W > 6$	Pure Composite	$p_1 = 2, p_2 = 3$. Composite
Condition 2	$p_1 < \sqrt{W}$	$p_2 < \sqrt{W}$	General Composite	$p_1$composite, p_2 is prime
	$p_1 < \sqrt{W}$	$p_2 > \sqrt{W}$	General Composite	$p_1$composite, p_2 is prime
Condition 3	$p_1 < \sqrt{W}$	$p_2 < \sqrt{W}$	Gap Base 2, 4	(p_1, p_2) is prime
	$p_1 > \sqrt{W}$	$p_2 < \sqrt{W}$	Gap Base 2	(p_1, p_2) is prime
Condition 4	$p_1 < \sqrt{W}$	$p_2 > \sqrt{W}$	Gap Base 4	(p_1, p_2) is prime
	$p_1 > \sqrt{W}$	$p_2 < \sqrt{W}$	Gap Base 4	(p_1, p_2) is prime

Table 232

The condition can be applied by considering (p_1, p_2) in the context of the value of x_2 which drives the factorization process. Now from condition 3, we know that for gap base 2, then $p_1 > \sqrt{W}$ and $p_2 < \sqrt{W}$. We can ignore the first part of condition 3 since both are less than $\sqrt{W}$. What this condition states is that when $p_2 < \sqrt{W}$ is found, then logically from the structure $p_1 > \sqrt{W}$ must be found before the factor switch. The factor switch can be described in terms of x_2. This then establishes a relationship between the three constructs x_2, $\sqrt{W}$, and (p_1, p_2).

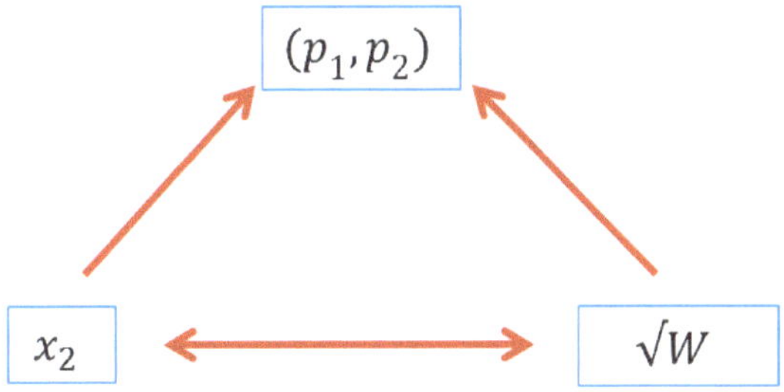

Figure 115

Some numbers are chosen from gap base 2 and 4, and a cutoff point is established as to when all prime factors can be found. For gap base 2, the cut-off point is established when the factor switch occurs, and for gap base 4 it is established by looking for the greatest value of x_2 or k_3 since there is no symmetry. In the table, this is shown as actual. Then $Int(\sqrt{W})$, $Int(\frac{\sqrt{W}}{4})$ and $Int(\sqrt{W}+1)/6)$ are found to see how they compare with the actual value in terms of accuracy where we have the result (p_1, p_2).

W	**Prime Factors**	**Actual**	$INT(\sqrt{W})$	$INT(\frac{\sqrt{W}}{4})$	$INT(\sqrt{W}+1)/6)$
678727907	11,13,17,23,61,199	3975	26052	6513	4342
678727913	167, 359, 11321	1887	26052	6513	4342
678727919	29, 31, 754981	150	26052	6513	4342
678727931	7, 601, 161333	701	26052	6513	4342
678727937	19,151, 236573	478	26052	6513	4342
678717981	31, 41, 379, 1409	2590	26052	6513	4342
678727987	7, 47, 2063003	55	26052	6513	4342
678727993	467, 547, 2657	443	26052	6513	4342
678727999	23, 829, 35597	3178	26052	6513	4342
678728005	5, 127, 4159, 257	3466	26052	6513	4342
678728011	13, 52209847	2	26052	6513	4342
678728017	11, 53, 1164199	97	26052	6513	4342
77221	31, 47, 53	9	277	69	46
77227	29, 2663	5	277	69	46
325	5, 5, 13	2	18	4	3
49	7, 7	1	7	1	1

Table 233

Looking at the table, we can choose either of two equations to use, that is $Int(\frac{\sqrt{W}}{4})$ and $Int(\sqrt{W}+1)/6)$ are to give close enough to give us the required accuracy of operation to establish (p_1, p_2). An alternative would be to take the average of the two to derive;

$$Maximum\ x_2 = (5\sqrt{W}+2)/24$$

The challenge here really is that there is no direct relationship; only a heuristic approach can be used. For example for the number 678728011, then there are two prime numbers that are found when $x_2 = 2$. However the previous number has $x_2 = 3466$.

Square Root Factor Theorem

For $W \in D_t$, there exists (p_1, p_2) such that $x_2 < Int(\frac{\sqrt{W}+1}{6})$, where p_1 is the greatest prime in factor-1 and p_2 the greatest prime in factor-2.

The application of this is that it is not really necessary for x_2 to be that large for the factorization process. For gap base 2, this has the effect of stopping at the factor switch in looking for factor pairs. Consequently factor pairs found this way are called efficient pairs.

Definition

Factor pairs that derive (p_1, p_2) are called efficient pairs.

An example is shown with 1111111111.

$$W = 1111111111$$

Number is gap base 2

Only these factors are necessary to determine factors, they are efficient pairs (rows $X2$ = 2 to 1852)

Don't need these factor pairs (rows $X2$ = 16667 to 16835017)

X2	K1	K2	Factor-1	Factor-2
2	16835017	14245014.08	101010101	11
7	4516712	4306632.047	27100271	41
45	688420.9294	683340	4100041	271
75	412439.3318	410610	2463661	451
497	62122	62080.01609	372731	2981
1515	20374.81571	20370	122221	9091
1852	16667	16663.66715	100001	11111
16667	1852	1851.629631	11111	100001
20370	1515.358128	1515	9091	122221
62122	497	496.6640008	2981	372731
410610	75.33339435	75	451	2463661
683340	45.33335537	45	271	4100041
4516712	7	6.666666162	41	27100271
16835017	2	1.66666663	11	101010101

Now let us consider searching for factor pairs of $W = 1111111111$. There will be 14 pairs, and this agrees with the formula calculation for factor pairs. Furthermore, we note that it takes up to more than the value 16 835 017 to establish all these pairs. Now the prime factors of W in this case are 11, 41, 271 and 9091. Hence we determine the maximum value of x_2, that is $INT((33333 + 1)/6) = 5556$. This implies the value of x_2 need not be more than this, a far cry from 16 835 017.

It can be noted from the table that the value of $x_2 = 1515$ for the number 9091 and that the count ends at $x_2 = 1852$, after that it is more than 5556 and we begin to get duplicate pairs. This means we can compare the expectations of the two systems assuming $W = 1\,111\,111\,111$, a gap base 2 number.

Base Result	**Standard Pair Maximum x_2**	**Efficient pairs Maximum x_2**	**Advantage of Efficient pairs**	x_2 smaller than W by
GB2	37 037 037	5556	6 666 times	199 984 times
GB2	26 455 026	5556	4 762 times	199 984 times

Table 234

This presents a greater efficiency of processing as we can see in the example, as much as 6 666 times compared to the standard factor pairs. When we run the computer program, we get the following interesting result by using efficient pairs.

$For\ W\ =\ 1111111111$

Using all factor pairs

Prime Factor	Time (ms)
11	6266
41	6266
9091	14
271	14

Using efficient pairs

Prime Factor	Time (ms)
11	4
41	4
9091	0
271	0

Time to find 2 463 661

Using all factor pairs

6280 milliseconds

Using efficient pairs

4 milliseconds

Table 235

From $W = a \cdot b$, we get 11 and 41 which leads to the number 2 463 661 that is factorized to get 271 and 9091.

A slightly different approach is used for gap base 4 because it produces unique pairs already. It is called a double barrel approach because it uses two competing variables to establish certainty in factorization. These are the variables k_3 and x_2 where these are unique with respect to each other unlike in gap base 2. Secondly because of the uniqueness feature, it implies we produce the same number of pairs, but we change the counting system to do so in a shorter time.

In the example of $W = 26\ 161\ 157$, x_2 will produce 9 factor pairs and k_3 will produce the remaining 6. However, in using k_4, it implies the program will jump between an expression for k_3 and x_2 seemlessly.

W = 26 161 157

W is gap base 4

These factor pairs will be produced by X2

X2	K3	Factor-1	Factor-2
2	396381	2378287	11
7	106346	638077	41
12	61411	368467	71
35	20862	125173	209
79	9218	55309	473
130	5597	33583	779
225	3232	19393	1349
294	2473	14839	1763
509	1428	8569	3053
1498	485	2911	8987
5337	136	817	32021
5583	130	781	33497
9668	75	451	58007
101400	7	43	608399
229484	3	19	1376903

These factor pairs will be produced by K3

Figure 116

The maximum valued need to produce all these pairs for certain is $x_2 = 853$. This creates efficiency in the factor pair production. Conceptually how do we do this?

- The first barrel is k_3. That is, x_3 is contant, but we can vary the values of x_2 to get k_3. This we are already doing.
- The other option is to use the expression of calculating x_2, that is, vary k_3 in order to get x_2. The value of x_3 will still be a constant. This is the second barrel.

K3 = (X3 – X2)/(6X2 – 1)
X3 is constant, we vary X2 to get K3

X2 = (X3 +K3)/(6K3 +1)
X3 is constant, we vary K3 to get X2

Figure 117

The only way to make it to work is to vary, x_2 and k_3 simultaneously. In both cases the maximum value is $x_2 = (\sqrt{W} + 1)/6$ or $k_3 = (\sqrt{W} + 1)/6$. As a result, two new variables are used, say k_4 and k_5 to represent x_2 and k_3 respectively. Therefore the equations become;

a) $k_4 = (x_3 - x_2)/(6x_2 - 1)$. This is the normal equation we have been using for k_3;
b) $k_5 = (x_3 + x_2)/(6x_2 - 1)$. We substitue the k_3 value with the x_2 value, and define a new variable k_5.

Hence we seek an integer value of k_4 and k_5, and x_2 becomes the common value. For example, in $W = 26\,161\,157$, when $x_2 = 3$ hence it implies $k_4 = 256481.7647$, and $k_5 = 229484.00$. Hence we have a factor pair produced from k_5.

W	**26,161,157**		
X3	**4,360,193**		
		GB2	GB4
X2	3	Factor-1	Factor-2
K4	256481.7647	1538891.588	17

		GB2	**GB4**
X2	3	Factor-1	Factor-2
K5	229484.00	19.0000	1376903.000

Table 236

Hence the major difference is the value of x_2, but the pairs are the same. An example is made from $W = 26161187$. For standard process the pairs are found after 335 400, and only 368 under the efficient process.

Standard

X2
8
102
152
1974
335400

Efficient

X2
2
8
102
152
368

Figure 118

Standard GB4 Pairs

X2	K3	Factor-1	Factor-2
8	92770	556621	47
102	7136	42817	611
152	4786	28717	911
1974	368	2209	11843
335400	2	13	2012399

GB4 Efficient Pairs

X2	K4	K5	Factor-1	Factor-2
2	396381.45	335400	13	2012399
8	92770	88983.796	556621	47
102	7136	7113.0506	42817	611
152	4786	4775.8488	28717	911
368	1975.4554	1974	2209	11843

The diagram illustrates the program approach towards factorization.

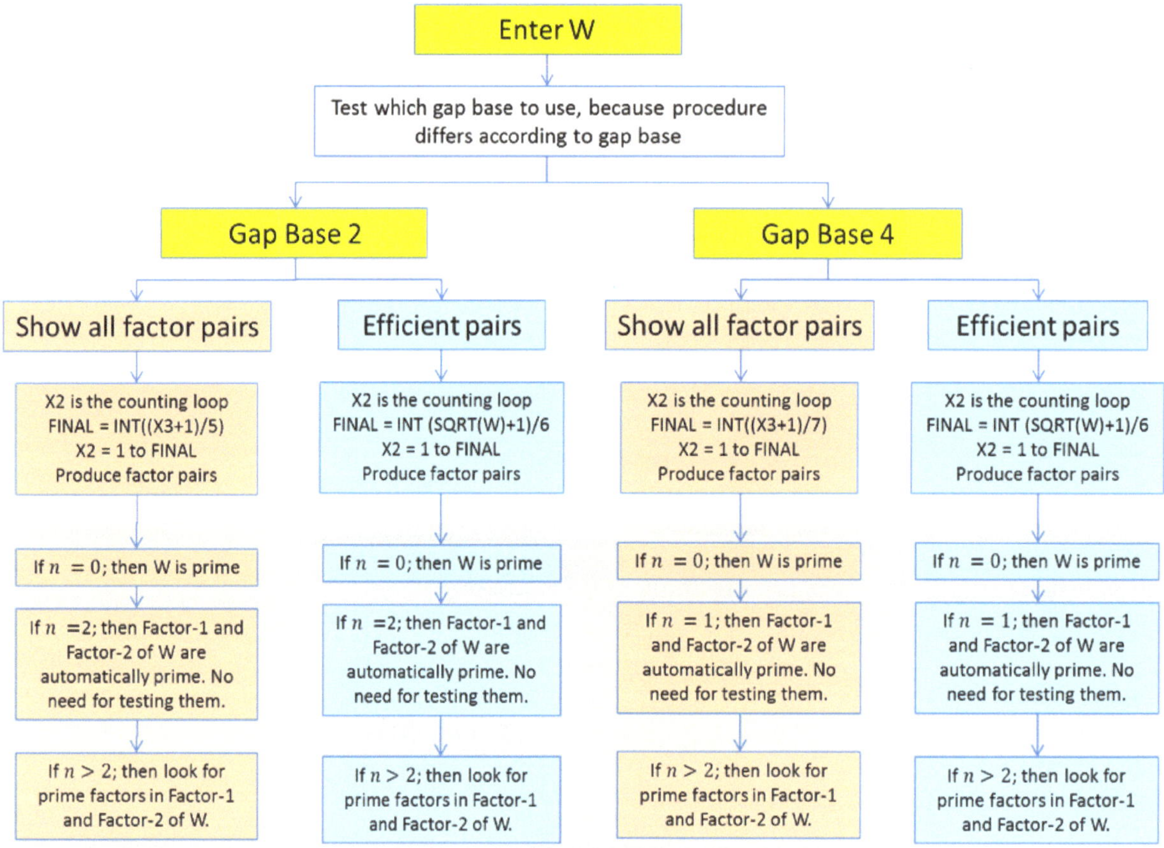

Figure 119

11.3.11 Symmetry and Gap Base 4 Factor Pairs

For factorization, another method may be used to create efficiency. All the factor pairs are unique. This we can clearly see from the number 26161157. Now $(\sqrt{W}+1)6$ would imply x_2 does not count beyond 852, or from 1498 downward where this would leave some of our critical factor pairs out like 19 and 43. That is we would be looking for a factor pair from $11 \cdot 41 = 451$, and this would not exist in the given range.

X2	K1	K2	K3	Factor-1	Factor-2
2	396381.3333	335399.2821	396381	2378287	11
7	106346.3333	101399.6667	106346	638077	41
12	61411.33333	59728.50228	61411	368467	71
35	20862.33333	20664.25434	20862	125173	209
79	9218.333333	9179.186667	9218	55309	473
130	5597.333333	5582.666667	5597	33583	779
225	3232.333333	3227.215149	3232	19393	1349
294	2473.333333	2470.197545	2473	14839	1763
509	1428.333333	1427.06503	1428	8569	3053
1498	485.3333333	484.8920533	485	2911	8987
5337	136.3333333	135.9914957	136	817	32021
5583	130.3333333	129.9922286	130	781	33497
9668	75.33333333	74.99740845	75	451	58007
101400	7.333333333	6.999976441	7	43	608399
229484	3.333333333	2.9999954	3	19	1376903

Table 237

The first factor-2 is always the smallest number possible in gap base 4, which implies that the first factor-1 will be the biggest pseudoprime possible, and therefore should contain 19 and 43.

Hence an advantage can be taken of this structural set-up for the GB-4 product. Since Factor-1 will always be a GB2 number, and this is symmetric, therefore we take the first factor pair of the GB4 product, and set it as the new value for W. This allows us now to apply the rule developed for GB2 factor pairs, and this is fast because we do not exhaust all the factor pairs of W in GB4. We take only the first pair because we are sure that it contains all the other factors when the GB4 factor has been identified.

X2	K1	K2	K3	Factor-1	Factor-2	Number
2	396381.3333	335399.2821	396381	**2378287**	11	**W=26161157**
3	23316.70588	20862	23316.37255	125173	19	**W=2378287**
7	9668	9218	9667.666667	58007	41	
12	5583	5429.712329	5582.666667	33497	71	
130	509	507.3636364	508.6666667	3053	779	
136	486.5239264	485	486.190593	2911	817	
225	294	293.2316802	293.6666667	1763	1349	
294	225	224.411898	224.6666667	1349	1763	
485	136.4269508	136	136.0936175	817	2911	
509	130	129.5816694	129.6666667	779	3053	
5583	12	11.66596018	11.66666667	71	33497	
9218	7.333592493	7	7.000259159	43	55309	
9668	7	6.666431071	6.666666667	41	58007	
20862	3.333383931	3	3.000050597	19	125173	

Table 238

We note that when $W = 2378287$, then $(\sqrt{W} + 1)6$ gives us $x_2 = 257$. Hence we get the first pair for $W = 26161157$ at $x_2 = 2$, and then get the rest of the pairs under the second iteration process counting only to 257. If we did this under gap base 4 only, then we would have counted to 229 484 before finding all the pairs we need. Indeed, this saves a lot of breath! Another example is illustrated from $W = 807811277$.

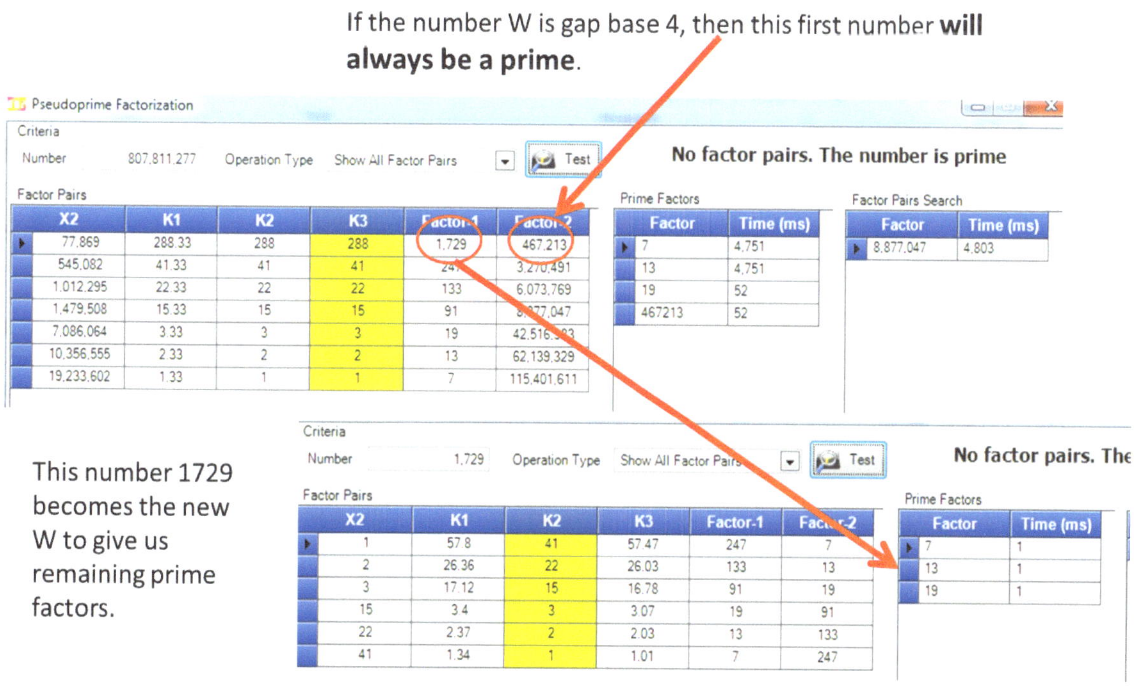

Figure 120

The limitation of this method is that it can only be used in factorization but would not improve efficiency in terms of developing a primality test.

11.4 Mbulawa Factor Pair Primality Test

It has been established that if $n = 0$, then it implies no factor pairs exist, which implies that the number is a prime. Secondly, there is an endpoint for searching for factor pairs, where these may be symmetric or non-symmetric. Now different types of primality tests have been defined so far based on the following conditions.

a) The algebraic sieve, based on knocking out pseudoprimes so that one remains with the prime numbers only.
b) The algebraic primality test that uses the algebraic sieve structure to define the primality condition $k_1 < 1$ and $k_2 < 1$.
c) The Mbulawa primality test based on the condition that condition $k_1 < 1$ and $k_2 < 1$ for a gap base 2 number and $k_3 < 1$ for a gap base 4 number.

The Mbulawa factor pair primality test is a quest for power time, that is, making it take shorter and shorter to establish if a number is prime no matter how large. Normally larger numbers chew up larger amounts of processing time.

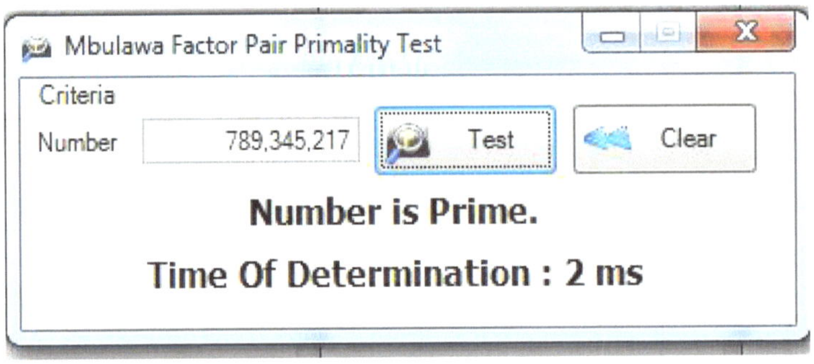

Figure 121

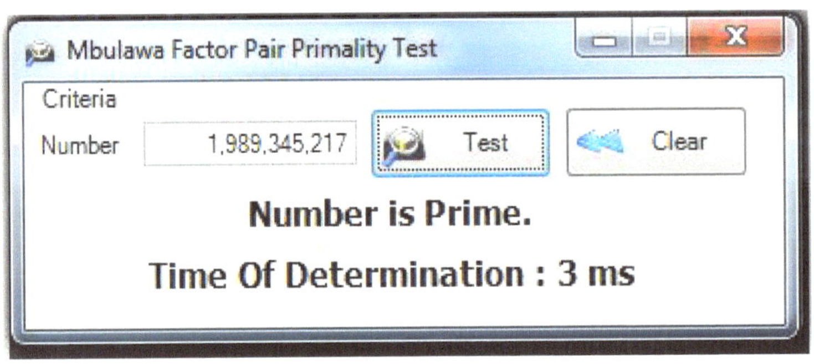

Figure 122

The argument of no primality factor pairs is useful because the search can be restricted further, and therefore can be used as a criterion for primality. For prime factors of W, it is known that the smaller prime number cannot be more than $\sqrt{W}$. Assuming the larger value in defining x_2, then the maximum value for x_2 in a prime number search should not go beyond $x_2 = (\sqrt{W} + 1)/6$ when using a factor pair search. Similarly if the number is gap base 4, then we also use $k_3 = (\sqrt{W} + 1)/6$ since both x_2 and k_3 are applied to generate factor pairs. In this case we have only one primality condition.

Factor Pair Primality Theorem

If for $W \in D_t$ *there is no factor pair at* $x_2 = Int\left(\frac{\sqrt{W}+1}{6}\right)$, *then* W *is a prime number.*

Consequently the primality test is the fastest so far because it can take less time compared to the other three tests. This test clearly illustrates the power of UNDT.

It is of interest to find out the relationship between x_2 and W. Since primality is determined at $x_2 = (\sqrt{W} + 1)/6$, then the ratio between the two is;

$$W/x_2 = 6W/(\sqrt{W} + 1)$$

Hence as an example assume that x_2 is increasing from 1 in steps of 100 000. This derives the simple equation $f(x_2) = 100000x + 1$, a straight line. However, the relationship between x_2 and W is not linear. Typically we expect that as W gets larger, the process of testing for primality becomes more and more difficult and unwieldy. The end result is that primality has to be done by high powered computers working over extensively long hours and days to determine primality or find factors. Therefore this algorithm presents an entirely different scenario, instead of increasing proportionately x_2 has a declining relative increase compared to W.

This can be generally expressed in the following format. Let W_r be the ratio of x_2 to W, where $p > 1$ and $0 < q < 1$ are constants, then generally;

$$W_r = px^q$$

The fact that $q < 1$ defines the observed slow rate, since the differential of this relationship is qpx^{q-1}. Similarly, $0 < q < 1$, then $q - 1 < 0$, and therefore x_2 has a slow change with respect to W. This can also be confirmed directly in defining the rate of change of x_2 through a differential of $(\sqrt{W} + 1)/6$, that is $1/12(W^{-0.5})$.

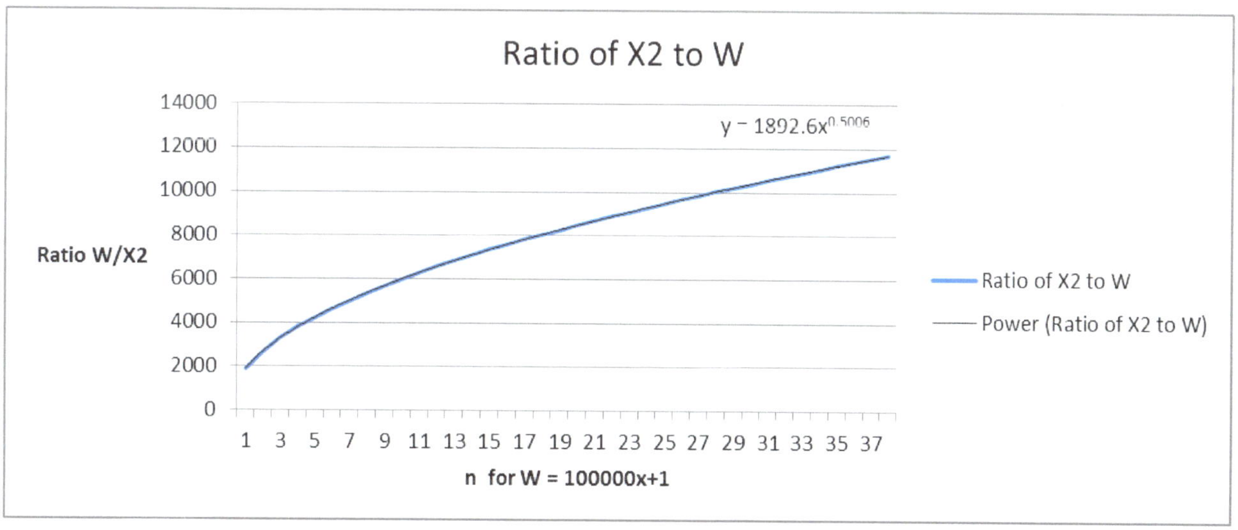

Figure 123

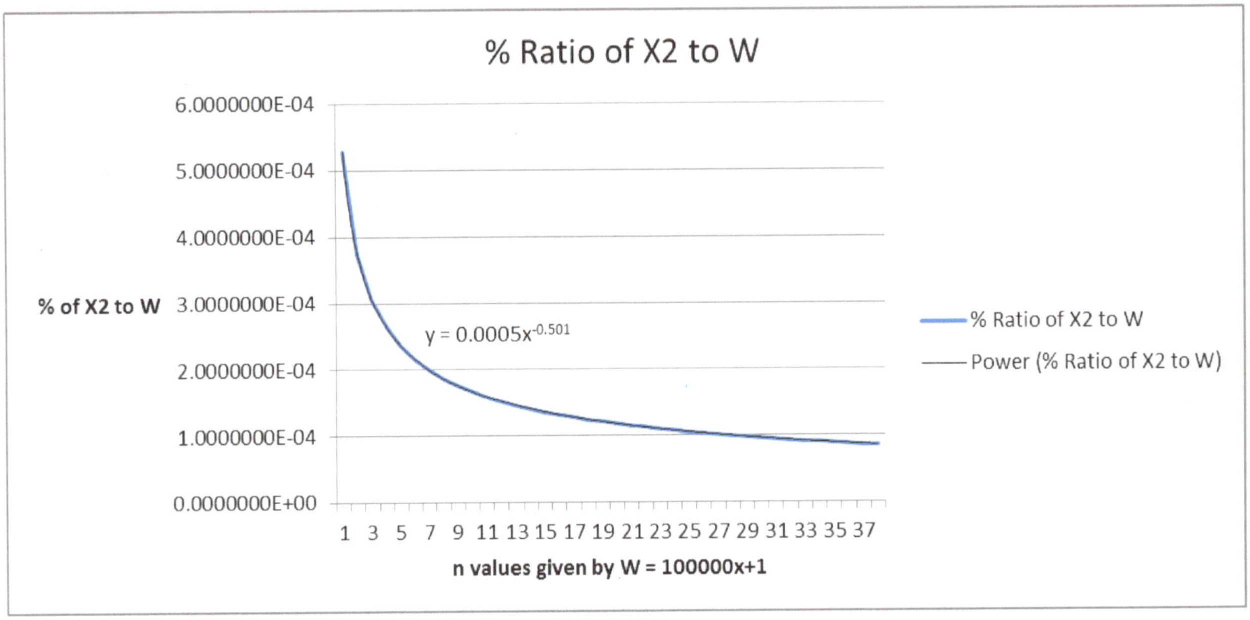

Figure 124

We can also express x_2/W as a percentage, and this has the general format, where $0 < u < 1$ and $0 < v < 1$.

$$W_p = ux^{-v}$$

In that case, having a big number W to determine primality or find factors should not be a major hurdle in using the factor pair theory.

11.5 Factor Pair Product Distribution

In order to establish what patterns the products take, we consider the options that are available. The assumption is that the prime numbers that are involved are from a symmetric and non-symmetric distribution. That is, they do not involve the asymmetric distribution.

11.5.1 Two Prime Factors

The basic idea is to establish the behavior of the factorization process and the structure in which it exists. This is done by finding out what options are actually there in terms of the product, and then working backwards to the factorization process. Hence the options then define possibilities in terms of products at given number of prime factors. The pattern is constructed in two ways in order to be able to deal with complexity.

a. The standard product option. In this option the number of group in a product structure is the same as the number of prime factors of W.
b. The extended product option. Here the number of groups in a product structure is more than the number of prime factors of W.

The standard option is easier to start with, hence the extended option will be introduced later. In the UNDT framework, a product is not just between two numbers, it is also between two types of numbers. Consequently this results in a product structure that has specific outcomes. The factorization is defined for a number W which is obviously a pseudoprime, where this number belongs to either gap base 2 or 4. Under the gap base, we have three sub-types for gap base 2 these being the $K1$, $K2$ and $K3$ numbers while gap base 4 only has $K3$ type.

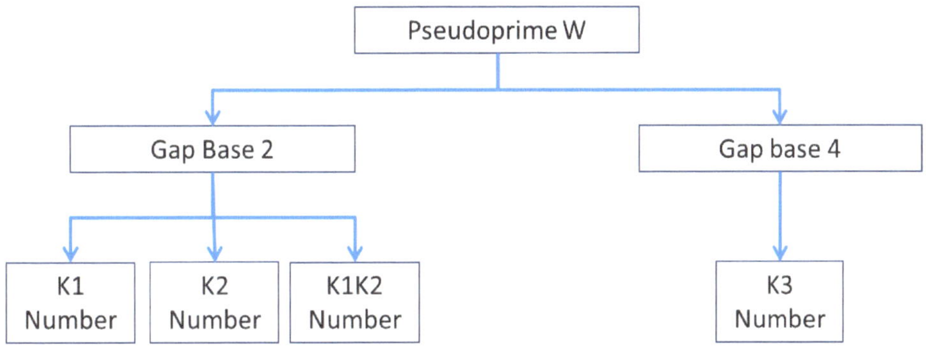

Figure 125

The product structure is also studied in terms of these characteristics. The product structure is based on a symmetric and non-symmetric distribution. It can also be done for the asymmetric case, where the values for the asymmetric case will arise mostly from a product of primes that results in numbers that have the same coordinate. For example 462465025 has 78 pairs instead of 79 because the factors $5 \cdot 5$ and 23 have the same coordinate.

The simplest product is that of two prime numbers. The possibilities are as shown in the table, we do not consider the product (Prime-2)(Prime-2) because it implies the same things as (Prime-1)(Prime-1). This means two prime numbers that are the same multiplying each other. Option 1 and 2 define the levels that are available given a certain number of prime factors for W. The levels together form a factorization category.

Options	Number of prime factors of W		Group
Option 1	Prime-1	Prime-1	(2,1)
Option 2	Prime-1	Prime-2	(1,0)

Table 239

In (2,1) the 2 stands for the maximum number of factor pairs in gap base 2 and 1 for the maximum number of factor pairs in gap base 4 for the group. Therefore if we multiply two numbers that are the same we will only have two factor pairs or one factor pairs respectively depending on the gap base.

Definition

a. *Every symmetric and non-symmetric distribution has product group categories* (g_1, g_2) *where* g_1 *and* g_1 *are the maximum number of factor pairs in gap base 2 and 4 respectively.*
b. *A factorization category is unique and denotedas P*x*-*y*, where* x *is the number of prime factors and* y *is the level within the category.*
c. *If* W *has* x *prime factors, then* $1 \leq y \leq x$.

As stated before, the groups together form a product category, and each group in the category defines a level. For example, when W has two prime factors that are not equal, then the maximum number of factor pairs is 1 in gap base 2, and 0 in gap base 4; hence we have the group (1,0). Now there are only two levels in the category, denoted as P2-1 and P2-2 for level 1 and 2 respectively. As indicated in the definition, there is no other factorization category P2, and its behavior is unique to itself. It is also possible to have more than one level in a group as will be shown later.

The next stage then demonstrates the power of UNDT, that products are indeed different. Note that the only difference in pattern between the first and second products is $K1$ and $K2$ because in both cases W is a GB2 number. As a result, in the compressed space, $K1$ and $K2$ play a significant result and are given an attribute name.

Definition

The variables $K1$, $K2$, and $K3$ are factorization identifiers of W.

In other words, they can be used to describe a particular factorization process. They have been used to identify a type of number, but this is now directly related to the process of finding prime factors of a number. It becomes useful as an identifier when related to the number of factor pairs in a product structure. When expanding the product options, the following is derived.

(2,1) Group

Prime-1	Prime-2	Result	Factor Pairs	Identifier
GB4	GB4	GB2	2	K1
GB2	GB2	GB2	2	K2
GB2	GB4	GB4	1	K3

	Prime-1	Prime-1	Result	Factor Pairs	Identifier
(1,0)K1 Group	GB4	GB4	GB2	1	K1
(1,0)K2 Group	GB2	GB2	GB2	1	K2

Table 240

Note that in the (1,0) group there is no $K3$ number, hence the zero. Also (1,0)K1 and (1,0)K2 can be represented simply as (1,0) group. This gives us the first group that has two levels due to K1 and K2. That is each level actually defines a sub-group that gives us the expected number of factor pairs. Therefore not only are products different, but they can also be grouped into specific product group for a given number of prime factors. This is a general observation is not only true for two prime numbers.

Consequently, the factor pair theory gives us an ability to study the distribution behavior of prime numbers according to prime factors of a given number. We can now use the number of factor pairs as a variable of study, including the number of prime factors and the identifier.

W	Largest X2	No Factor pairs	No of Prime factors	Identifier	Time in milliseconds
4,490,287	5713	6	3	K1, K2	30
83,020,543	82855	6	3	K1, K2	541
202,343,797	169467	6	3	K1, K2	1317
661,447,867	392318	6	3	K1, K2	4354

Table 241

For example, in the above numbers, we are looking at the product format $GB4 \cdot GB4 \cdot GB2$, and we can consider its characteristics. In studying these numbers, we have kept the three variables the same. In this example we note that:

- The larger the number, the larger the value of X2
- The larger the number, the longer it takes to factorize.

The last observation is to be expected, but what we can now do in the UNDT framework is to take numbers in the same range but have them with a K2 identifier and note the time it takes to factorize.

We can also study the product in terms of category, that is by looking at the product level behavior

GB2	Factor-1	Factor-2	P4-2
1854547	109091	17	Factor Pairs = 10 Group (10,5)
	GB4	GB4	

GB4	Factor-1	Factor-2	P3-2
109091	1859	59	Factor Pairs = 2 Group (4,2)
	GB2	GB4	

GB2	Factor-1	Factor-2	P2-2
1859	43	43	Factor Pairs = 1 Group (1,1)
	GB2	GB2	

Figure 126

For example consider factorizing the number $W = 1854547$, and we take only b_1 values in each case and iterate, then we notice that the factorization remains within the same level. In this case, this is level 2. Another example is W= 23515453. Here the factorization takes place along level 1.

GB2	Factor-1	Factor-2	P5-1
23515453	1808881	13	Group (30,15)
	GB2	GB2	

GB2	Factor-1	Factor-2	P4-1
1808881	78647	23	Group (14,7)
	GB4	GB4	

GB4	Factor-1	Factor-2	P3-1
78647	1333	59	Group (6,3)L
	GB2	GB4	

GB2	Factor-1	Factor-2	P2-1
1333	43	31	Group (2,1)
	GB2	GB2	

Figure 127

However, the difference is observed when we have a gap base 4 number. This time the level goes progressively down.

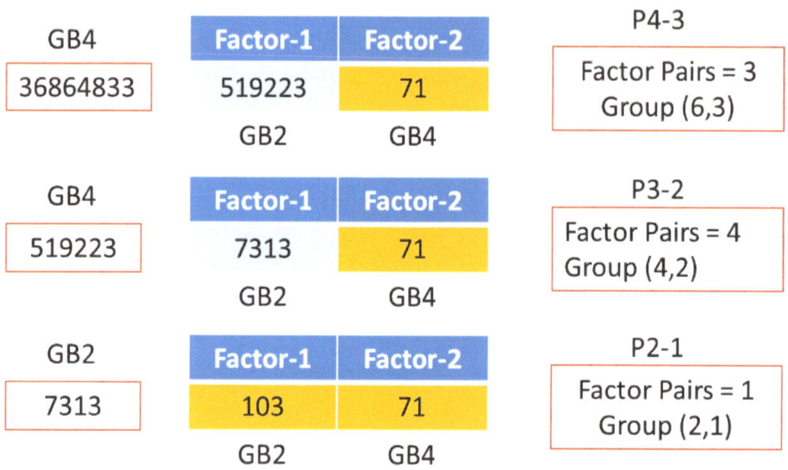

Figure 128

The rules that are applied for prime factors only can also apply to any pair of pseudoprimes to give the required answer. That is, if we have a product $GB2 \cdot GB2 = GB2$, this is true whether we using prime numbers only, pseudoprimes only, or even a mixture of them. However, the factorization process will differ depending on how the product is actually described in terms of its constitution. Consider for example the product $169 \cdot 217 = 36673$, the product is defined within the (2,1) group with identifier K2. That is, instead of (Prime1)(Prime2) we have (Pseudoprime1)(Pseudoprime2). The result of the product is in a different group though the identifier has remained the same.

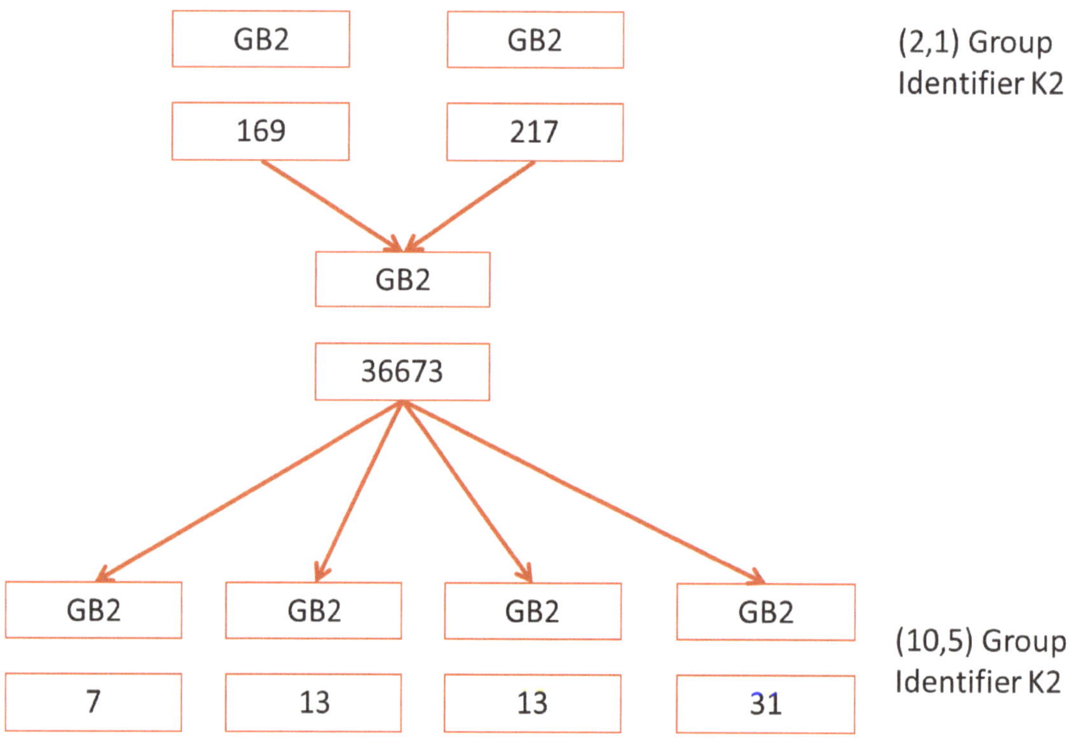

Figure 129

11.5.2 Three Prime Factors

For a three prime factor number, the following possibilities exist.

Prime-1	Prime-2	Prime-3
Prime-1	Prime-1	Prime-1
Prime-1	Prime-1	Prime-2
Prime-1	Prime-2	Prime-2
Prime-2	Prime-2	Prime-2

Table 242

However, because of commutativity, this comes down to three options;

Options	Number of prime factors of W			Groups
Option 1	Prime-1	Prime-2	Prime-3	(6,3)
Option 2	Prime-1	Prime-1	Prime-1	(2,0)
Option 3	Prime-1	Prime-1	Prime-2	(4,2)

Table 243

Hence we look for structures that define the above configuration. Using the limit factor pair theorem, such a structure will have 6 factor pairs or less. Secondly, we then consider the gap base configuration of the products as shown. The three options result in three product groups.

							Asymmetric	
(6,3) Group	Prime-1	Prime-2	Prime-3	Result	Factor Pairs	Identifier	Factor Pairs	Identifier
	GB2	GB2	GB2	GB2	6	K2	NONE	
	GB2	GB2	GB4	GB4	3	L	3	L
	GB2	GB4	GB4	GB2	6	K1,K2	5	K1,K2
	GB4	GB4	GB4	GB4	3	R	3	R
							Asymmetric	
(4,2) Group	Prime-1	Prime-1	Prime-2	Result	Factor Pairs	Identifier	Factor Pairs	Identifier
	GB2	GB2	GB2	GB2	4	K2	NONE	
	GB4	GB4	GB4	GB4	2	K3	2	K3
	GB2	GB2	GB4	GB4	2	K3	2	K3
	GB4	GB4	GB2	GB2	4	K1,K2	NONE	
							Asymmetric	
	Prime-1	Prime-1	Prime-1	Result	Factor Pairs	Identifier	Factor Pairs	Identifier
(2,0)K2 Group	GB2	GB2	GB2	GB2	2	K2	NONE	
(1,0)K1 Group	GB4	GB4	GB4	GB2	1	K1	NONE	

Table 244

Note that (1,0)K1 Group is the last of its type under symmetric or non-symmetric distribution. After this, no other values occur that have K1 as an identifier.

The table illustrates the fact that the maximum number of factor pairs is six for gap base 2 and 3 for gap base 4. Asymmetry has no effect on gap base 4 because it does not change the number of factor pairs. However, for gap base 2, it has an effect, but there is only one asymmetric product with 5 factor pairs. The rest do not qualify to have an asymmetric distribution because it can only be made for two numbers that are in different gap bases. The "L" and "R" indicate a left and right position of the prime factors. That is the prime factors are in the left or at position Factor-1 and right at position Factor-2.

11.5.3 Four Prime Factors

In a similar manner, we have four options that are available to determine the product structure for four prime factors in a symmetric and non-symmetric distribution.

Options	Number of prime factors of W				Groups
Option 1	Prime-1	Prime-2	Prime-3	Prime-4	(14,7)
Option 2	Prime-1	Prime-1	Prime-1	Prime-1	(3,3)
Option 3	Prime-1	Prime-1	Prime-1	Prime-2	(6,3)
Option 4	Prime-1	Prime-1	Prime-2	Prime-2	(7,0)
Option 5	Prime-1	Prime-1	Prime-2	Prime-3	(10,5)

Table 245

The four options result in four product groups. In line with the limit theory, the maximum number of factor pairs here is 14.

(14,7) Group
Option 1

Prime-1	Prime-2	Prime-3	Prime-4	Result	Factor Pairs	Identifier
GB2	GB2	GB2	GB2	GB2	14	k2
GB2	GB2	GB2	GB4	GB4	7	k3
GB2	GB2	GB4	GB4	GB2	14	K1, k2
GB2	GB4	GB4	GB4	GB4	7	K3
GB4	GB4	GB4	GB4	GB2	14	K1, K2

(10,5) Group
Option 5

Prime-1	Prime-1	Prime-2	Prime-3	Result	Factor Pairs	Identifier
GB2	GB2	GB2	GB2	GB2	10	k2
GB2	GB2	GB2	GB4	GB4	5	k3
GB2	GB2	GB4	GB4	GB2	10	K1,k2
GB4	GB4	GB4	GB4	GB2	10	K1,k2
GB4	GB4	GB4	GB2	GB4	5	K3
GB4	GB4	GB2	GB2	GB2	10	K1,k2

(7,0) Group
Option 4

Prime-1	Prime-1	Prime-2	Prime-2	Result	Factor Pairs	Identifier
GB2	GB2	GB2	GB2	GB2	7	K2
GB2	GB2	GB4	GB4	GB2	7	K1,K2
GB4	GB4	GB4	GB4	GB2	7	K1,K2

(6,3) Group
Option 3

Prime-1	Prime-1	Prime-1	Prime-2	Result	Factor Pairs	Identifier
GB2	GB2	GB2	GB2	GB2	6	K2
GB2	GB2	GB2	GB4	GB4	3	K3
GB4	GB4	GB4	GB4	GB2	6	K1,k2
GB4	GB4	GB4	GB2	GB4	3	K3

(3,0) Group
Option 2

Prime-1	Prime-1	Prime-1	Prime-1	Result	Factor Pairs	Identifier
GB2	GB2	GB2	GB2	GB2	3	K2
GB4	GB4	GB4	GB4	GB2	3	K1, k2

Table 246

This clearly shows that indeed factorization is highly structured in process.

11.5.4 Five Prime factors

A similar pattern for five prime factors is derived as shown. The options are:

Options	Number of prime factors of W					Groups
Option 1	Prime-1	Prime-2	Prime-3	Prime-4	Prime-5	(30,15)
Option 2	Prime-1	Prime-1	Prime-1	Prime-1	Prime-1	(4,2)
Option 3	Prime-1	Prime-1	Prime-1	Prime-1	Prime-2	(8,4)
Option 4	Prime-1	Prime-1	Prime-1	Prime-2	Prime-3	(14,7)
Option 5	Prime-1	Prime-1	Prime-2	Prime-2	Prime-3	(16,8)
Option 6	Prime-1	Prime-1	Prime-2	Prime-3	Prime-4	(22,11)

Table 247

This has five levels, but only the first level is shown from option one, that is the (30,15) group.

Prime-1	Prime-2	Prime-3	Prime-4	Prime-5	Result	Factor Pairs	Identifiers
GB2	GB2	GB2	GB2	GB2	GB2	30	K2
GB2	GB2	GB2	GB2	GB4	GB4	15	K3
GB2	GB2	GB2	GB4	GB4	GB2	30	K1,K2
GB2	GB2	GB4	GB4	GB4	GB4	15	K3
GB2	GB4	GB4	GB4	GB4	GB2	30	K1,K2
GB4	GB4	GB4	GB4	GB4	GB4	15	K3

Table 248

In summary, the groups for 5 prime factors are as follows.

(30,15) Group
Option 1

Factor Pairs	Identifiers
30	K2
15	K3
30	K1,K2
15	K3
30	K1,K2
15	K3

(22,11) Group
Option 6

Factor Pairs	Identifiers
22	K2
11	K3
22	K1,K2
11	K3
22	K1,K2
11	K3

(14,7) Group
Option 4

Factor Pairs	Identifiers
14	K2
7	K3
14	K1,K2
7	K3
14	K1,K2
7	K3

(8,4) Group
Option 3

Factor Pairs	Identifiers
8	K2
4	K3
8	K1,K2
4	K3

(4,2) Group
Option 2

Factor Pairs	Identifiers
4	K2
2	K3

(16,8) Group
Option 5

Factor Pairs	Identifiers
16	K2
8	K3
16	K1,K2
8	K3
16	K1,K2
8	K3

Table 249

This category does not have a K1 number, only K2, K3 and K1K2 numbers as shown by the identifiers. Similarly, it is observed that some products may not have certain identifiers.

Prime Factors	K1	K2	K1K2	K3
2	✓	✓	✓	✓
3	✓	✓	✓	✓
4		✓	✓	✓
5		✓	✓	✓

Table 250

For example a product of four and five prime factors is not defined by K1 numbers. This leads to a conjecture.

Conjecture

$K1$ numbers are defined for $n = 1$, where n is the number of factor pairs and P_1, P_2, P_3 are prime factors.

P_1	P_2	P_3	Condition
GB4	GB4	NONE	$P_1 \neq P_2$
GB4	GB4	NONE	$P_1 = P_2$
GB4	GB4	GB4	$P_1 = P_2 = P_3$

Table 251

The conjecture can be applied in the context of doing self-primality testing, that is, this is considered a reasonable framework by which to consider it. Besides this conjecture, the number of additional conditions that allows self-testing to be defined increases such that it is more practical to consider it as a consequence of this conjecture. After three prime factors, then for every GB4 product where $P_1 = P_i$, the result gives a $K1K2$ number, and this increases the complexity of self-testing.

11.5.5 The Factor Pair Triangle

The name factor pair triangle is derived from a reminder of Pascal's triangle of binomial co-efficient, only this time it involves factorization. Two types of development are suggested here. The first is the standard prediction. This consists of factor pairs that we are certain will be present, and the next is those that occur to "disturb" the standard pattern. Both of these patterns are based on a symmetric and non-symmetric distribution, it does not include the asymmetric distribution. If the asymmetric distribution is included, the table would be very unwieldy for bigger values of prime factors, and it would be difficult to develop a prediction tool. Therefore the table that includes asymmetric distribution is called an outlier table.

The standard pattern is derived by setting the number of factor pairs to relate to the number of options that are there. For example, if we are doing four prime factors, then the highest level number is also 4, and so on. In order to generate the standard prediction, option 1 is defined to consist of unique prime factors of W, and option 2 to consist of the same prime factor. Then the other options are derived by permuting other prime factors until we reach [prime 1, prime 1].

Option	Number of prime factors of W							Group
Option 1	Prime-1	Prime-2	Prime-3	Prime-4	Prime-5	Prime-6	Prime-7	(126,63)
Option 2	Prime-1	Prime-1	Prime-1	Prime-1	Prime-1	Prime-1	Prime-1	(6,3)
Option 3	Prime-1	Prime-1	Prime-1	Prime-1	Prime-1	Prime-1	Prime-2	(12,6)
Option 4	Prime-1	Prime-1	Prime-1	Prime-1	Prime-1	Prime-2	Prime-3	(22,11)
Option 5	Prime-1	Prime-1	Prime-1	Prime-1	Prime-2	Prime-3	Prime-4	(38,19)
Option 6	Prime-1	Prime-1	Prime-1	Prime-2	Prime-3	Prime-4	Prime-5	(62,31)
Option 7	Prime-1	Prime-1	Prime-2	Prime-3	Prime-4	Prime-5	Prime-6	(94, 47)

Table 252

The missing option here is the one indicated below as option 8. It consists of at least two pairs of prime factors being the same. Under the UNDT context, it can give several types of results. The group is (52,26)

Prime-1	Prime-1	Prime-2	Prime-2	Prime-3	Prime-3	Prime-7	Factor Pairs	Type
GB2	GB2	GB2	GB2	GB2	GB2	GB2	52	K2
GB2	GB2	GB2	GB2	GB2	GB2	GB4	26	K3
GB2	GB2	GB2	GB2	GB4	GB4	GB2	52	K1,K2
GB2	GB2	GB2	GB2	GB4	GB4	GB4	26	K3
GB2	GB2	GB4	GB4	GB4	GB4	GB2	52	K1,K2
GB2	GB2	GB4	GB4	GB4	GB4	GB4	26	K3

Table 253

The standard prediction would be as follows without option 8. Each level presents a product option, and between each level is a difference row. The color shows how the pattern is derived. Level 1 is derived using the limit theorem, while level two we add a difference that is doubling as the number of prime factors increases. For example from 4 prime factors to 5, it doubles from 4 to 8; that is (14,7) derives (10,5) and consequently (30,15) derives (22,11). Since 30-8=22, then from the limit theorem the number of pairs for the $K3$ number will be half of that, which is 11. Hence we have (22,11).

	Factor pairs for a given number of prime factors of W					
No. of Prime Factors	2	3	4	5	6	7
Level 1	(2,1)	(6,3)	(14,7)	(30,15)	(62,31)	(126, 63)
	1	2	4	8	16	32
Level 2	(1,1)	(4,2)	(10,5)	(22,11)	(46,23)	(94,47)
		2	4	8	16	32
Level 3		(2,1)	(6,3)	(14,7)	(30,15)	(62,31)
			3	6	12	24
Level 4			(3,3)	(8,4)	(18,9)	(38,19)
				4	8	16
Level 5				(4,2)	(10,5)	(22,11)
					5	10
Level 6					(5,5)	(12,6)
						6
Level 7						(6,3)

Table 254

The numbers add diagonally so we have the series;

- 1, 2, 3, 4, 5, 6, ...
- 2, 4, 6, 8, 10, ...; where this is double the above series
- 4, 8, 12, 16, ...; which is also double the above series
- 8, 16, 24, ... ; which is double the above series

The is the pattern of the standard prediction, which is derived on the basis of exclusion of two prime factors being the same except for [prime 1]. After the standard prediction has been made, then the adjustment that includes the missing options can be put in place. This is shown for 7 and 8 prime factors of W.

Note that we have factor pair number such as (1,0), (7,0), (25,0) and (79,0). These are regarded as anomaly occurrences because they do not subscribe to the rule of dividing by 2, as this would give a K3 number with a decimal number of factor pairs, which is not realistic. The adjustments and the anomalies make the series pattern that we observed earlier to be less obvious. Infact it would have been very difficult for one to observe the pattern if the final table was to be derived directly.

	Number of factor pairs of W						Prediction	Actual
No. of Prime Factors	2	3	4	5	6	7	8	8
Level 1	(2,1)	(6,3)	(14,7)	(30,15)	(62,31)	(126,63)	(254,127)	(254,127)
	1	2	4	8	16	32	64	64
Level 2	(1,0)	(4,2)	(10,5)	(22,11)	(46,23)	(94,47)	(190,95)	(190,95)
		2	3	6	12	24	64	64
Level 3		(1,0)	(7,0)	(16,8)	(30,15)	(62,31)	(126, 63)	(126,63)
		(2,0)						
			1	2	5	10	48	47
			(6,3)	(14,7)	(25,0)	(52,26)	(78, 39)	(79,0)
			3	6	7	14	32	1
Level 4			(3,0)	(8,4)	(18,9)	(38,19)	(46, 23)	(78,39)
				4	8	16	20	32
Level 5				(4,2)	(10,5)	(22,11)	(26, 13)	(46,23)
					5	10	12	20
Level 6					(5,0)	(12,6)	(14, 7)	(26, 13)
						6	7	12
Level 7						(6,3)	(7, 0)	(14, 7)
								7
Level 8								(7, 0)

NO K3 NUMBER (pink)

Table 255

However, the system and concept still needs more improvement.

Optimizing Factorization

It is always a must to look for ways to shorten the factorization of a number. Hence unlocking the distributive behavior of numbers in terms of a factorization space is an interesting subject to explore. The integer classification theory has led to the concept of defining a factorization space in contrast to just forming an algorithm.

We either compress the time or the space, or both.

12 The Compressed Factorization Space (CFS)

This consists of two types of approach;

i. The algebraic method based on specific conditions that result in self testing primality. The method uses the factor two theorem in the context of standard factor pairs for the conditions.

ii. Factor Two Theorem in the context of efficient pairs. Only one factor pair is used per operation where $W = F\{W_1, W_2, \cdots W_n\}$.

Therefore the basic concept behind compressed factorization is the use of self-test primality combined with factor-2 theorem. Here the aim is to fight for micro-seconds of processing time, the Mbulawa factor pair primality test was fighting for milliseconds. As a result, self-testing becomes a critical tool to create a short-cut in processing time.

The UNDT environment has developed different types of space for primality and factorization. However, these can be described in two basic ways.

i. The positional framework. This takes into consideration the arrangement of the gap base 2 numbers and gap base 4 numbers according to a single variable x. The numbers then form a matrix consisting of either prime numbers or pseudoprimes. The consequence of this is a framework for determining primality since a number is either a psuedoprime or a prime. The framework then makes use of two variables k_1, k_2 to determine if a number is a pseudoprime or prime. These varibales cannot be used for factorization.

ii. The product framework. This is based on the product law $GB2 \cdot GB2 = GB2$, $GB4 \cdot GB4 = GB2$ and $GB2 \cdot GB4 = GB4$. Each of these products define the variables k_1, k_2, k_3 where these are used for two purposes; to define primality as well as to carry out factorization.

Positional Framework

x → X, k_1 → Gap Base 2, k_2 → Gap Base 4

X	Gap Base 2	Gap Base 4
18	115	119
19	121	125
20	127	131
21	133	137
22	139	143
23	145	149
24	151	155
25	157	161
26	163	167

↓ Algebraic Sieve

Product Framework

$$k_3 = \frac{x_3 - x_2}{6x_2 - 1}$$ Gap Base 4

$$k_2 = \frac{x_2 + x_3}{6x_2 - 1}$$ Gap Base 2

$$k_1 = \frac{x_3 - x_2}{6x_2 + 1}$$ Gap Base 2

↓ Factor Pair Theory

Figure 130

The positional framework results in the concept of the algebraic sieve, and the product framework results in the factor pair theory. For these to form a holistic approach, the foundational conditions had to assume that there are three types of numbers; the composite, the pseudoprimes and the primes. The similarity between the two frameworks is that if the variables are less than 1, then we have a prime number and if the variables are integers then we have a pseudoprime. Secondly, both frameworks are based on a counting approach. A variable x or x_2 will vary from 1 to a given range to determine if a number is prime or pseudoprime, or to find factors of W. Efficiency is then determined from shortening the counting cycle as much as possible.

The compressed space is therefore based on trying to use the number of factor pairs n in the factorization process. They key in the factorization has been primarily the counting; if we have counted to a certain level we are sure that no other factor can be found, or that the number is a prime. We therefore create a set of conditions that define either primality or factorization, where such conditions are linked to the value of n. This then means we must study n and how it relates to providing opportunity for faster processing time. The approach is as follows.

a. The method factorizes by considering the number of prime factors a number should have, starting with 2, then 3, 4 and 5. One can go beyond, but focus here is limited to five for the compressed space. In this approach, when the program comes to $n = 3$, it assumes the number is not a two prime factor number, when it come to $n = 4$, it knows the number is not a 3 factor number and so on. Note that when $n = 1$ or $n = 2$, both of these result in a two prime factor outcome.
b. It involves the use of the number of factor pairs of the number to determine what formulas to use. For example, the number may have three prime factors, but their gap base combination as a product determines outcome.

Different types of space may be defined to described the various techniques that have been developed.

a) The efficient standard space. This uses the variables k_1, k_2, k_3 to find primality; that is, the Mbulawa primality test. The test depends on one of these variables being less than 1 for primality. The alternative is using $x_2 = Int\left\{\frac{W+7}{30}\right\}$ to reduce the search count, hence creating efficiency in the space. This is the best that can be done for efficiency in this space. That is, when x_2 reaches the defined level, we are certain the number is either a prime or not a prime number.
b) The standard factor pair space. The development of factor pair theory for factorization followed, which has the added advantage of providing primality testing. Hence factor pairs were produced under the condition $x_2 = Int\left\{\frac{W+7}{30}\right\}$. This produced all possible factor pairs, and we are certain that at this value of x_2 all pairs will be produced.
c) Efficient factor pair space. This space was developed as a quest for higher speeds of generating factor pairs and hence deriving prime factors of a given number. The combination of factor pairs and the limitation of the search count to $(\sqrt{W} + 1)/6$ produced faster results.
d) The compressed factorization space. This is the ultimate space that combines all the methodologies produced in the previous space descriptions. That is, using factor pairs, $(\sqrt{W} + 1)/6$, the factor-2 theorem and the self-testing primality findings. The word compressed implies the time element is so compressed that it should take a far less time to find prime factors than compared to the other spaces.

As indicated before, the general approach has been to reduce the count on x_2 so as to create greater efficiency. This meant we count in a shorter cycle. Now the aim here is to make the factorization space as small as possible by reducing two things; firstly the counting and secondly the space in which the counting takes place. In order to create such a space, we recognize that:

- Using factor pairs creates a transient database, and that this database is driven by a coordinate engine.
- That $(\sqrt{W}+1)/6$ gives us a fairly good estimate that reduces the count thousands of times the size of the number we are dealing with, and it does so proportionally. The bigger the number, the bigger the reduction. Hence it implies that as the number gets bigger, this does not necessarily increase the proportionality in the difficulty of getting factor pairs.
- Also $(\sqrt{W}+1)/6$ gives us the probabilistic certainty that all factor pairs have been derived.
- If $n=0$ implying no factor pairs, then W is a prime number. In this case, the count will always reach $(\sqrt{W}+1)/6$ to establish W as prime.
- For gap base 4, the double barrel approach was used to reduce the count on x_2. This produces the same number of factor pairs as the standard factor space but at far reduced timeframes.
- Now factor-2 theorem tells us that the first factor-2 will always be the smallest prime number if the number is gap base 2, and that it will always be prime if it is a gap base 4 number.
- The factor-2 theorem tells us that factor-1 will always contain the rest of all other prime factors of W.

However, the new approach in the compressed space is twofold, and is based on self-testing primality. In other words, we gain time by not doing primality testing. Instead knowing what structure is used determines which numbers are prime.

Input a number:		☐ Show repeated factors as p^k	Run time:
9900785544332110097	Factors:	9900785544332110097	00:00:00.050
9900785544332110079	Factors:	43*47*1187*4127172500177	00:00:00.033
1700937621009745675	Factors:	5*5*47*1303*1110979651547	00:00:00.037
8888777755553333111	Factors:	2791*2312621*1377138901	00:00:00.051
5379021907352187697	Factors:	520974523*10324923139	00:00:19.127
6743106854081178643	Factors:	7129*5134039*184235053	00:00:00.087
41098674532186797833	Factors:	139*63377*432613*10784047	00:00:00.022
39000959321785432199	Factors:	17*163*14074687593571069	00:00:00.043

Figure 131: Factorization from JavaScripter.net (Feb 2013)

12.1 Self-Testing Primality

What is interesting about the factor pair theory is the fact that one can produce a factor pair that immediately gives you a prime factor without testing for primality. This is referred to as self-testing primality, and it can be very advantageous in terms of time efficiency for extremely big numbers. The test is defined for gap base 2 and gap base 4, though the conditions differ.

Definition

Self-testing primality derives prime factors and verifies primality structurally.

Self-testing primality is a function of the number of prime factors that one is looking for. The minimum number of prime factors is obviously 2, then the next is 3, 4 and so on. In each of such cases, some prime factors repeat. For example, there can be two prime factors $11 \cdot 17$ as unique prime factors, and $17 \cdot 17$ as repeat factors. Each of these can then be represented by different structures for derivation.

Self-Test Primality Theorem

Let W have a factor pair (a_1, b_1) and n is the number of factor pairs, then a_1 and b_1 are prime if;

a. $n = 1$, and $W \in GB4$
b. $n = 2$, and $W \in GB2$

This is the change in approach in terms of dealing with primality, now the number of factor pairs of W is used to determine primality. This gives us a framework of a very simple primality test algorithm, a self-testing primality test for the element b_1. The basis of the theorem is that Factor-1 will be the largest possible pseudoprime or prime, which by implication means that Factor-2 will be the smallest possible number that divides W. If it is the smallest possible number, then it must be prime.

26161157	Factor-1	Factor-2	
Pseudoprime	2378287	11	Prime
Pseudoprime	638077	41	Prime
Pseudoprime	368467	71	Prime
Pseudoprime	125173	209	Pseudoprime
Pseudoprime	55309	473	Pseudoprime
Pseudoprime	33583	779	Pseudoprime
Pseudoprime	19393	1349	Pseudoprime
Pseudoprime	14839	1763	Pseudoprime
Pseudoprime	8569	3053	Pseudoprime
Pseudoprime	2911	8987	Pseudoprime
Pseudoprime	817	32021	Pseudoprime
Pseudoprime	781	33497	Pseudoprime
Pseudoprime	451	58007	Pseudoprime
Prime	43	608399	Pseudoprime
Prime	19	1376903	Pseudoprime
	GB2	GB4	

2378287	Factor-1	Factor-2	
Pseudoprime	125173	19	Prime
Pseudoprime	58007	41	Prime
Pseudoprime	33497	71	Prime
Pseudoprime	3053	779	Pseudoprime
Pseudoprime	2911	817	Pseudoprime
Pseudoprime	1763	1349	Pseudoprime
Pseudoprime	1349	1763	Pseudoprime
Pseudoprime	817	2911	Pseudoprime
Pseudoprime	779	3053	Pseudoprime
Prime	71	33497	Pseudoprime
Prime	43	55309	Pseudoprime
Prime	41	58007	Pseudoprime
Prime	19	125173	Pseudoprime
	GB2	GB2	
	GB4	GB4	

22184437

Factor-1	Factor-2	Base	Type
2016767	11	GB4	Prime
715627	31	GB2	Prime
331111	67	GB2	Prime
65057	341	GB4	Pseudoprime
30101	737	GB4	Pseudoprime
22847	971	GB4	Prime
10681	2077	GB2	Pseudoprime
2077	10681	GB2	Pseudoprime
971	22847	GB4	Pseudoprime
737	30101	GB4	Pseudoprime
341	65057	GB4	Pseudoprime
67	331111	GB2	Pseudoprime
31	715627	GB2	Pseudoprime
11	2016767	GB4	Pseudoprime
GB2	GB2		
GB4	GB4		

216209	Factor-1	Factor-2	
Pseudoprime	469	461	Prime
Prime	67	3227	Pseudoprime
Prime	7	30887	Pseudoprime

1333	Factor-1	Factor-2	
Prime	43	31	Prime
prime	31	43	Prime

2489	Factor-1	Factor-2	
Prime	19	131	Prime

Table 256

The above table illustrates the general statement of the theorem. It shows the factors, and their type whether prime or pseudoprime, and at the bottom gives the group classification of the factors. The number 26161157 shows a typical gap base 4 number pattern, where all factor-2 numbers are gap base 4. Then 2378287 gives a gap base 2 result with partial symmetry, where 43 disturbs the complete symmetry of the pattern. Note that 971 breaks the pattern of pseudoprimes in the number 22184437, indicating that the type pattern can be broken. Similarly, if 11 is GB4, then 2016767 is GB2, if 31 is GB2 then 715627 is GB4.

12.1.1 Gap Base 2

The self-testing primality test arises from the condition (a_1, b_1) consists of prime numbers if $n = 2$, and $W \in GB2$ in the theorem. What is important to note here is the two constraints that bind the development of factor pairs in gap base 2. That is;

a) Gap base 2 develops its factor pairs in a symmetric manner
b) The first factor-2 is always prime.

Therefore the smallest number of factor pairs for gap base 2 is 2 pairs (a_1, b_1) and (a_2, b_2). Because of the symmetric condition, it means $a_1 = a_2$ and $b_1 = b_2$. Now if b_1 is prime according to the factor-2 theorem, and no further factor pairs can be derived, then a_1 must be prime. This condition implies that when $n = 2$, then the factor pairs have the format;

Factor-1	Factor-2
a_1	b_1
b_1	a_1

Table 257

Technically, this is the first factor switch in gap base 2. We can then expand the number of factor pairs for gap base 2 by allowing a pseudoprime and assuming that a_1 and b_1 remain prime – this is the constraint.

GB2 Two Factor Theorem

Let $W \in GB2$, then if n=2 and $a_1 = b_2$, then a_1 and b_1 are prime numbers.

This is a self-testing primality condition. There is no need for searching for any new factor pairs, nor doing any primality test.

Now assume that we have a pseudoprime for factor-1 denoted p_1, and since gap base 2 is symmetric, this is what will reflect in the factor pair expansion. Hence if there is p_1, then there must be p_2 to preserve symmetry.

Factor-1	Factor-2
p_1	b_1
p_2	a_1
a_1	p_2
b_1	p_1

Table 258

An example of this structure is for $W = 325$. This consists of 4 factor pairs.

X2	K1	K2	Factor-1	Factor-2
1	11		65	5
2		4	25	13
4		2	13	25
11	1		5	65

Table 259

As a general structure, this may be expressed as follows, provided $a_1 > b_1$. The structure gives 4 pairs with 2 prime factors in gap base 2, but one of the factors is repeated to give 3 prime factors. Hence the structure assumes that;

a. $p_1 = a_1 \cdot b_1$
b. $p_2 = b_1 \cdot b_1$

for any $a_1 > b_1$, where $a_1 \in GB2$ and $b_1 \in GB4$, and $W = a_1 \cdot b_1 \cdot b_1$.

Factor-1		Factor-2	
$a_1 \cdot b_1$		b_1	GB4
$b_1 \cdot b_1$		a_1	GB2
a_1	GB2	$b_1 \cdot b_1$	
b_1	GB4	$a_1 \cdot b_1$	

Table 260

Consider adding another row for factor-1 with factor p_3. Now p_1 and p_2 are pseudoprimes, and we are tracing the development of this expansion, hence we keep the notation of a_1 and b_1 in order to understand what is happening. The major difference is that now b_1 is related to $p_1$and a_1 to p_2 but they are still prime according to the original assumption.

The structure gives us 6 factor pairs with 3 prime factors. Hence we can obtain 3 prime factors from 4 or 6 factor pairs.

Factor-1	Factor-2
p_1	b_1
p_2	a_1
p_3	b_3
b_3	p_3
a_1	p_1
b_1	p_2

or

Factor-1	Factor-2
p_1	b_1
p_2	a_1
p_3	b_3
b_3	p_3
a_1	p_1
b_1	p_2

Table 261

The expansion can be continued with another addition of a factor pair (p_3, b_3), where p_3 is a pseudoprime. Let us assume that W is a $K1$ or $K2$ number, then the result will be symmetric. Hence we expect six factor pairs and a factor switch. The coloring shows that we expect either b_3 or p_3 to be a prime number.

The structure shows that for factor-2, then b_1 and b_2 remain as prime numbers, and this leads to the following corollary.

Corollary

If $n \geq 2$, then $b_1$and b_2 are prime factors of W provided $W \in GB2$.

a_1	b_1
a_2	b_2
...	...
b_2	a_2
b_1	a_1

Table 262

Now $W = a \cdot b \cdot c$, hence $c = b_1 \cdot b_2$. The consequence of this corollary is that if we follow the pattern of the expansion and the fact that W is gap base 2, then we can infer that $b_1 = a_n$ and $b_2 = a_{n-1}$ from the symmetry property.

We can now apply this to understand pseudoprime generation within the factor pairs. An example is given for the number 2378329, where we use the format that was applied in the expansion. It is then clear that the prime factors of 2378329 are 61 and 127. Now note that, for pseudoprime generation, then:

$$p_1 = 127 \cdot 307$$

$$p_2 = 61 \cdot 307$$

$$p_3 = 61 \cdot 127$$

X2	K2	Factor-1	Factor-2
10	6498	$p_1 = 38989$	$b_1 = 61$
21	3121	$p_2 = 18727$	$b_2 = 127$
51	1291	$p_3 = 7747$	$b_3 = 307$
1291	51	$b_3 = 307$	$p_3 = 7747$
3121	21	$b_2 = 127$	$p_2 = 18727$
6498	10	$b_1 = 61$	$p_1 = 38989$

Table 263

This clearly shows that whenever we have 6 factor pairs, then b_3 has the probability to be prime. Now b_1 and b_2 are already prime, so this will give us three prime numbers; b_1, b_2 and b_3 and the number is a $K2$ number. This implies the product is of the form $GB2 \cdot GB2 \cdot GB2$.

However, the factor switch introduces a dimension where p_3 can also be prime. This is due to the reflective behavior of the factor switch. In this case b_3 is not prime, rather b_4 is a prime number.

X2	K1	K2	Factor-1	Factor-2
4	17238		$p_1 = 103427$	$b_1 = 23$
10	6720		$p_2 = 40319$	$b_2 = 59$
226		292	$p_3 = 1753$	1357
292		226	$p_1 = 1357$	$p_3 = 1753$
6720	10		$b_2 = 59$	$p_2 = 40319$
17238	4		$b_1 = 23$	$p_1 = 103427$

Table 264

This time we can have the prime numbers b_1, b_2 and b_4, and this s a $K1K2$. That is, it has the product form $GB2 \cdot GB4 \cdot GB4$. This leads to a further corollary.

GB2 Three Factor Theorem

If $n = 6$*, and* $W \in GB2$*, hence;*

iii. $b_1 \cdot b_2 = b_3$, *then* b_4 *is prime, and* W *is a* $K1K2$ *number*

iv. $b_1 \cdot b_3 = b_4$, *then* b_3 *is prime, and* W *is a* $K2$ *number.*

For the case $n = 6$, then b_1, b_2, b_3 *or* b_4 are prime, this is another conditional self-testing primality test. Conditionality is brought about by the factor switch. It is generally observed that numbers in a factor switch can all be pseudoprime when $n > 6$, hence this is a particular observation for $n = 6$. It further leads to the following construction. If these are the three primes of W then it implies;

$$p_1 = b_2 \cdot b_3$$

as this will give us the biggest pseudoprime. Hence the next biggest pseudoprime is given by;

$$p_2 = b_1 \cdot b_3$$

The smallest pseudoprime will be;

$$p_3 = b_1 \cdot b_2$$

12.1.2 Gap Base 4

Now we consider development in a gap base 4 environment. For gap base 4, there is no symmetric distribution of factor pairs. This is the first major difference. For $n = 1$, we will have the following format.

Factor-1	Factor-2
a_1	b_1

Table 265

Since there is no symmetry, it then implies a_1 and b_1 are prime numbers – that is the only way for them to be unique. If they can be factored again, then the result will yield more pairs, but they cannot be factored further, so they must be prime.

GB4 Two Factor Theorem

If $x_2 = Int(\sqrt{W} + 1)/6)$ *and n=1, then* a_1 *and* b_1 *are prime factors.*

The fact that the maximum value of x_2 has been obtained and we have only one factor pair is a primality test, with the beauty that no further testing is necessary.

The next addition of factor pairs then gives;

Factor-1	Factor-2
a_1	b_1
a_2	b_2

Table 266

However, the number of factors pairs in gap base 4 must be odd, as demonstrated by the factor pair limit theorem. Therefore this result is defined as a non-standard distribution in gap base 4, or just like we do not expect a result that gives us one factor pair in gap base 2. An example is the gap base 4 number 2378411.

W = 2378411

X2	K3	Factor-1	Factor-2
8090	8	49	48539
56629	1	7	339773

Table 267

Another example is given from $W = 52559$.

X2	K3	Factor-1	Factor-2
52	28	169	311
674	2	13	4043

Table 268

There is no way to discover or establish this except through pattern methodology. Non-standard distributions always occur for the case where one of the prime number repeats itself. Assuming that a_2 is a prime number from GB2 and b_1 is a prime number from GB4, the pattern for this non-standard distribution is;

Factor-1		Factor-2	
$a_1 = a_2 \cdot a_2$		b_1	GB4
a_2	GB2	$b_2 = a_2 \cdot b_1$	

Table 269

That is, $W = a_2 \cdot a_2 \cdot b_1$. It gives us a self-testing primality structure. Another format is given from the example $W = 6647$.

X2	K3	Factor-1	Factor-2
3	65	391	17
4	48	289	23

Table 270

This has a different structure, where $W = b_2 \cdot b_1 \cdot b_1$. This also has a self-testing attribute, since b_1 and b_2 will be prime numbers. In this case b_2 is from GB4.

Factor-1		Factor-2	
$a_1 = b_2 \cdot b_1$		b_1	GB4
$b_1 \cdot b_1$	GB2	b_2	GB4

Table 271

Now $GB \cdot GB4 = GB4$ and this satisfies the structure format. However $G4 \cdot GB4 = GB2$, this does not satisfy the format. Hence product format has a bearing on the final result output of the structure.

The next result is three factor pairs, where the rest of the factor pairs are a repetition.

Factor-1	Factor-2
a_1	b_1
a_2	b_2
a_3	b_3
b_3	a_1
b_2	a_2
b_1	a_3

Table 272

From the factor-2 theorem, then $b_1$is a prime number. Now $W = a_1 \cdot b_1$, but if b_1 is prime, then $a_1$cannot be prime since the number of factor pairs is more than two, or the basic assumption is that there must be three prime factors of which b_1 is one of them. Two conditions apply in terms of relationship between a_1 and b_1, that is;

a. $a_1 > b_1$, then $a_3 > b_1$
b. $a_1 < b_1$, then $a_3 < b_1$

Factor-1	Factor-2
13330903	$b_1 = 5$
310021	215
$a_3 = 43$	1550105

Factor-1	Factor-2
91	$b_1 = 732467$
13	5127269
$a_3 = 7$	9522071

Table 273

This is assumed from the fact that $b_1 < b_3$, and $a_1 > a_3$, where this is a structural condition for the factor product. If it is always true that $a_1 > a_3$, then if $a_1 < b_1$it implies $a_3 < b_1$.

If a_1 is a pseudoprime, then its product must be defined from one or more of the factor pairs. That is, we assume;

a. $a_1 = a_2 \cdot b_2$
b. $a_1 = a_2 \cdot a_3$
c. $a_1 = a_3 \cdot b_3$
d. $a_1 = b_2 \cdot b_3$

However, from pattern methodology, the following is established.

a. $a_1 = a_2 \cdot a_3$
b. $a_1 = b_2 \cdot b_3$

Hence from deduction we have the following scenarios, assuming that W will always have three prime factors;

a. Either b_2 is prime, then b_3 is prime, implying a_2 and a_3 are not prime
b. Either b_2 is prime, then a_2 or a_3 is prime
c. Either b_3 is prime, then b_2 is not prime, and a_2 or a_3 is prime
d. Neither b_2 or b_3 are prime, hence a_2 and a_3 must be prime.

In the same manner, from pattern methodology, we establish that provided b_1 is prime and a_3 is prime or pseudoprime;

a. $a_2 = b_1 \cdot b_3$, where b_2 and b_3 are prime factors
b. $a_3 = b_1 \cdot b_2$, where b_2 and b_3 are prime factors
c. $b_2 = b_1 \cdot a_3$, where a_2 and a_3 are prime factors
d. $b_3 = a_2 \cdot b_1$, where a_2 and a_3 are prime factors

It is evident that the structure is influenced by the product law, that is $GB2 \cdot GB2 \cdot GB4 = GB4$ and also $GB4 \cdot GB4 \cdot GB4 = GB4$.

Factor-1		Factor-2	
$a_1 = a_2 \cdot a_3$		b_1	GB4
a_2	GB2	$b_2 = b_1 \cdot a_3$	
a_3	GB2	$b_3 = a_2 \cdot b_1$	

Factor-1	Factor-2	
$a_1 = b_2 \cdot b_3$	b_1	GB4
$a_2 = b_1 \cdot b_3$	b_2	GB4
$a_3 = b_1 \cdot b_2$	b_3	GB4

Table 274

This is illustrated by the example below.

X2	K3	Factor-1	Factor-2
1	2221817	13330903	5
36	51670	310021	215
258351	7	43	1550105

W = 66654515

X2	K3	Factor-1	Factor-2
122078	15	91	732467
854545	2	13	5127269
1587012	1	7	9522071

W=66654497

X2	K3	Factor-1	Factor-2
18	103823	622939	107
23	81088	486529	137
758	2443	14659	4547

W=66654473

X2	K3	Factor-1	Factor-2
2	1009916	6059497	11
3	653475	3920851	17
59407	31	187	356441

W=66654467

Table 275

Therefore we conclude that this is a self-test primality condition because a_1 always has two prime factors either a_2 and a_3 or b_2 and b_3. It is independent of the permutation of the structure.

Two other variations are given for a 3 factor pair number. The first involves an asymmetric example for $W = 181589$. The pattern is similar as in the case of $W = 66654515$, the difference is the relationship between b_1 and a_3. Both numbers share the same SFM coordinate, hence the asymmetric nature.

X2	K3	Factor-1	Factor-2
7	738	4429	41
294	17	103	1763
704	7	43	4223

Table 276

The general pattern is shown below, giving a self-primality testing pattern.

Factor-1		Factor-2	
$a_1 = a_2 \cdot a_3$		b_1	GB4
a_2	GB2	$b_2 = b_1 \cdot a_3$	
$a_3 = b_1 + 2$	GB2	$b_3 = a_2 \cdot b_1$	

Table 277

A different pattern of 3 factor pairs can also be defined with 4 prime factors where 3 of them are the same.

X2	K3	Factor-1	Factor-2
3	4965	29791	17
88	160	961	527
2723	5	31	16337

Table 278

In this case $W = a_3 \cdot a_3 \cdot a_3 \cdot b_1$, with the following structure. The factor-1 part is quite interesting since all the numbers there are expressed in terms of only one variable a_3.

Factor-1		Factor-2	
$a_1 = a_3 \cdot a_3 \cdot a_3$		b_1	GB4
$a_2 = a_3 \cdot a_3$	GB2	$b_2 = b_1 \cdot a_3$	
a_3	GB2	$b_3 = a_2 \cdot a_3 \cdot b_1$	

Table 279

We can keep the same pattern $W = a_3 \cdot a_3 \cdot a_3 \cdot b_1$but change the gap base. An example is from $W = 152\ 303$. This derives the following pattern.

X2	K3	Factor-1	Factor-2
3	1493	8959	17
88	48	289	527
819	5	31	4913

Table 280

The structure of the product is given as follows, where $W = a_3 \cdot b_1 \cdot b_1 \cdot b_1$.

Factor-1		Factor-2	
$a_1 = a_3 \cdot b_1 \cdot b_1$		b_1	GB2
$a_2 = b_1 \cdot b_1$	GB2	$b_2 = b_1 \cdot a_3$	
a_3	GB4	$b_3 = b_1 \cdot b_1 \cdot b_1$	

Table 281

The next is 5 factor pairs, with examples as shown. According to the limit theorem, this should have a maximum of 4 prime factors.

X2	K3	Factor-1	Factor-2
7089	302	1813	42533
49622	43	259	297731
262287	8	49	1573721
347353	6	37	2084117
1836008	1	7	11016047

X2	K3	Factor-1	Factor-2
38	6887	41323	227
1173	222	1333	7037
1627	160	961	9761
36358	7	43	218147
50432	5	31	302591

Table 282

Using the similar approach, it is observed that the pattern for constructing the pairs is as follows. This also provides a framework for self-test primality. As usual $b_1$is a prime number, and in this structure a_4 and a_5 are also prime numbers.

Factor-1	Factor-2	
$a_5 \cdot a_5 \cdot a_4$	b_1	GB4
$a_4 \cdot a_5$	$a_5 \cdot b_1$	
$a_5 \cdot a_5$	$a_4 \cdot b_1$	

a_4	GB2	$a_5 \cdot a_5 \cdot b_1$
a_5	GB2	$a_4 \cdot a_5 \cdot b_1$

Table 283

In this structure we do not have $GB4 \cdot GB4 \cdot GB4 \cdot GB4 = GB2$, so this would not be a gap base 4 number. Instead we have the product structure defined by $GB2 \cdot GB2 \cdot GB2 \cdot GB4 = GB4$. However, it is not clear if this is self-testing primality because of the possibility of permutation in structure. Generally, if there are two consecutive prime numbers in gap base 2 such as p, q, and any prime r in gap base 4, then;

$$W = p^2 \cdot q \cdot r$$

is a five factor pair number with 3 prime factors provided $p < q < r$.

Different structures exists for $n = 7$. For example the structure for $n = 7$ is given as follows, provided $b_1 < b_2 < b_5$ and a_7 is some number in gap base 2.

Factor-1		Factor-2	
$a_7 \cdot b_2 \cdot b_5$		b_1	GB4
$a_7 \cdot b_1 \cdot b_5$		b_2	GB4
$b_2 \cdot b_5$		$a_7 \cdot b_1$	
$b_1 \cdot b_5$		$a_7 \cdot b_2$	
$a_7 \cdot b_1 \cdot b_2$		b_5	GB4
$b_1 \cdot b_2$		$a_7 \cdot b_5$	
a_7	GB2	$b_1 \cdot b_2 \cdot b_5$	

Table 284

The product structure is defined by $GB2 \cdot GB4 \cdot GB4 \cdot GB4 = GB4$. Another format can be derived from the following examples.

Factor-1	Factor-2
15422491	5
2203213	35
211267	365
30181	2555
511	150905
73	1056335
7	11016065

W= 77112455

Factor-1	Factor-2
7010227	11
1001461	77
218449	353
31207	2471
27181	2837
3883	19859
7	11016071

W=77112497

Table 285

With many different types of formats, this is not defined as a self-testing primality structure.

12.1.3 Conditions for Self-primality Testing

Hence it may be observed that the more factor pairs open up the space through higher number of prime factors, the higher number of permutations for product structures. This then makes it sensible to limit the process to a reasonable number of factor pairs, which is six for gap base 2 and 3 for gap base 4. However, the rest of the pairs in gap base 2 are a repeat, hence three is the standard that is set with some exceptions here and there.

Self-Primality Rule

Self-testing primality occurs for $n \leq 3$.

The condition $n \leq 3$ is applicable for both gap base 2 and 4 since automatically there is no $n = 3$ for gap base 2, only $n = 2$. However, $n = 1$ and $n = 3$ apply for gap base 4. The condition implies that numbers more than three factors are not classified for the self-testing concept in primality, or $n > 3$ means the standard approach must be used of producing factor pairs and then testing them for primality accordingly. The reason is that the number of equations needed to process the self-testing may require the same input as the standard approach. Three prime factors are derived from 6 factor pairs in gap base 2, and from 3 factor pairs in gap base 4. Two prime factors are derived from one factor pair in gap base 4, and from 2 factor pairs in gap base 2. However an exception is made to the above rule for $n = 3$ for four prime factors provided three of the factors are the same.

For both gap base 2 and 4, $b_1$is always a prime number and $a_1$has the remaining prime factors of W if it is pseudoprime. This is an important observation. In looking into product behavior, we can also consider the aspect of self-testing primality. The model is as shown in the diagram.

- $n = 0$ represents a prime number, that is W is prime.
- $n = 1$ represents a gap base 4 product of two prime numbers
- $n = 2$ represents a gap base 2 product of two prime numbers
- $n = 3$ represents a product of three prime numbers
- $n > 3$ implies that W has more than three prime numbers.

The approach has been to make the cut off point using counting on the basis of x_2. Now the cut off point is determined by the number of factor pairs. That is, by establishing how many factor pairs there are, we determine primality as well as factorization patterns.

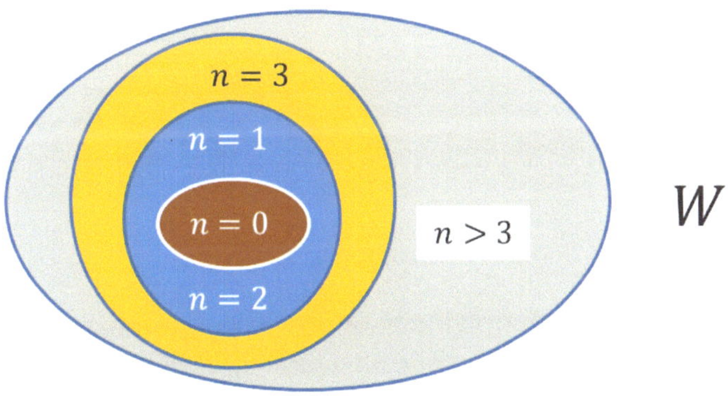

Figure 132

The concept of self-testing primality is structural. For example it is clear that there are five types of products to give us a pattern of two prime factors, an interesting feature and consequence of UNDT. The prime factors are then broken down into products in terms of gap base and the number of factor pairs. In the beginning of the factorization process, the first step is identifying the gap base of W. After that other processes kick in to determine the next procedure. For example if $n = 2$ and $K1 = 0$, then the number is a $K2$ number, meaning a product of two GB2 numbers that are prime. The argument is settled, no further operations are necessary. Similarly if $n = 1$ and $K2 = 0$, then the number is a $K1$ or $K3$ number. However if we have one factor pair, then the two prime numbers are from different gap bases. This is the self-testing structure for two prime factors of W. Note that 1 and 2, and 4 and 5 are exactly the same in format and gap base, the difference is the identifier. Therefore the test conditions are;

a. If $a_1 = b_1$ then $n = 1$ and a_1 is a prime number, for GB2 or GB4.
b. If $a_1 = b_2$ or $a_2 = b_1$ then $n = 2$ and a_1 and b_1 are prime numbers for GB2 or GB4.

This is a slightly different but significant presentation from the GB2 two factor primality theorem which states, *"if n=2 and $a_1 = b_2$, then a_1 and b_1 are prime numbers"* . In this theorem, one has to first of all establish that n =2 by doing a complete search for factor pairs. However, for the compressed space, it is not necessary to do so, the focus is the structural behavior that If $a_1 = b_1$ then $n = 1$ and a_1 is a prime number, or If $a_1 = b_2$ or $a_2 = b_1$ then $n = 2$ and a_1 and b_1 are prime numbers.

Factor-2 theorem says $b_1$is always prime, hence this derives the above two conditions, and any other conditions that we may derive. The self-testing is then based on three main conceptual points;

i. The fact that $b_1$is always prime
ii. The fact that we have conditions that determine primality
iii. The fact that the conditions limit the counting that is necessary to a certain number of factor pairs less than six.
iv. Efficient pairs are not applicable since this may exclude certain conditions form being met.

For example, consider the number $W = 19\,422\,631$. If $b_4 = b_1 \cdot b_3$ then b_2 and b_3 are prime numbers is a condition that is applied, which implies that the program will count only to 9329. This condition combined with the product $a_1 = b_2 \cdot b_3$ tells us that we have $n = 6$ for factor pairs. There is now no need to say we have to count to $x_2 = Int\left\{\frac{W+7}{30}\right\} = 647\,421$ in a standard operation. We cannot use efficient pairs since we observe that this derives $x_2 = Int\left(\frac{\sqrt{W}+1}{6}\right) = 734$ and at this value the condition $b_4 = b_1 \cdot b_3$ cannot be determined. The conditions determine how far we go or when we have to stop.

X2	Factor-1	Factor-2
37	87097	223
42	77381	251
58	55973	347
9329	347	55973
12897	251	77381
14516	223	87097

Table 286

Conditions can be derived for two prime factors in the following table. The UNDT framework provides an elaborate description in how products related to define the self-primality testing environment. The identifier tells us which gap base constructs the product because it is uniquely defined for each product.

Level	Prime-1	Prime-1	Factor Pairs	Identifier	Type
1	a_1	a_1	1	K2	GB2 Numbers

	Prime-1	Prime-1	Factor Pairs	Identifier	Type
2	a_1	a_1	1	K1	GB4 numbers

	Prime-1	Prime-2	Factor Pairs	Identifier	Type
3	a_1	b_1	1	K3	GB2 & GB4 numbers

	Prime-1	Prime-2	Factor Pairs	Identifier	Type
4	a_1	b_1	2	K1	GB4 numbers

	$a_2 = b_1$	$b_2 = a_1$

	Prime-1	Prime-2	Factor Pairs	Identifier	Type
5	a_1	b_1	2	K2	GB2 numbers
	$a_2 = b_1$	$b_2 = a_1$			

Table 287

Now here is the beauty of this theory; if $n > 2$, then the structure is saying the number of prime factors of W cannot be two. Immediately there is exclusion, hence now the procedure then checks to see if the number of primes is three factors. This power of application arises from the factor pair limit theorems.

Theorem

If $a_1 \neq b_1$ *or* $a_1 \neq b_2$ *then* $n > 2$.

The theorem gives us attest to see if the number of factor pairs is greater than two or not.

A similar approach is used for three prime factors. This derives the following self-testing structure. Note that if the three prime numbers are different, then it's very easy to declare primality without referring to the identifiers. The prime factors have the same position for the different identifiers. However for the case of three factor pairs, we have to bring in an additional condition to determine whether we have a left or right situation. The condition is;

- If $a_1 = a_2 \cdot a_3$, then it is an L-product, hence a_2 and a_3 are primes in GB2, and $b_1 \in$ GB4 a prime.
- If $a_1 = b_2 \cdot b_3$, then it is an R-product, hence b_2 and b_3 are primes in GB4, and $b_1 \in$ GB4 a prime.

L			
Factor-1		Factor-2	
$a_1 = a_2 \cdot a_3$		b_1	GB4
a_2	GB2	$b_2 = b_1 \cdot a_3$	
a_3	GB2	$b_3 = a_2 \cdot b_1$	

R		
Factor-1	Factor-2	
$a_1 = b_2 \cdot b_3$	b_1	GB4
$a_2 = b_1 \cdot b_3$	b_2	GB4
$a_3 = b_1 \cdot b_2$	b_3	GB4

Table 288

Therefore now we do not rely on x_2, once the two conditions above are satisfied, we know that we have three prime factors and the factor pairs will be six. Note that in every case we have b_1 as a prime factor, hence the idea is to identify the other two prime factors in the structure without doing any primality testing. Note that the following two conditions are exactly the same because the structure is the same even though the identifiers are different.

- If $a_1 = b_2 \cdot b_2$ and $a_2 = b_1 \cdot b_2$, or $a_1 = b_1 \cdot b_2$ and $a_2 = b_1 \cdot b_1$, then $W = b_1 \cdot b_1 \cdot b_2$ or $W = b_1 \cdot b_2 \cdot b_2$ respectively and $n = 4$. This is for a $K2$ or $K1K2$ number, the same equations apply.
- If $a_1 = b_2 \cdot b_2$ and $a_2 = b_1 \cdot b_2$ or $a_1 = b_1 \cdot b_2$ and $a_2 = b_1 \cdot b_1$, then $W = b_1 \cdot b_1 \cdot b_2$ or $W = b_1 \cdot b_2 \cdot b_2$ respectively and $n = 2$. This is for a $K3$ number.

Therefore the condition is not repeated in the conditions for three prime numbers. The order of the conditions is in terms of elimination. Hence if the first condition is true, then the rest are ignored and a solution is stated.

i. If $a_1 = b_1$ then $W = b_1 \cdot b_1 \cdot b_1$. The prime number is a GB4 number, and $n = 1, b_1 \in GB4$.
ii. If $a_1 = b_1 \cdot b_1$ and $a_2 = b_1$ then $W = b_1 \cdot b_1 \cdot b_1$. The prime is a GB2 number, and $n = 2, b_1 \in GB2$.
iii. If $a_1 = b_2 \cdot b_2$ and $a_2 = b_1 \cdot b_2$ or $a_1 = b_1 \cdot b_2$ and $a_2 = b_1 \cdot b_1$, then $W = b_1 \cdot b_1 \cdot b_2$ or $W = b_1 \cdot b_2 \cdot b_2$ respectively and $n = 2 \ or \ 4$. This is for a K2, K1K2 and K3 number.
iv. If $a_1 = a_2 \cdot a_2$ then $W = a_2 \cdot a_2 \cdot b_1$and $n = 3$. This is a repeat prime factor.

v. If $a_1 = b_1 \cdot b_2$ then $W = b_1 \cdot b_1 \cdot b_2$ and $n = 3$. This is a repeat prime factor.
vi. If $a_1 = a_2 \cdot a_3$, hence a_2 and a_3 are primes, and $n = 3$. This is a left product, $W = b_1 \cdot a_2 \cdot a_3$.
vii. If $a_1 = b_2 \cdot b_3$, hence b_2 and b_3 are primes, and $n = 3$. This is a right product, $W = b_1 \cdot b_2 \cdot b_3$.
viii. If $b_3 = b_1 \cdot b_2$ then b_2 and b_4 are prime numbers and $n = 6$.
ix. If $b_4 = b_1 \cdot b_3$ then b_2 and b_3 are prime numbers, and $n = 6$.

Level	Prime-1	Prime-2	Prime-3	Factor Pairs	Identifier	Type
1	b_1	b_2	b_3	6	K2	GB2 only
	b_1	b_2	b_3	6	K1,K2	GB2*GB4*GB4

2	Prime-1	Prime-2	Prime-3	Factor Pairs	Identifier	Type
	b_1	$b_1 = b_2$ or $b_2 = b_3$	b_3	4	K2	GB2 only
	b_1	$b_1 = b_2$ or $b_2 = b_3$	b_3	4	K1, K2	GB2*GB4*GB4

3	Prime-1	Prime-2	Prime-3	Factor Pairs	Identifier	Type
L	a_2	a_3	b_1	3	K3	GB2*GB2*GB4
R	b_1	b_2	b_3	3	K3	GB4*GB4*GB4

4	Prime-1	Prime-1	Prime-2	Factor Pairs	Identifier	Type
	b_1	b_1	b_2	2	K3	GB4 only
	a_2	a_2	b_1	2	K3	GB2*GB2*GB4

5	Prime-1	Prime-1	Prime-1	Factor Pairs	Identifier	Type
	$a_2 = b_1$	b_1	b_1	2	K2	GB2 only

6	Prime-1	Prime-1	Prime-1	Factor Pairs	Identifier	Type
	$a_1 = b_1$	a_1	b_1	1	K1	GB4 only

Table 289

We then combine the conditions for two and three prime factors to define the self-testing primality.

Pair Value	Condition	Result	Comment
$n = 1$	$a_1 = b_1$	$W = b_1 \cdot b_1$	GB2 product or GB4 product
	$a_1 = b_1$	$W = b_1 \cdot b_1 \cdot b_1$	GB4 numbers only (K3 number)
$n = 2$	$a_1 = b_1 \cdot b_1$ and $a_2 = b_1$	$W = b_1 \cdot b_1 \cdot b_1$	GB2 numbers only (K2 number)
(but includes n=4 values)	$a_1 = b_2$ and $a_2 = b_1$	$W = b_1 \cdot b_2$	GB2 numbers only or GB4 numbers only
	$a_1 = b_2 \cdot b_2$ and $a_2 = b_1 \cdot b_2$ or $a_1 = b_1 \cdot b_2$ and $a_2 = b_1 \cdot b_1$	$W = b_1 \cdot b_1 \cdot b_2$ or $W = b_1 \cdot b_2 \cdot b_2$	K1, K1K2, K3 numbers. You need only go as far as $n = 2$ for condition to hold, but it includes pairs for $n = 4$.
$n = 3$	$a_1 = a_2 \cdot a_3$	$W = b_1 \cdot a_2 \cdot a_3$	Left product
	$a_1 = b_2 \cdot b_3$	$W = b_1 \cdot b_2 \cdot b_3$	Right product
	$a_1 = a_2 \cdot a_2$	$W = a_2 \cdot a_2 \cdot b_1$	Two repeat factors in GB2
	$a_1 = b_1 \cdot b_2$	$W = b_1 \cdot b_1 \cdot b_2$	Two repeat factors in GB4
$n = 6$	$b_3 = b_1 \cdot b_2$	$W = b_1 \cdot b_2 \cdot b_4$	K1K2 number
	$b_4 = b_1 \cdot b_3$	$W = b_1 \cdot b_2 \cdot b_3$	K2 number

Table 290

The above table is called a product mapping matrix because it defines the conditions and type of products for W. Another condition can be added as an exception because we can have $n = 3$ which is a reasonable number of factor pairs but for 4 prime factors. The structure is $W = a_3 \cdot a_3 \cdot a_3 \cdot b_1$ and $W = a_3 \cdot b_1 \cdot b_1 \cdot b_1$. This looks the same, but there is a difference in the classification of b_1 and a_3.

b_1 is GB4

<table>
<tr><th colspan="2">Factor-1</th><th colspan="2">Factor-2</th></tr>
<tr><td colspan="2">$a_1 = a_3 \cdot a_3 \cdot a_3$</td><td>$b_1$</td><td>GB4</td></tr>
<tr><td>$a_2 = a_3 \cdot a_3$</td><td>GB2</td><td colspan="2">$b_2 = b_1 \cdot a_3$</td></tr>
<tr><td>a_3</td><td>GB2</td><td colspan="2">$b_3 = a_2 \cdot a_3 \cdot b_1$</td></tr>
</table>

Table 291

b_1 is GB2

<table>
<tr><th colspan="2">Factor-1</th><th colspan="2">Factor-2</th></tr>
<tr><td colspan="2">$a_1 = a_3 \cdot b_1 \cdot b_1$</td><td>$b_1$</td><td>GB2</td></tr>
<tr><td>$a_2 = b_1 \cdot b_1$</td><td>GB2</td><td colspan="2">$b_2 = b_1 \cdot a_3$</td></tr>
<tr><td>a_3</td><td>GB4</td><td colspan="2">$b_3 = b_1 \cdot b_1 \cdot b_1$</td></tr>
</table>

Table 292

This changes the $n = 3$ row to be inclusive for 3 and 4 factors.

$n = 3$ Three factors	$a_1 = a_2 \cdot a_3$	$W = b_1 \cdot a_2 \cdot a_3$	Left product
	$a_1 = b_2 \cdot b_3$	$W = b_1 \cdot b_2 \cdot b_3$	Right product
	$a_1 = a_2 \cdot a_2$	$W = a_2 \cdot a_2 \cdot b_1$	Two repeat factors in GB2
	$a_1 = b_1 \cdot b_2$	$W = b_1 \cdot b_1 \cdot b_2$	Two repeat factors in GB4
	$a_1 = b_2 \cdot b_3$	$W = b_1 \cdot b_2 \cdot b_3$	Right product
$n = 3$ Four factors	$a_1 = a_3 \cdot a_3 \cdot a_3$	$W = a_3 \cdot a_3 \cdot a_3 \cdot b_1$	W is gap base 4
	$a_1 = a_3 \cdot b_1 \cdot b_1$	$W = a_3 \cdot b_1 \cdot b_1 \cdot b_1$	W is gap base 4

Table 293

12.1.4 CRAWL Function

From this we can define a CRAWL function, where this is a program that looks for prime factors in terms of a multiple of three factors or less. The word CRAWL is conceptual, that is, we imagine a program that crawls through the sieve to look for numbers with three prime factors, and uses a three prime factor cycle to break down those with more than three factors. This implies factors more than three are broken down further, for example 4 prime factors is 3 prime factors and 1 prime factor to make 4. This is as illustrated.

Category	No. of Prime Factors	Breakdown			
0a	2	2			
1b	3	3			
1a	4	3	1		
1b	5	3	2		
2a	6	3	3		
2b	7	3	3	1	
2c	8	3	3	2	
3a	9	3	3	3	
3b	10	3	3	3	1

Table 294

A category is defined by (number of prime factors)/3, and this has a class of *a, b* and *c*. Therefore a category of *3b* for W means there will be three cycles for a ten factor number. A computer program would then be instructed for example to look for numbers with 8 prime factors in a given range.

Consider the number $W = 12\,907\,783$, the first three prime factors as follows.

X2	K2	Factor-1	Factor-2
1	307328	1843969	7
3	113226	679357	19
6	58143	348859	37

Table 295

Therefore the remaining number to factorization is $\frac{W}{4921} = 2623$. This is then entered as the new value of W to derive the following result.

X2	K2	Factor-1	Factor-2
7	10	61	43
10	7	43	61

Table 296

This shows the 3-2 configuration of the factorization process, from 30 factor pairs to 5 factor pairs. Another example is given from $W = 493\,589\,657$.

X2	K3	Factor-1	Factor-2
2	7478631	44871787	11
5	2836722	17020333	29
12	1158661	6951967	71

Table 297

Therefore the remaining number to factorization is $\frac{W}{22649} = 21793$. This is then entered as the new value of W to derive the following result.

X2	K2	Factor-1	Factor-2
3	191	1147	19
5	117	703	31
6	98	589	37

Table 298

This is a 3-3 configuration.

Furthermore, we can add a condition that we are dealing with unique prime factors of W in order to create a band. The theory is that there must be a band in which a certain number of prime factors begin to exist, and the CRAWL function would establish that. As an example, if we assume no repetition of prime factors;

- 1729 is the first number to have 3 prime factors in gap base 2
- 935 is the first number to have 3 prime factors in gap base 4
- 385 is the smallest number to have 3 primes numbers that includes both gap base 2 and 4.

A similar exercise is done for 4 prime factors.

- 53599 is the first number to have 4 prime factors in gap base 2
- 21505 is the first number to have 4 prime factors in gap base 4
- 5005 is the smallest number to have 4 primes numbers that includes both gap base 2 and 4.

Therefore;

a) From 1729 to 53599, there are no numbers with 4 prime factors. This is a 3-factor band for gap base 2.
b) From 935 to 21505, there is no number with 4 prime factors. This is a 3-factor band for gap base 4.
c) From 385 t0 5005, there are no numbers with 4 prime factors where this includes both gap base 2 and 4. This is an inclusive 3-factor band for GB2 & GB4.

This is consolidated through a definition.

Definition

Two patterns of prime factors are defined:

a) Unique factors when W has two or more factors that are not the same,
b) Repeat factors when W has two or more factors some of which are the same.

This leads to the following theorem in regard to the unique factor pattern.

Theorem

There exist minimum occurrence factorization bands for gap base 2 and 4 for unique factors.

For example, what the theorem implies is that we do not expect to find five factors less than 1983163 in gap base 2, hence the term minimum occurrence. Similarly one can map the distribution of such bands along the number line in the context of UNDT, that is noting that 2 and 3 are not prime numbers. Allowing a repetition would give slightly lower results values, and a more complex arrangement. For example the smallest pseudoprime in gap base 2 is 49, and 25 in gap base 4 for two prime numbers, but if we do not allow repetition, then the numbers are 91 and 55.

Number of Unique Factors	2	3	4	5	6	7	8
GB2	91	1,729	53,599	1,983,163	85,276,009	5,201,836,549	348,523,048,783
GB4	55	935	21,505	623,645	25,569,445	1,201,763,915	63,693,487,495

Table 299

The gap base 2 band is in multiples of 7 while the gap base 4 band is in multiples of 5. Using the CRAWL function, one may then reduce the factorization from the six band to the three band in two moves. This is a dramatic reduction. The interesting observation is that gap base 4 numbers have smaller bands compared to gap base 2 numbers. Note that without the gap base approach or the UNDT environment, defining a band in which prime factors of a certain number occur would be a major challenge but this environment makes it achievable. A factorization band can also be derived for repeat factors. Note that GB2 evolves much faster than GB4 in the unique factor pattern.

12.2 Factor-2 Factorization

Self-testing primality focused on using up to a maximum of three factor pairs in order to determine primality of the factorization process. It used standard pairs under certain conditions to look for prime factors. Factor-2 factorization uses efficient pairs, and is based on using only one factor pair, the one defined by the factor 2 theorem. The whole process of factorization is then based on this, as a result it is an iteration operation. This is done by defining W in a different context as consisting of factors that are key to factorization. These factors are referred to as key factors. Therefore;

$$W = F\{W_1, W_2, \cdots W_n\}$$

where each W_i is represents a factor 2 process and F denotes that W can be broken down into key factors.

Definition

For any $W \in D_t$, then $F\{W_1, W_2, \cdots W_n\}$ is called a compressed factorization space, where W_n is a prime number and $W_1, W_2, \cdots W_{n-1}$are pseudoprime .

As an example of the compressed space, let us consider the number $W = 799788119$. If we factor this using the standard approach in applying all the factor pairs, we derive the following result. The time it takes to find all the factor pairs is 5194 milliseconds, which implies it would take even shorter to find the first factor pair. Now we know that at $x_2 = Int\left\{\frac{W+7}{30}\right\}$ we are sure that all the factor pairs have been established for standard pairs, where this is referred to as the verification level. However at around 38% of this level, all the pairs for W are found. Normally the program will continue to look for factor pairs until the verification level, where this is a 62% of wasted effort.

Verification Level	Largest X2	%
26,659,604	10,253,694	38%

X2	K3	Factor-1	Factor-2
604	36792	220753	3623
7850	2830	16981	47099
10253694	2	13	61522163

Table 300

The compressed space uses two approaches to deal with this wasted effort, the first being self-testing primality. Now in using self-primality testing, we do not need to reach the verification level but consider the number of factor pairs and specific conditions. This is a left product and if $a_1 = a_2 \cdot a_3$ then $W = b_1 \cdot a_2 \cdot a_3$. This would give us the required answer immediately. The second approach is applying the factor 2 theorem for a single pair only, where we compress the operation further as shown in the diagram.

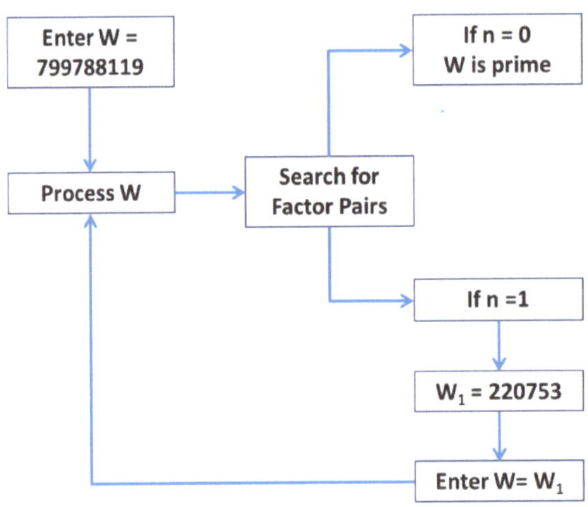

Figure 133

We note that *n=1* occurs at $x_2 = 604$, and this derives $W_1 = 220753$. Factor 2 will be a prime number in gap base 4. Then we process W_1 and this gives us $W_2 = 16981$, and factor 2 is 13 in gap base 2. Then 16981 is a prime number.

W	Verification level	Standard Pairs - Largest X2	Efficient Pairs - FPLT	Compression
799,788,119	26,659,604	10,253,694	4713	604
220,753	7,358	2,830	78	2

16,981	566	-	21	21
Additive Weight	26,667,528	10,256,524	4812	627

Table 301

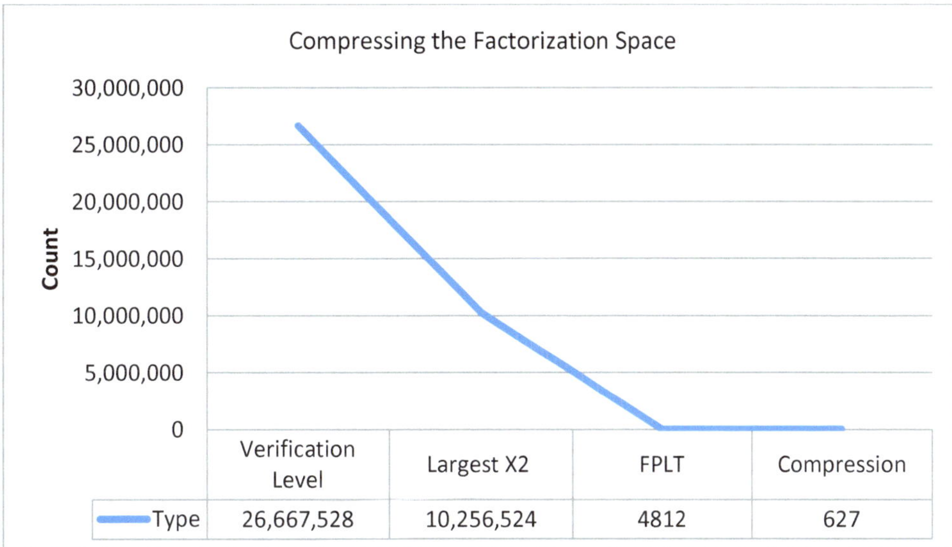

Figure 134

The additive weight is a way of establishing comparative measure in order to conceptualize the level of compression of the factorization space. The total of the X2 values are called an X2 additive weight. When the total time is divided by the X2 additive weight, this is defined as the compression rate of W.

			Compressed pairs	Factor-1	Factor-2	
	W	**Time**	**X2 Value**	W_i	**Prime**	**Type**
n=1	799,788,119	5,091	604	220,753	3,623	GB4
n=2	220,753	2	2	16,981	13	GB2
n=3	16,981	1	0	0	16,981	GB2
		5,094	606			
	Compression rate	8.41				

Table 302

Hence the compression rate is another measure that divides the total time by the total count to find how many seconds each unit of counting took. This enables us to compare between two or more factorization spaces. Consider $W = 799887179$, where this is almost the same size as the previous number.

Verification Level	Largest X2	%	FPLT	Compression
26,662,906	19,044,193	0.714258	4713	3
1,568,406	1,120,290	0.714286	1143	1
224,058	86,176	0.384615	432	2
17,235	658	0.038178	120	22
131	-	0	10	0

Table 303

Here the statistic are very different in terms of behavior, as we note that the largest X2 is quite large compared to the verification level. The first number had three prime factors, this one has five prime factors. We note that the compression rate is much higher than the first number though the additive weight for compression is 606 and 28 respectively.

			Compressed pairs	Factor-1	Factor-2	
	W	Time	X2 Value	W_i	Prime	**Type**
n=1	799,887,179	5,174	3	47,052,187	17	GB4
n=2	47,052,187	309	1	6,721,741	7	GB2
n=3	6,721,741	45	2	517,057	13	GB2
n=4	517,057	4	22	3,947	131	GB2
n=0	3,947	1	0	0	3,947	GB4
		5,533	28			
	Compression rate	197.607				

Table 304

The compressed space is a great advantage for situations when the number of factor pairs is very large. The advantage may not be so obvious for small numbers or when the factors pairs of W are probably less than ten or so. According to the limit theorem, we can draw up the following table looking at compression rates and number of factors.

W	Number of factor pairs	Number of prime factors	Time	Compression rate	Additive X2 Weight
799,788,119	3	3	5,091	8.41	606
799,887,179	15	5	5,174	197.607	28
799,887,127	14	4	5,225	3.824	108
554,827	16	5	12	2.333	6

Table 305

The other issue that counts is whether the number is gap base 2 or 4, since factor pairs repeat in gap base 2.

12.3 Turbo Charged Factorization

As the two words imply, here the aim is to develop a compressed space that has an element of being turbo charged to do the factorization faster.

We first begin by developing a turbo charged factorization space, and then compressing it to define a compressed turbo charged space. It has been demonstrated that the ditation set can also be produced from $12x \pm y$, where $y = \pm 1, and\ y = \pm 5$. This gives us the following product results.

GB2	**GB4**
$12x + 1$	$12x - 1$
$12x - 5$	$12x + 5$

Table 306

As a result we can have a different factorization engine besides $6x \pm 1$. This gives us more power in the factorization process, hence the use of the term turbo charged. The turbo charging is defined by x, whereas in using $6x + 1$, we could perhaps change it using the constant to $6x + 7$ and $6x + 13$. However this would not increase the power of the factorization process. The table below shows the difference in the number of elements that are generated by $12x \pm y$ compared to $6x \pm 1$. That is from $x = 0$ to 10, $12x \pm y$ has generated 44 elements whereas $6x \pm 1$ has only generated 22, that is half the number.

These are described as the "6" and "12" factorization engines, and the other is a turbo because it has twice as much generating power. This has implication in reducing the time it takes to process factor pairs.

	Turbo Charge				**Normal**	
	12x-5	**12x-1**	**12x+1**	**12x+5**	**6x-1**	**6x+1**
x	GB2	GB4	GB2	GB4	GB4	GB2
0	-5	-1	1	5	-1	1
1	7	11	13	17	5	7
2	19	23	25	29	11	13
3	31	35	37	41	17	19
4	43	47	49	53	23	25
5	55	59	61	65	29	31
6	67	71	73	77	35	37
7	79	83	85	89	41	43
8	91	95	97	101	47	49
9	103	107	109	113	53	55
10	115	119	121	125	59	61

Table 307

Using the previous approach, we can then define the equations for the factorization process. Note that 1 up to 4 defines the commutative group since the same number multiplies itself, that is $x_1 = x_2$. Hence some of the numbers that are a square in terms of prime factors will have the product definition of $12x_3 + 1$. This already simplifies the factorization process since once classification is determined as being in this category, then we know immediately that we have a number that has two factors which are a square or two different factors.

For example $17 \cdot 17 = 289$, and $12x_3 + 1 = 289$ gives us $x_3 = 24$. Other examples are shown, where we can divide a GB2 product result $12x + 1$ into three parts up to x = 24.

Square $x_1 = x_2$		Prime x_1		$x_1 \neq x_2$	
x	12x+1	x	12x+1	x	12x+1
2	25	1	13	7	85
4	49	3	37	11	133
10	121	5	61	12	145
14	169	6	73	17	205
24	289	8	97	18	217
		9	109	21	253
		13	157	22	265
		15	181		
		16	193		
		19	229		
		20	241		
		23	277		

Table 308

However, other product results represent the result $x_1 \neq x_2$, or a prime number and hence a primality test!

Eq No.	Product	Type	X1	Result	Definition of k values
Gap Base 2 Product Result – 6 altogether					
1	$(12x_1+1)(12x_2+1)$	GB2xGB2	$x_1=(x_3-x_2)/(1+12x_2)$	$12x_3+1$	k_1
2	$(12x_1-5)(12x_2-5)$	GB2xGB2	$x_1=(x_3+5x_2-2)/(12x_2-5)$	$12x_3+1$	k_2
3	$(12x_1+5)(12x_2+5)$	GB4xGB4	$x_1=(x_3-5x_2-2)/(12x_2+5)$	$12x_3+1$	k_3
4	$(12x_1-1)(12x_2-1)$	GB4xGB4	$x_1=(x_3+5x_2)/(12x_2-1)$	$12x_3+1$	k_4
5a	$(12x_1-5)(12x_2+1)$	GB2xGB2	$x_1=(x_3+5x_2)/(12x_2+1)$	$12x_3-5$	k_5
5b	$(12x_1+1)(12x_2-5)$	GB2xGB2	$x_1=(x_3-x_2)/(12x_2-5)$	$12x_3-5$	k_6
Gap Base 4 Product Result – 10 altogether					
6a	$(12x_1+5)(12x_2-1)$	GB4xGB4	$x_1=(5x_2-x_3)/(1-12x_2)$	$12x_3-5$	k_7
6b	$(12x_1-1)(12x_2+5)$	GB4xGB4	$x_1=(x_2+x_3)/(5+12x_2)$	$12x_3-5$	k_8
7a	$(12x_1-5)(12x_2+5)$	GB2xGB4	$x_1=(x_3+5x_2+2)/(12x_2+5)$	$12x_3-1$	k_9
7b	$(12x_1+5)(12x_2-5)$	GB4xGB2	$x_1=(x_3-5x_2+2)/(12x_2-5)$	$12x_3-1$	k_{10}
8a	$(12x_1+1)(12x_2-1)$	GB2xGB4	$x_1=(x_3-x_2)/(12x_2-1)$	$12x_3-1$	k_{11}
8b	$(12x_1-1)(12x_2+1)$	GB4xGB2	$x_1=(x_3+x_2)/(12x_2+1)$	$12x_3-1$	k_{12}
9a	$(12x_1-5)(12x_2-1)$	GB2xGB4	$x_1=(x_3+5x_2)/(12x_2-1)$	$12x_3+5$	k_{13}
9b	$(12x_1-1)(12x_2-5)$	GB4xGB2	$x_1=(x_3+x_2)/(12x_2-5)$	$12x_3+5$	k_{14}
10a	$(12x_1+5)(12x_2+1)$	GB2xGB4	$x_1=(x_3+5x_2)/(12x_2+1)$	$12x_3+5$	k_{15}
10b	$(12x_1+1)(12x_2+5)$	GB4xGB2	$x_1=(x_3-x_2)/(12x_2+5)$	$12x_3+5$	k_{16}

Table 309

This derives a total of sixteen equations that will find the required factor pairs to factorize a given number. However the testing does not involve all 16 equations, only four equations will apply for a given W.

An example is given for looking for prime factors of $W = 34908263$. As usual, the process first of all tests to find out the result type in terms of gap base. In the example, the product result is a gap base 4 number, and it is under the category of $12x-1$. That is, we have a product that results in an answer that is gap base 4, where this is classified by $12x-1$. Therefore even though there are sixteen equations, this immediately narrows the area of search activity to only four equations starting from k_9 to k_{12}. This means the process will not waste time checking other equations.

W	34908263	
GB2	2909022.33	12x-5
GB2	2909021.83	12x+1
GB4	2909022	12x-1
GB4	2909021.5	12x+5

Table 310

The transient factorization database created has the factors as shown. Note that x2 varies only from 1 to 8. Both x1 and x2 must be integers for a solution to exist.

Equation Product Result	x2	x1	Factor-1	Factor-2
12x-1	1	223771	13	2685251
12x-1	2	153106	19	1837277
12x-1	3	93839	31	11609
12x-1	4	61894	742729	47
12x-1	8	29990	97	359879

Table 311

The database is created from the following. For every x2, there is a $GB2 \cdot GB4$ product that results in $12x - 1$ shown by ***R1***. In this case there are actually 8 as shown in the last row. Each of these derives a Factor-1 & Factor-2 pair. Hence for example when x2 = 1, then we derive the result as shown.

GB4	W	34908263				
	GB2	2909022.33	12x-5			
	GB2	2909021.83	12x+1			
	GB4	2909022	12x-1	GB2	GB4	
	GB4	2909021.5	12x+5	12x+1	12x-1	EQ 8
				Factor-1	Factor-2	
	x2	1		3173478.5	11.0	
8a	x1	264456.455		13	2685251	
8b	x1	223771				

Table 312

The implication is that this uses $(12x_1 - 1)(12x_2 + 1) = 12x - 1$ since we have integers for x1 at the bottom, that is at 8b or equation k_{12}. The next one is x2 = 2, which uses equation 7b.

GB4	W	34908263				
	GB2	2909022.33	12x-5			
	GB2	2909021.83	12x+1			
	GB4	2909022	12x-1	GB2	GB4	
	GB4	2909021.5	12x+5	12x-5	12x+5	EQ 7
				Factor-1	Factor-2	
	x2	2		1203733	29	Integer
7a	x1	100311.517		19	1837277	Integer
7b	x1	153106				

Table 313

The table below presents a summary from valid x2 values starting from 1 to 8, the answer being A or B.

	GB2	GB4	GB2	GB4	GB2	GB4	GB2	GB4	GB2	GB4
EQ	12x+1	12x-1	12x-5	12x+5	12x-5	12x+5	12x+1	12x-1	12x+1	12x-1
	Factor-1	Factor-2	Factor-1	Factor-2	Factor-1	Factor-2	Factor-1	Factor-2	Factor-1	Factor-2
A	3173478.5	11	1203733	29	851421	41	742729	47	367455.4	95
B	13	2685251	19	1837277	31	1126073	49	7344.469	97	359879
X2	**1**		**2**		**3**		**4**		**8**	

Table 314

Since $12x - 1$ is in gap base 4, then we have equation A and B to test for decimal or integer as shown in the above two examples. But observe that in as much as 1203733 and 29 are integers for x2 = 2, but x1 is not an integer, so the result is not selected. The same procedure applies to other results. The table shows the gap base of the product as well as the equations involved to derive a result that is $12x - 1$.

Now we can compare this to the standard pair database. Under standard pairs the count gets to over 400 thousand and under efficient pairs it gets to 760. The number of factor pairs is the same because this is a gap base 4 number. In a gap base 2 number, the efficient pairs are half the standard process.

We note that by the time the efficient space arrives at 16, it has actually found all the necessary factor pairs. Therefore it is evident that the turbo charge has the same characteristic of self-testing primality, provided you know the number has four or five factors. The main advantage of the turbo charge is that it only derives the needed factor pairs in direct relation to the exact number of prime factors. As a result, it becomes very efficient.

Standard Factor Pairs					Efficient Factor Pairs			
X2	K3	Factor-1	Factor-2		K4	K5	Factor-1	Factor-2
8	123788	742729	47		2	447542	13	2685251
102	9522	57133	611		3	306213	19	1837277
149	6515	39091	893		5	187679	31	1126073
243	3993	23959	1457		8	8	742729	47
760	1276	7657	4559		16	59980	97	359879
1935	501	3007	11609		41	23555	247	141329
3157	307	1843	18941		67	14437	403	86621
4614	210	1261	27683		98	9878	589	59267
9878	98	589	59267		102	102	57133	611
14437	67	403	86621		149	149	39091	893
23555	41	247	141329		210	4614	1261	27683
59980	16	97	359879		243	243	23959	1457
187679	5	31	1126073		307	3157	1843	18941
306213	3	19	1837277		501	1935	3007	11609
447542	2	13	2685251		760	760	7657	4559

Table 315

Let us now consider $W = 16221043$. This is a gap base 2 number with equation 12x-5.

Equation	x2	x1	Factor-1	Factor-2
12x-5	1	79515	17	954179
12x-5	69	1631	19567	829
12x-5	96	1174	14093	1151

Table 316

The standard pairs will count to 159030, while the efficient pairs will end at 192.

X2	K1	K2	Factor-1	Factor-2	
3	159030	142289.6842	954179	17	Efficient Pairs
138	3269.220073	3261	19567	829	
192	2349	2344.592368	14093	1151	
2349	192	191.6394466	1151	14093	
3261	138.3474572	138	829	19567	
159030	3	2.666660728	17	954179	

Table 317

12.3.1 Compressing the Turbo Charged Factorization Space

Using these observations, we can now begin to compress the turbo charged space. The issue is to look for a suitable verification level that has a certainty of deriving all factors of W. In the turbo charged space, it appears by observation that we can confidently consider $\sqrt{x_3}$ rather than the $\sqrt{W}$ as a compression factor. Secondly, this can be further reduced by a factor of four since the values for the turbo charged space are comparatively low using examples that have been studied. Therefore we use $0.25\sqrt{x_3}$ as the final trial and error compression factor, it is possible that other values may be used. At first a factor of 10 was used to reduce $\sqrt{x_3}$, it was fine for some values but was lower than actual for one value in the case of 13474139. The value that is chosen has to have a reasonable level of certainty. A table is constructed to test this hypothesis.

a) x_3 ; the highest verification level
b) $\sqrt{W}$; the next lower verification level
c) $\sqrt{x_3}$; an assumed verfiication level
d) $0.25\sqrt{x_3}$; an adjustment on the assumed verification level

			Turbo		Maximum k Value			
W	x_3	$\sqrt{W}$	$\sqrt{x_3}$	$0.25\sqrt{x_3}$	STD	EFF	Turbo	EFF Pairs
34908263	2909022	5908.322	1705.58553	426.3964	447542	760	8	15
16221043	1351754	4027.536	1162.64956	290.6624	159030	192	96	3
76652489	6387707	8755.141	2527.391343	631.8478	190678	195	98	3
3480992563	290082714	58999.94	17031.81476	4257.954	N/A	N/A	37	N/A
271440061	22620005	16475.44	4756.049306	1189.012	2381053	1302	9	15
13474139	1122845	3670.714	1059.643808	264.911	320813	311	156	3

Table 318

The above table shows that at time the count from efficient pairs is lower than the turbo charge count, and this happens when the number of factor pairs is small, such as 3. An example is the number 76652489, the efficient pair count is much lower at 195 compared to the turbo charge count of 632.

Even though the values for 3480992563 are not available because the computer cannot do ten figures, we can use the equations to calculate the needed count that should give us a certainty that the test has been exhaustive in searching for factor pairs and the prime factors. We need only count to 37 in this case!

W = 3480992563				
Prime Factors	79	227	389	499
x2 values	7	19	32	37

Table 319

The following comparison can therefore be made, where starting with the primitive space the gains in compression are enormous!

$W = ax_3 + b$	Primitive Factorization Space	Standard Factorization Space	Compressed Factorization Space	
	Trail pairs	**Standard Pairs**	**Efficient Pairs**	**Turbo Charged Pairs**
	$x_2 < x_3$	$x_2 = Int\left\{\frac{W+7}{30}\right\}$	$x_2 = Int\left(\frac{\sqrt{W}+1}{6}\right)$	$0.25\sqrt{x_3}$
3,480,992,563	290,082,714	116,033,085	9,833	4,257

Table 320

The factor pairs have a maximum value of 37 which is far less than 4257. Note that the bigger the number, then the bigger is $\sqrt{W}$ but the turbo charge count does not relate in exactly the same way.

W	$0.25\sqrt{x_3}$	STD	EFF	Turbo
34908263	426.3964	447542	760	8
16221043	290.6624	159030	192	96

Table 321

The first value of W is bigger, but if we consider the efficient pair values and the turbo charge values the bigger number has a lower turbo charge value. However 192 relates to a bigger turbo charge value. What the figures means is that it took a quicker time to find the factor pairs of 34908263 compared to 16221043 even though it is bigger. The gap base 4 is faster and gap base 2 is slower.

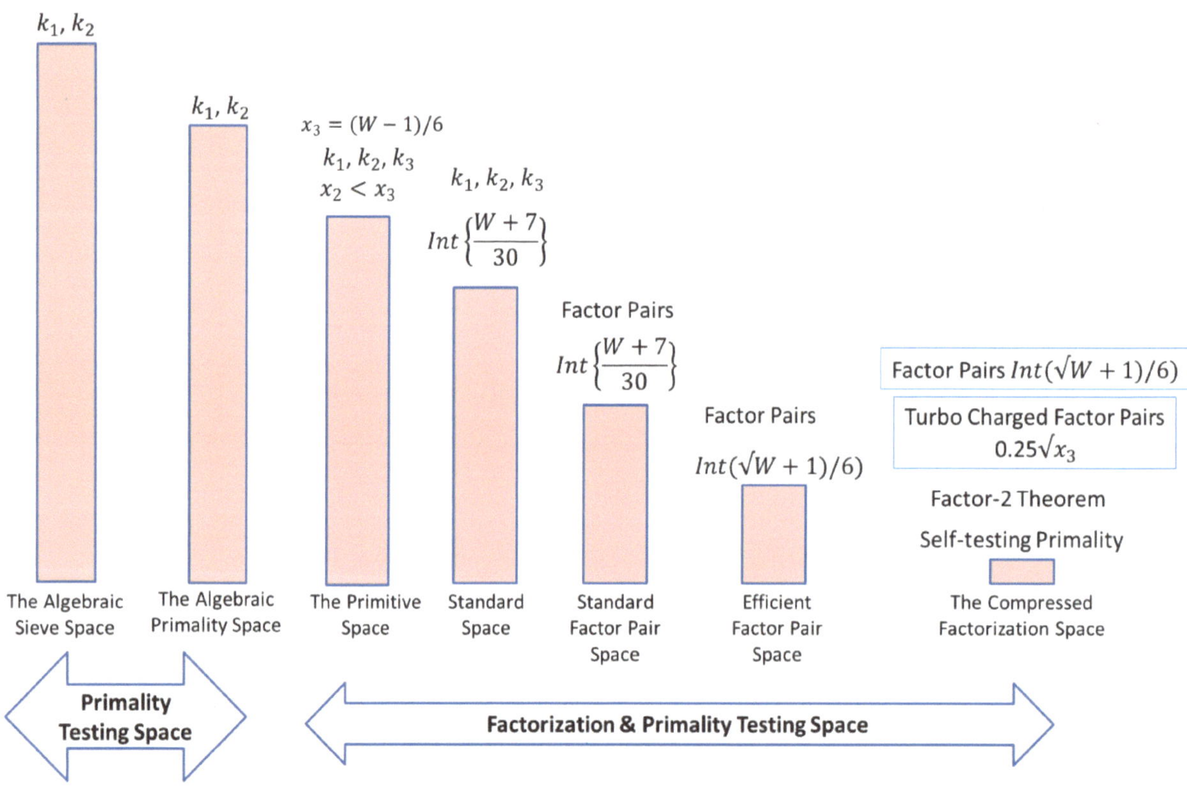

Figure 135

The quest for the microseconds can continue as more testing takes place. The compression will be very useful for extremely large numbers since there is no trial division and the approach structured.

12.3.2 Turbo Charged Primality Test

The primality condition for a turbo charged test is as follows.

Definition

If $0.25\sqrt{x_3} < x_2$, *the* W *is a prime number.*

As an example consider testing 1201 for primality. The number is gap base 2, and falls under $12x+1$. This implies that only equation 1 to 4 need be referred to in order to check for primality, the rest of the equations are redundant.

GB2	W	1201				
	GB2	100.5	12x-5			
	GB2	100	12x+1			
	GB4	100.1667	12x-1	GB2	GB2	
	GB4	99.66667	12x+5	12x+1	12x+1	EQ 1
				Factor-1	Factor-2	
	x2	2		48.04	25	
	x1	3.92				

x2	2		GB2	GB2	
x1	5.684211		12x-5	12x-5	EQ 2
			Factor-1	Factor-2	
			63.21053	19	

x2	2		GB4	GB4	
x1	3.034483		12x+5	12x+5	EQ 3
			Factor-1	Factor-2	
			41.41379	29	

x2	2		GB4	GB4	
x1	4.434783		12x-1	12x-1	EQ 4
			Factor-1	Factor-2	
			52.21739	23	

Table 322

Since $x_3 = 100$, therefore $0.25\sqrt{x_3} = 2.5$ which means we do not have to count beyond 3 for x_2.

	Eq 1	**Eq 2**	**Eq 3**	**Eq 4**
x2	1	1	1	1
x1	7.615385	14.71429	5.470588	9.181818
x2	2	2	2	2
x1	3.92	5.684211	3.034483	4.434783
x2	3	3	3	3
x1	2.621622	3.645161	2.02439	2.942857
x2	15	15	15	15
x1	0.469613	0.988571	0.124324	0.642458

Table 323

However practically we find that we are certain there are no prime factors when $x_2 = 15$ because that is when $x_1 < 1$, the general condition for the primality test. Consider for example $W = 1207$, which is not a prime number.

GB4	W	1207				
	GB2	101	12x-5			
	GB2	100.5	12x+1			
	GB4	100.666667	12x-1	GB4	GB4	
	GB4	100.166667	12x+5	12x+5	12x-1	EQ 6
				Factor-1	Factor-2	

x2	1
x1	8.72727273
x1	6

109.7273	11
17	71

Table 324

We have only gone as far as $x_2 = 1$ to find the prime factors of the number, we have not gone beyond 3, which is what turbo charged factorization is about. If after $x_2 \leq 3$ there is no other integer values, the algorithm assumes none will exist after that, and that primality testing verification is achieved. This is a structural decision based on factor pairs since it does four pairs at a time, and 3x4=12, which is about 15.

12.4 SFM for Composites- Factorisation Paths

The UNDT framework has three families; hence we now have to consider factorization structure for other classes as well. Basically there are only two groups to factorize;

a) The set of pseudoprimes, and a procedure for this has been developed already.
b) The set of composites. The good news here is that there are only two factors to consider for factorization, 2 and 3. The main issue is a mechanism for finding the power.

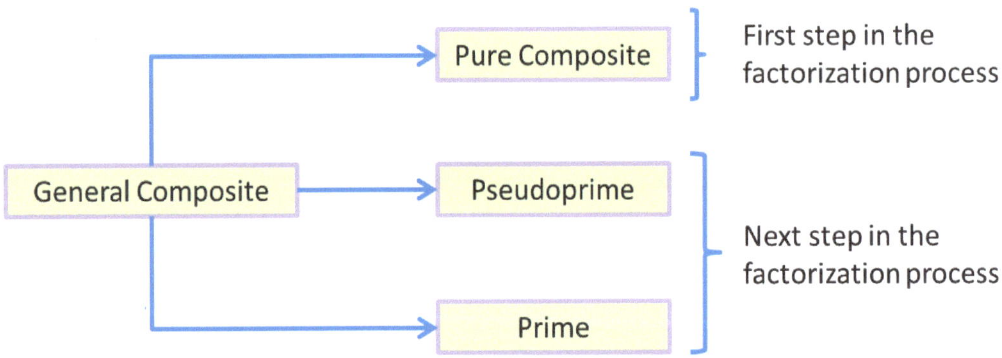

Figure 136

This actually gives us an alternative construction of a composite. The first definition that was made was algebraic, where a composite is any number that was not an element of GB2 or GB4. Now it is conceptual as the definition is in terms of arithmetic. That is, a general composite is a product of a pure composite with a prime or pseudoprime.

That means the number can have factors consisting of 2, or consisting of 3, or both 2 and 3. This again clearly defines the role of 2 and 3 in the number structure under UNDT. The number 192 is an example of a pure composite, there is nothing in the product that is defined from the pseudoprime. This results in a transfer principle if a composite is not pure.

We can remind ourselves of the integer product law. The integer product law finds the relationship in products between the three type of numbers; the primes, the pseudoprimes and the composites. It also states that the result of a product is either in the composite group or is a pseudoprime.

Class	Class	Resulting Product Class
Class 1A	Class 1A	Class 1A
Class 1A	GB-2	Class 1A
Class 1A	GB-4	Class 1A

Table 325

This implies that the product can be traced as to what type of numbers defines it, where this is called a factorization path. Class 1A for example consists of odd numbers, and this can only result from the following product classes.

In the composite family, the Class 1A product is closed and therefore unique. Class 2B also has the same characteristic as Class 1A. That is, if we take the composites only, any product that results in Class 2B is between Class 1B and 2B. The general product distribution shows that Class 2A product answer can come from several classes. For example for the number 24, this can be the result of three product paths.

	Factor	Factor	Product
Class 2A x 1B	6	4	24
Class 1A x 2B	3	8	24
Class 2A x 2B	2	12	24

Table 326

Infact over 50% of products are either in Class 1B or Class 2A. From the product table by class, we can compile a probability of occurrence for each product class.

	1A	1B	2A	2B	GB2	GB4	
Occurrence	3	4	8	3	2	1	21
Probability	14%	19%	38%	14%	10%	5%	

Table 327

Class GB4 is unique in a special manner because it can only be formed from a GB4 family.

Consider for example the number 533547. When we punch it into the program, $533547 = 567 \times 941$, 567 is in Class1A, 941 in Gap Base 4, and the answer 533547 is in Class 1A.

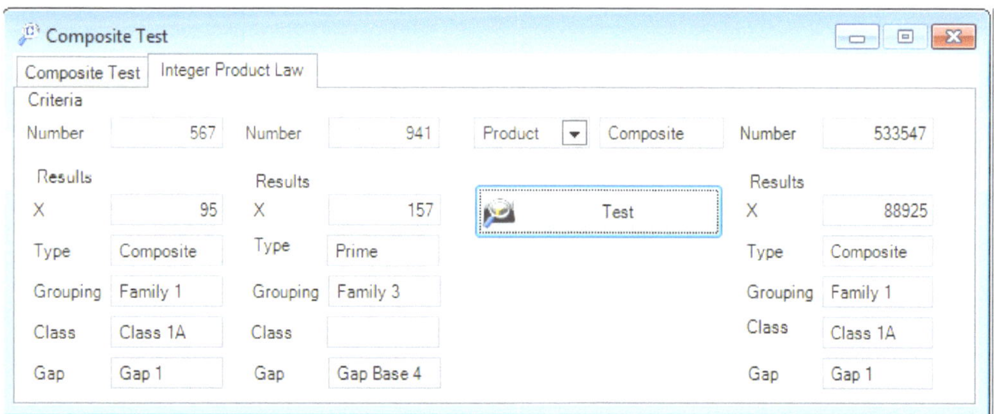

Figure 137

We also recognize that there is a one-to-one relationship between the variable X and the numbers that are being multiplied through the relationship $f(x_1) \cdot f(x_2) = f(x_3)$. This means that we can multiply two numbers without knowing their actual values, and better still, it provides a method of relating numbers in the way they multiply. This becomes useful for developing the technique for factorization.

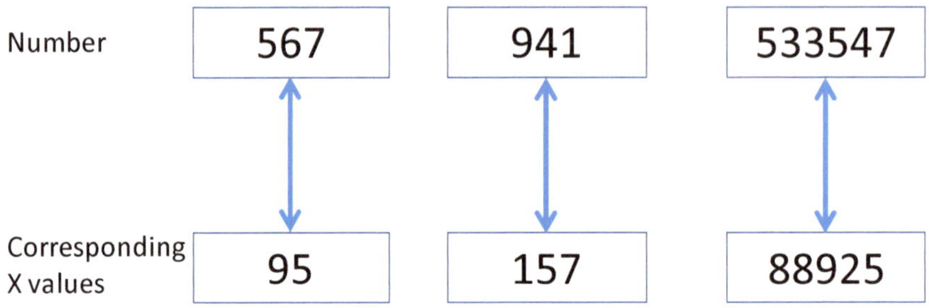

Figure 138

Each class has a set of equations that describe how factorization occurs, this is the power of SFM. The system works in two ways, through a control equation and a calculation equation to define a factorization path.

Theorem

Every integer has a unique algebraic factorization path

This theorem gives us a deterministic approach to factorization and is the cornerstone for the development of SFM. The essence of this theorem has been demonstrated for the pseudoprimes, and will now include the composites. For example, let us consider the number 9. First of all we have to know the classification of the number in order to determine how to deal with it. The point of entry then is making use of the inverse of the classifying equations.

Classifying equations →

Test equations →

x	6x-3	6x-2	6x-1	6x	6x+1	6x+2
2	9	10	11	12	13	14
W	(W+3)/6	(W+2)/6	(W+1)/6	W/6	(W-1)/6	(W-2)/6
9	2	1.833333	1.666667	1.5	1.333333	1.166667

Figure 139

When $x = 2$, then 9 occurs under $6x - 3$ which is class 1A. Similarly, when we use the control or test equations, if we enter $W = 9$, then we get $x = 2$ as the only integer result, indicating that it belongs to Class 1A. This identifies the starting point of the path, and this will always be uniquely the same.

Definition

Every factorization path initiates with the integer W and terminates with 1 for composite factorization.

Therefore the class is clearly defined and should be consistent. Now each class has its control equation and calculating equations that specify factorization procedures. What happens is that the procedure will derive the sequence 9,3,1. You divide 9 by 3 to get 3, the 3 by 3 to get 1. Each class has a structure that determines which number you divide by. In this case the structure is $3^n[P]$ for class 1A. That is, a number consist of a key and a base, and the key is given by 3^n and $[P]$ is the base. Hence we always start by looking for the factorization key, then the base. If the base is one, that is $[P] = 1$, then it implies that number is made up of the key only, that is 3^n.

How do we identify the factorization paths. We use the technique of pattern methodology. Consider the graphic of class 1A elements where these are divided by 3. Using class 1A, we recognize that:

1. If you divide by 3 any element of Class A, there two possibilities; getting consecutive elements of [P], or an element of class A again.
2. The blue bar ends with an element of gap base 2, that is 1,7,13,19 and so on. This is a factorization path.

CLASS 1A

3	1			
9	3	1		
15	5			
21	7			
27	9	3	1	
33	11			
39	13			
45	15	5		
51	17			
57	19			
63	21	7		
69	23			
75	25			
81	27	9	3	1
87	29			
93	31			
99	33	11		
105	35			
111	37			
117	39	13		
123	41			
129	43			
135	45	15	5	
141	47			
147	49			
153	51	17		
159	53			
165	55			
171	57	19		
177	59			
183	61			
189	63	21	7	
195	65			
201	67			
207	69	23		
213	71			
219	73			
225	75	25		
231	77			
237	79			
243	81	27	9	3
249	83			
255	85			
261	87	29		
267	89			
273	91			
279	93	31		
285	95			
291	97			

Figure 140

CLASS 1B

4	2	1				
10	5					
16	8	4	2	1		
22	11					
28	14	7				
34	17					
40	20	10	5			
46	23					
52	26	13				
58	29					
64	32	16	8	4	2	1
70	35					
76	38	19				
82	41					
88	44	22	11			
94	47					
100	50	25				
106	53					
112	56	28	14	7		
118	59					
124	62	31				
130	65					
136	68	34	17			
142	71					
148	74	37				
154	77					
160	80	40	20	10	5	
166	83					
172	86	43				
178	89					
184	92	46	23			
190	95					
196	98	49				
202	101					
208	104	52	26	13		
214	107					
220	110	55				
226	113					
232	116	58	29			
238	119					
244	122	61				
250	125					
256	128	64	32	16	8	4

Figure 141

3. Another bar (dark orange) the same length as the blue bar ends with elements of gap base 4, that is 5,11,17,23 and so on. This forms a factorization path.
4. The yellow bar ends with elements of [P], that is, both gap base 2 and 4. This forms a factorization path.
5. The light green bar ends with elements of [P], and of class A. This is a repetition of number 1, but we recognize it as a factorization path.

To form the idea, we define the following.

Definition

a) Each factorization path has a factorization key for W.
b) A number that divides into a composite according to the factorization key is called a root.
c) Roots that are from the same class as W are major.
d) Roots from a different class from W are minor.

Let us use the example of the number 243 in class 1A to illustrate the concept of a factorization path. According to the definition, 243 is a major root, and the factorization key is 3. Therefore 243/3 gives 81, and 81 is a root by definition. Similarly, 81/3 is 27, hence 27 is a root.

3				
9	**3**			
27	**9**	**3**		
81	**27**	**9**	**3**	
243	**81**	**27**	**9**	**3**

Figure 142

The factorization path is 243,81,27,9,3,1. The surprising finding is that class 1A has only major roots 3, 9, 27, 81, 243 as they are all in the same class! The operation is closed.

Now let us consider Class 1B. From the definition, each class has a particular divisor called the factorization key. For Class 1B this is 2. We observe that;

a) The first division by 2 gives us elements of gap base 4 and elements of Class 2B. That is 2,8,14, 20 and so on are in class 2B.
b) We can further group this occurrence into two, as 2-14-26... and 8-20-32.... Both of these are in class 2B.
c) The 2-14-26 group terminates with an element of gap base 2.
d) The 8-20-32 group is diverse as each has its own subgroup. That is, we have subgroups 8-56-104, 20-44-68, and 32-80-128 in the pattern.
e) The subgroup 20-44-68 ends with an element of gap base 4.

The graphic below illustrates class 1B characteristics which are rather complex compared to class 1A because of the subgroups.

Class 1B

4	2	1				
10	5					
16	8	4	2	1		
22	11					
28	14	7				

34	17					
40	20	10	5			
46	23					
52	26	13				
58	29					
64	32	16	8	4	2	1

Figure 143

If we use as an example 256 in Class 1B, we realize that 4, 16, 64,and 256 are the major roots, and 2,8,32,128 are the minor roots.

4	2					
16	8	4	2			
64	32	16	8	4	2	
256	128	64	32	16	8	4

Figure 144

The factorization path is given as 256, 128, 64,32,16,8,4,1. The two examples that have been given here for 243 and 256 shows the regularity and power of the UNDT. It links numbers together in a way that makes it easy for us to find factors using algebraic methodologies.

We now illustrate using Class 2B, and show the transfer that takes place in a factorization path. Consider $W = 512$, which is a 2B element. Then this gives us the following relationships between Class 2B and Class 1B.

2							
8	4	2					
32	16	8	4	2			
128	64	32	16	8	4	2	
512	256	128	64	32	16	8	4
2B	1B	2B	1B	2B	1B	2B	1B

Figure 145

The factorization path and transfer between classes is then shown by the graphic below.

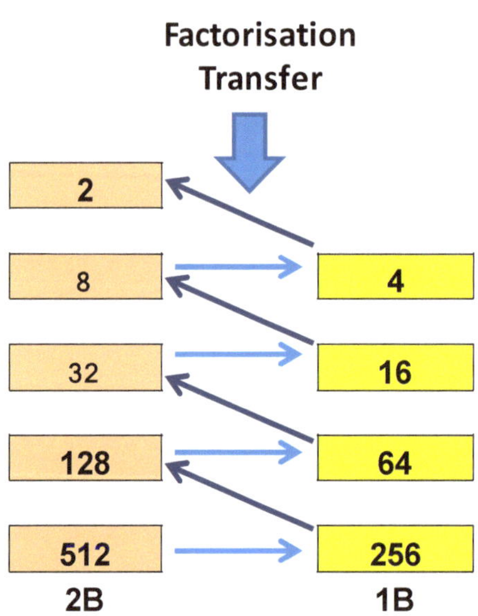

Figure 146

We then derive the control equations by looking for an algebraic description of the path, because each class consists of a set of paths. When W is entered, the control equation determines which path should be used to factorize the number.

12.4.1 Deriving Control Equations

These equations are found on the basis of the definition of the set and the pattern it derives.

12.4.1.1 *Class 1A*

An example is made by considering Class 1A. Since this is a class consisting of odd numbers, we divide by 3.

UNDT Equation	Class	Structure	Examples
$6x-3$	1A	$3^n[P]$, Odd	1. $14067 = 3\cdot 3\cdot [521]$ 2. $3399 = 3\cdot [11\cdot 103]$

Table 328

The examples show [P], but since this has been covered, our interest now lies in establishing the method for Class 1A only.

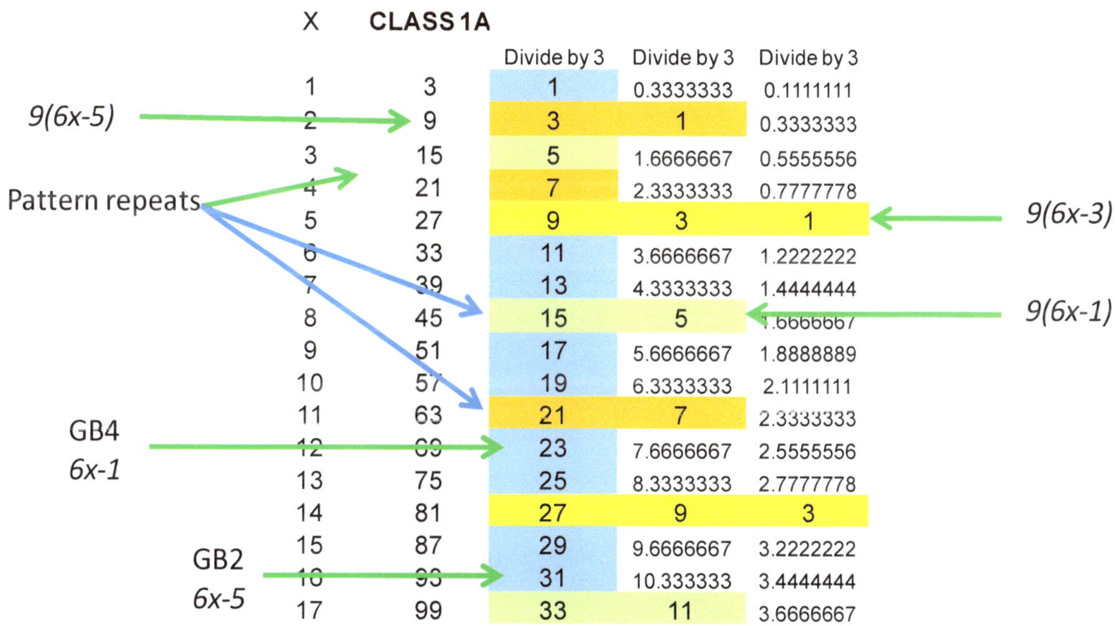

Figure 147

When we divide by 3, we get a pattern that is structured around five equations.

a) $54x - 45 = 9(6x - 5)$.
b) $18x - 3 = 3(6x - 1)$
c) $18x + 3 = 3(6x + 1)$
d) $54x - 27 = 9(6x - 3)$
e) $54x - 9 = 9(6x - 1)$

Except for the fourth equation, the last entry always ends with a prime or pseudoprime, which implies looking for factors is transferred to the [P] procedure. These equations work as a system to break the number into its constituent factors. This is done as follows. The format of the structure is to divide the input into a Control and a Calculation procedure.

a) The control tells us which equation to use. If the number is an element of Class 1A, then one of the five equations will fit.
b) Then knowing the equation to use implies we can now calculate, so we transfer the figure to the calculation procedure.

Consider the number $W = 243$.

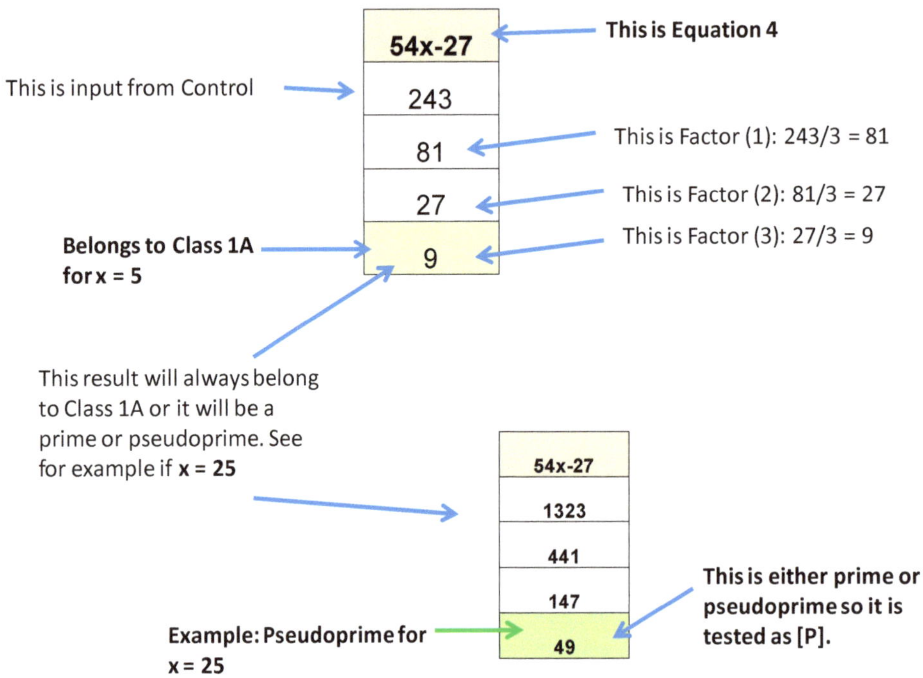

Figure 148

POWER COUNT

Suppose you are looking for factors of **243**
The control equation gives us **x = 5** and **x = 1** to use in the equations **54x - 27** and **54x-45** respectively.

X	54x-27	RESULT
5	243	
	81	3
	27	3
	9	3

Power count = 3

X	54x-45	Result
1	9	
	3	3
	1	3

Power count = 2

So $243 = 3^5$

Figure 149

Control

Input	54x-45	18x-3	18x+3	54x-27	54x-9
243	5.3333333	13.666667	13.333333	5	4.666667

Calculation

X	54x-45	18x-3	18x+3	54x-27	54x-9
5	225	87	93	243	261
	75	29	31	81	87
	25			27	29
				9	

Control

Input	54x-45	18x-3	18x+3	54x-27	54x-9
9	1	0.6666667	0.3333333	0.666667	0.333333

Calculation

X	54x-45	18x-3	18x+3	54x-27	54x-9
1	9	15	21	27	45
	3	5	7	9	15
	1			3	5
				1	

Answer:

243	81	27	9	3	1
	3	3	3	3	3

243= 3⁵

Figure 150

The equation to use is $54x - 27$. Our first factors are now 81, 27 and 9, and this gives us the last entry of 9. Therefore we take the last entry and hold 81 and 27. We transfer 9 back to the control, and it tells us that we should use the equation $54x - 45$ which gives us the factors 9, 3, 1. The answer then is given as *243 = 3*^{5}.

12.4.1.2 Class 2A

For example the set Class 2A has the following definition.

UNDT Equation	Class	Structure	Examples
$6x$	2A	$2^n3^m[P]$	a) $3402 = 2 \cdot 3 \cdot 3 \cdot 3 \cdot 3 \cdot 3 \cdot [7]$ b) $464070 = 2 \cdot 3 \cdot [5 \cdot 31 \cdot 499]$

Table 329

Therefore we expect the format 2^n3^m for such elements. We therefore divide by 2 and by 3. From this we derive numbers which we graph to get a straight line equation. The numbers 6,18 and 30 are in the same group, and 12, 24 and 36 the same applies. We connect them through an equation. In this case the equations are $12x - 6$ and $12x$. These are the control equations for Class 2A.

$12x - 6$	$12x$
6	12
18	24
30	36

Table 330

Class 2A			
0	Div by 2	Div by 3	Div by 2
6	3	1	0.5
12	6	2	1
18	9	3	1.5
24	12	4	2
30	15	5	2.5
36	18	6	3
42	21	7	3.5
48	24	8	4
54	27	9	4.5
60	30	10	5
66	33	11	5.5
72	36	12	6
78	39	13	6.5
84	42	14	7
90	45	15	7.5

12x-6	12x
6x-3	6x
2x-1	2x
	x

x	12x-6	6x-3	2x-1
1	6	3	1
2	18	9	3
3	30	15	5
4	42	21	7
5	54	27	9
6	66	33	11
7	78	39	13
8	90	45	15

Figure 151

When $12x - 6$ is divided by 2 and by 3, we get $6x - 3$ and $2x - 1$ where these form the calculation equations. These relationship are illustrated by the graphic. The equation control equation $12x$ will derive the other set of numbers which will also cascade as $12x, 6x, 2x$ and x as calculation equations.

x	12x	6x	2x	x
1	12	6	2	1
2	24	12	4	2
3	36	18	6	3
4	48	24	8	4
5	60	30	10	5
6	72	36	12	6
7	84	42	14	7
8	96	48	16	8

Table 331

For $12x - 6$, the transfer will be define by $2x - 1$ which is class 1A, and for $12x$, the transfer will be defined by x. Since x produces all possible numbers, it implies Class 2A will transfer to any class depending on the value of x. In terms of the factorization process, what will happen is that the last number of the x variable is transferred to the relevant set structure for further factorization. So some of them will go back to Class 2A where this will derive more factors of two and three again.

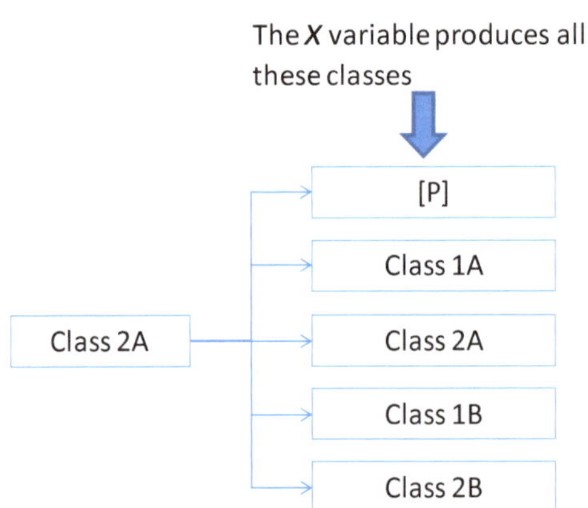

Figure 152

Each class has a different transfer definition depending on its equation structure.

12.4.1.3 Class 1B

We note the following pattern numbers to derive three straight lines, giving the control equations for Class 1B. That is, 4-28-52, 10-22-34 and 16-40-64 all form the same group which we can define by a straight line.

$24x - 20$	$12x - 2$	$24x - 8$
4	10	16
28	22	40
52	34	64

Table 332

This class transfers to GB4, GB2, and GB2 and Class 2B for the equations $24x - 20$, $12x - 2$ and $24x - 8$ respectively.

X	Class 1B	Div by 2	Div by 2	Div by 2
1	4	2	1	0.5
2	10	5	2.5	1.25
3	16	8	4	2
4	22	11	5.5	2.75
5	28	14	7	3.5
6	34	17	8.5	4.25
7	40	20	10	5
8	46	23	11.5	5.75
9	52	26	13	6.5
10	58	29	14.5	7.25
11	64	32	16	8
12	70	35	17.5	8.75
13	76	38	19	9.5
14	82	41	20.5	10.25
15	88	44	22	11
16	94	47	23.5	11.75
17				
18				
19				
20				

24x-20	12x-2	24x-8
12x-10	6x-1	12x-4
6x-5		6x-2
		3x-1
GB4	GB2	GB2/2B

x	24x-20	12x-10	6x-5
1	4	2	1
2	28	14	7
3	52	26	13
4	76	38	19
5	100	50	25
6	124	62	31
7	148	74	37
8	172	86	43
9	196	98	49
10	220	110	55
11	244	122	61
12	268	134	67
13	292	146	73

Table 333

12.4.1.4 Class 2B

We apply the same procedure to get calculating equations for class 2B. In this class we divide by 2 only.

12x-10	48x-40	24x-4	48x-16
2	8	20	32
14	56	44	80
26	104	68	128

Table 334

The groups are 2-14-26, 8-55-104, 20-44-68, and 32-80-128. These are many compared to the other classes. Here we transfer to GB2, GB2, GB4 and GB2/2B for the equations $12x - 10$, $48x - 40$, $24x - 4$ and $48x - 16$ respectively.

X	Class 2B	Div by 2	Div by 2	Div by 2	Div by 2
0	2	1	0.5	0.25	0.125
1	8	4	2	1	0.5
2	14	7	3.5	1.75	0.875
3	20	10	5	2.5	1.25
4	26	13	6.5	3.25	1.625
5	32	16	8	4	2
6	38	19	9.5	4.75	2.375
7	44	22	11	5.5	2.75
8	50	25	12.5	6.25	3.125
9	56	28	14	7	3.5
10	62	31	15.5	7.75	3.875
11	68	34	17	8.5	4.25
12	74	37	18.5	9.25	4.625
13	80	40	20	10	5
14	86	43	21.5	10.75	5.375
15	92	46	23	11.5	5.75
16	98	49	24.5	12.25	6.125
17	104	52	26	13	6.5
18	110	55	27.5	13.75	6.875
19	116	58	29	14.5	7.25
20	122	61	30.5	15.25	7.625
21	128	64	32	16	8
22	134	67	33.5	16.75	8.375
23	140	70	35	17.5	8.75
24	146	73	36.5	18.25	9.125
25	152	76	38	19	9.5

12x-10	48x-40	24x-4	48x-16
6x-5	24x-20	12x-2	24x-8
	12x-10	6x-1	12x-4
	6x-5		6x-2
			3x-1

	2B	1B	2B	1B	2B
X	**48x-16**	**24x-8**	**12x-4**	**6x-2**	**3x-1**
1	32	16	8	4	2
2	80	40	20	10	5
3	128	64	32	16	8
4	176	88	44	22	11
5	224	112	56	28	14
6	272	136	68	34	17
7	320	160	80	40	20
8	368	184	92	46	23
9	416	208	104	52	26
10	464	232	116	58	29
11	512	256	128	64	32
12	560	280	140	70	35
13	608	304	152	76	38
14	656	328	164	82	41
15	704	352	176	88	44
16	752	376	188	94	47
17	800	400	200	100	50
18	848	424	212	106	53

Table 335

We can summarize this result by creating a factorization path matrix that puts all the equations together.

12.5 The Factorization Path Matrix

Each equation in the matrix is defined in terms of the class and the factor class abbreviated FC.

The interesting thing about the factorization matrix is that in reality we do not even have to calculate the individual figures in the process of factorization because it is based on specific parameters. These are;

a) The factorization key, which is the number being multiplied
b) The power, which gives us the power on the key
c) The transfer, which tells us where to continue the process in which again we look for the key and the power.

For example, let there be a number in Class 2B. If the calculating equation is $48x - 40$, then;

a) The key is 2
b) The power is 3
c) And we transfer to GB2. In other words, there is no going back to the composite process, whatever the number!

What is key to know is the value of x. For example, 37 832 is 2^3 multiplying the base in GB2, and $x = 789$. This implies our number in GB2 is given by 6(789)-5 = 4729. Upon testing, we find this is a prime, so the answer is $2^3 \cdot 4729$.

Control Equations	Factor Class	Factor	Factor	Factor	Factor	Key	Power	Transfer to
Class 1A		Calculating Equations						
$54x-45$	FC-1	$18x-15$	$6x-5$			3	2	GB2
$18x-3$	FC-2	$6x-1$				3	1	GB4
$18x+3$	FC-3	$6x+1$				3	1	GB2
$54x-27$	FC-4	$18x-9$	$6x-3$	$2x-1$		3	3	GB2/GB4/1A
$54x-9$	FC-5	$18x-3$	$6x-1$			3	2	GB4
Class 1B								
$24x-20$	FC-1	12x-10	6x-5			2	2	GB4
$12x-2$	FC-2	$6x-1$				2	1	GB2
$24x-8$	FC-3	$12x-4$	$6x-2$	$3x-1$		2	3	GB4/2B
Class 2A								
$12x-6$	FC-1	$6x-3$	$2x-1$			2,3	1	GB2/GB4/1A
$12x$	FC-2	$6x$	$2x$	x		2,3	2,1	All Classes
Class 2B								
$12x-10$	FC-1	$6x-5$				2	1	GB2
$48x-40$	FC-2	$24x-20$	$12x-10$	$6x-5$		2	3	GB2
$24x-4$	FC-3	$12x-2$	$6x-1$			2	2	GB4
$48x-16$	FC-4	$24x-8$	$12x-4$	$6x-2$	$3x-1$	2	4	GB4/2B
Class [P]		**Base**	**Base**					
$x_3 = 6x_1x_2-(x_1-x_2)$		GB2	GB4					[P]
$x_3 = 6x_1x_2-(x_1+x_2)$		GB4	GB4					[P]
$x_3 = 6x_1x_2+(x_1+x_2)$		GB2	GB2					[P]

Table 336

The approach is illustrated by the following example for $W = 353\ 647$ in class 1A. No other calculations are necessary.

Enter W		**352647**				
	x=	58775	**Key**	**Power**	**Transfer**	
	FC-1	6531.333	3	2	39183	GB2
	FC-2	19591.67	3	1	117549	GB4
	FC-3	19591.33	3	1	117549	GB2
	FC-4	6531	3	3	13061	GB2/GB4/1A
	FC-5	6530.667	3	2	39183	GB4

Table 337

This can make computer programming very fast. But what is interesting is that no matter the size of the number, some of them will transfer only to [P]. The key is 3^3 and the base is 13061 in GB4.

12.5.1 Examples SFM Approach

Example 1

There are three types of equations used for the SFM approach;

a) the classifying equations
b) the control equations
c) and the calculating equations.

Since for composites we use factorization paths, we do not make use of the classifying equations, as the example shows that the control equations are enough to classify a number as well as determine its factorization path.

Consider $W = 848$. Only one control equation will give an integer value, and that indicates the class of the number. In this case it is Class 2B, and corresponds to the classifying equation. That is, we get 141 to indicate that the number is in class 2B, or 18. The advantage of 18 is that it also indicates the calculating equation. This means we can actually skip the classifying equations if necessary, it is just that then one has to test all the control equations and they are many. It is probably faster to include the classifying equation.

Classifying Equations

		Class 1A	Class 1B	Class 2A	Class 2B	GB4	GB2
		6x-3	6x-2	6x	6x+2	6x-1	6x+1
W	848	141.8333	141.667	141.3333	141	141.5	141.167

W value	848

Control Equations

Class 1A	54x-45	18x-3	18x+3	54x-27	54x-9
	16.53704	47.27778	46.9444	16.2037	15.87037

Class 1B	24x-20	12x-2	24x-8
	36.16667	70.83333	35.6667

Class 2A	12x-6	12x
	71.16667	70.66667

Class 2B	12x-10	48x-40	24x-4	48x-16
	71.5	18.5	35.5	18

GB2	6x+1
	141.1667
GB 4	6x-1
	141.5

Calculating Equations

848	48x-16	Power
424	24x-8	2
212	12x-4	2
106	6x-2	2
53	3x-1	Transfer GB4

Table 338

Example 2

Consider the example $W = 4432$ in class 1B.

Control Equations

W	12x-2	24x-8	24x-20
4432	369.5	185	185.5

Calculating Equations

x	12x-2	24x-8	24x-20
185	2218	4432	4420
	1109	2216	2210
		1108	1105
		554	
	GB4	GB2/2B	GB4

x	24x-8	Power
185	4432	2
	2216	2
	1108	2
	554	
	GB2/2B	

Table 339

The power is 2^3. This is then transferred to Class 2B as follows using 554.

Control Equations

W	12x-10	24x-4	48x-16	48x-40
554	47	23.25	11.875	12.375

Calculating Equations

x	12x-10	24x-4	48x-16	48x-40
47	554	1124	2240	2216
	277	562	1120	1108
		281	560	554
			280	277
			140	
	GB2	GB4	GB2/2B	GB2

x	12x-10	Power
47	554	2
	277	GB2

Table 340

The power is 2^1. Another transfer takes place to gap base 2 of the number 277. At x_2 = 10, this proves to be a prime number as x_1 is less than one.

x_1	0.949152542	GB2
	0.590163934	GB2
	0.615819209	GB4

Table 341

This ends our factorization process, and we have; $2^4 \times 277 = 4432$.

Example 3

Let $W = 4698$, which is in Class 2A.

Control Equation

W	12x-6	12x
4698	392	391.5

Calculating Equation

x	12x-6	12x
392	4698	4704
	2349	2352
	783	784
		392
Transfer	1A/1B/2A/2B/GB2/GB4	1A/1B/2A/2B/GB2/GB4

Table 342

The power is shown as follows.

x	12x-6	Power
392	4698	2
	2349	3
	783	Transfer
	1A	

Table 343

We then transfer 783 to Class 1A equations.

Control Equation

W	54x-45	18x-3	18x+3	54x-27	54x-9
783	15.33333	43.66667	43.33333	15	14.66667

Calculating Equation

X	54x-45	18x-3	18x+3	54x-27	54x-9
15	765	267	273	783	801
	255	89	91	261	267
	85			87	89
				29	
	GB2	GB4	GB2	GB4	GB4

Table 344

The power is derived as follows.

X	54x-27	Power
15	783	3
	261	3
	87	3
	29	Transfer

Table 345

Since 29 is prime, therefore $4698 = 2 \times 3^3 \times 29$.

Each time the process is repeated, the same path will be followed, hence the concept of factorization paths. As one will notice, the procedure is exact and precise, there is no guess work.

Predictive Factorization

The section covers further application of the UNDT framework to describe other ways of factorization.

13 The SFM Tree

There are three types of product operations; those that involve the set [P], the set [C], and the set [C]x[P]. There are only two types of numbers to factorize, a composite or a pseudoprime. The set [C_0] is a composite only factorization, while the set [P] is a pseudoprime factorization, and [C_0]x[P] is a combination. So even in this case we have a structured factorization methodology.

Factor	Factor	Product	Set Type
Composite	Composite	Composite	[C_0]
Prime/Pseudoprime	Prime/Pseudoprime	Pseudoprime	[P]
Composite	Prime/Pseudoprime	Composite	[C_0]x[P]

Table 346

The first factorization approach concerned the set [P], and the factorization paths involved the set [C_0]. By using the method of transfer, one can factor a number that belongs to [C_0]x[P] by using the technique for [C_0] followed by the technique for [P]. The method of transfer also takes place within the set [C_0].

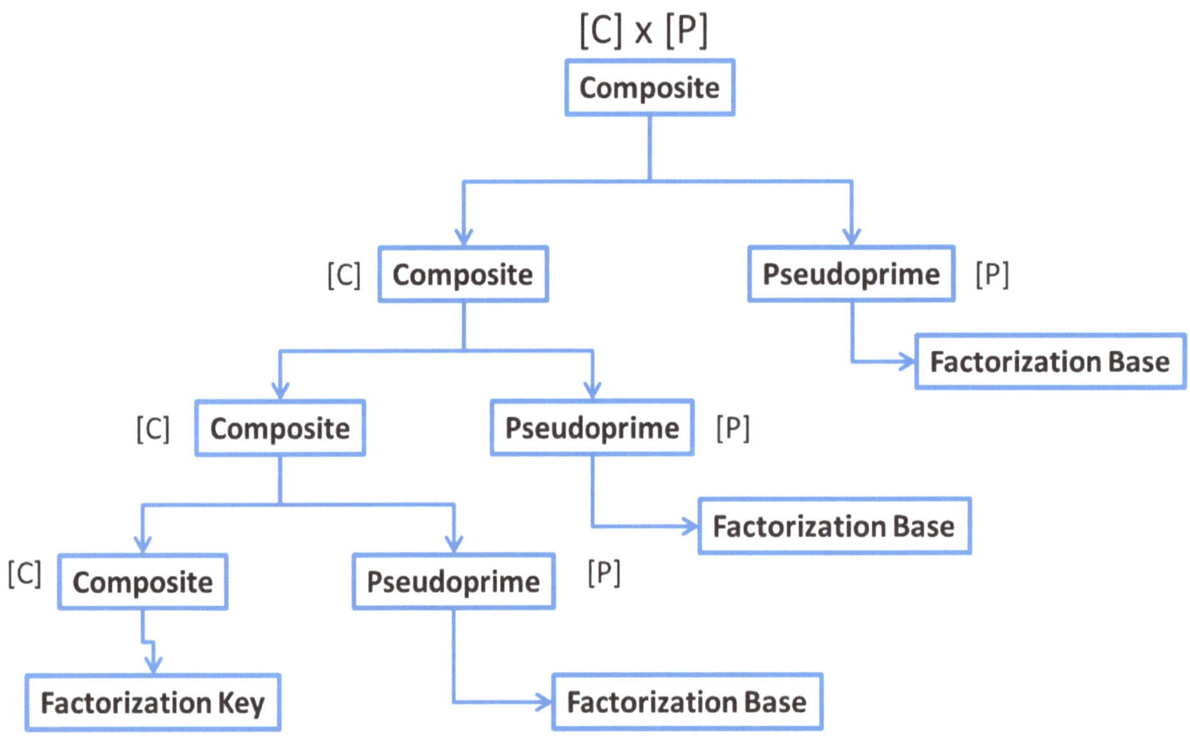

Figure 153

The aim in all these operations is to derive the factorization key and factorization base. In considering the set [C_0]x[P] = [C_0], a procedure can be developed in the form of an SFM tree as shown by the diagram. In this case we develop a structure that derives two types of factors, a composite from [C_0] and a pseudoprime from [P]. We can keep on breaking down an element of [C_0]x[P] in a repeated from until we get the prime factors that we seek. This should create more efficiency in the search for factors.

We use the same technique that was applied in the set [P] when looking for factors. The difference is that we now combine the set [P] and [C_0] to form equations that perform a trace. We define the set [C_0] to take Factor-1 and the set [P] the Factor-2 by default and determine the class of W. As usual, the first step is to classify the number and determine the value of x_3. The same approach is used as in the case of the set [P], where we vary the variable x_2 to get x_1. The value of x_1 must be an integer, the same being true for Factor-1 and Factor-2.

Factor-1	**Factor-2**	**W**	
Class	**Gap Base**	**Class Result**	**Equation for X_1**
Class 1A	GB-2	Class 1A	$x_1 = (x_3 + 3x_2)/(6x_2 + 1)$
Class 2A	GB-2	Class 2A	$x_1 = (x_3)/(6x_2 + 1)$
Class 1B	GB-2	Class 1B	$x_1 = (x_3 + 2x_2)/(6x_2 + 1)$
Class 2B	GB-2	Class 2B	$x_1 = (x_3 - 2x_2)/(6x_2 + 1)$
Class 1A	GB-4	Class 1A	$x_1 = (x_3 + 3x_2 - 1)/(6x_2 - 1)$
Class 2A	GB-4	Class 2A	$x_1 = (x_3)/(6x_2 - 1)$
Class 2B	GB-4	Class 1B	$x_1 = (x_3 - 2x_2)/(6x_2 - 1)$
Class 1B	GB-4	Class 2B	$x_1 = (x_3 + 2x_2)/(6x_2 - 1)$

Table 347

In a similar manner we can determine the maximum value of x_2 when searching for factors by setting $x_1 = 1$. This does not create the same synergy as in the previous set [P] since here for example $x_1 = (x_3 + 3)/7$ for the case Class 1A and GB2.

Factor-1	**Factor-2**	**W**		
Class	**Gap Base**	**Class Result**	**Equation for maximum X_2**	**Approximate value of W/X_2**
Class 1A	GB-2	Class 1A	$x_2 = (x_3 - 1)/3$	18 times
Class 2A	GB-2	Class 2A	$x_2 = (x_3 - 1)/6$	36 times
Class 1B	GB-2	Class 1B	$x_2 = (x_3 - 1)/4$	24 times
Class 2B	GB-2	Class 2B	$x_2 = (x_3 - 1)/8$	48 times
Class 1A	GB-4	Class 1A	$x_2 = x_3/9$	54 times
Class 2A	GB-4	Class 2A	$x_2 = (x_3 + 1)/6$	36 times
Class 2B	GB-4	Class 1B	$x_2 = (x_3 + 1)/8$	48 times
Class 1B	GB-4	Class 2B	$x_2 = (x_3 + 1)/4$	24 times

Table 348

Class 1A is very fast when a product involves GB4, but very slow when a product involves GB2. Class 1B is slower with GB2 but fairly fast with GB4, while class 2B is slow with GB4 but faster with GB2. Only class 2A has the same speed with both gap bases.

So how would this work practically? It would do much the same way as for factoring elements of [P].

a) First of we classify the number by deriving x_3
b) Then we vary the value of x_2 starting at 1.
c) Then we look for integer values of x_1. This will define our factor find.

For example, take the number 400. We note that it is in class 1B, and $x_3 = 67$.

	Class 1A	Class 1B	GB-4	Class 2A	GB-2	Class 2B
x	6x-3	6x-2	6x-1	6x	6x+1	6x+2
W	(W+3)/6	(W+2)/6	(W+1)/6	W/6	(W-1)/6	(W-2)/6
400	67.16667	67	66.83333	66.66667	66.5	66.33333

Table 349

We then substitute this into the equation matrix. We get our hit when x_2 =4, then x_1 is an integer. Our maximum value of x_2 is 22, where this is approximately (400/18) = 22.22.

X3	X2
67	4

Factor-1	Factor-2	W	X1	Factor-1	Factor-2	X2 Check
Class 1A	GB-2	Class 1A	3.16	15.96	25.06266	4.010443
Class 2A	GB-2	Class 2A	2.68	16.08	24.87562	3.97927
Class 1B	GB-2	Class 1B	3	16	25	4
Class 2B	GB-2	Class 2B	2.36	16.16	24.75248	3.958746
Class 1A	GB-4	Class 1A	3.391304	17.34783	23.05764	4.009607
Class 2A	GB-4	Class 2A	2.913043	17.47826	22.88557	3.980929
Class 2B	GB-4	Class 1B	2.565217	13.3913	29.87013	5.145022
Class 1B	GB-4	Class 2B	3.26087	21.56522	18.54839	3.258065

Table 350

The structure achieves our aim which is to separate a composite into a composite and prime or pseudoprime right from the word go as to create efficiency. Note the last column that does a check for consistency by deriving x_2 again. As shown by the diagram, this has a tree approach, hence the name SFM tree.

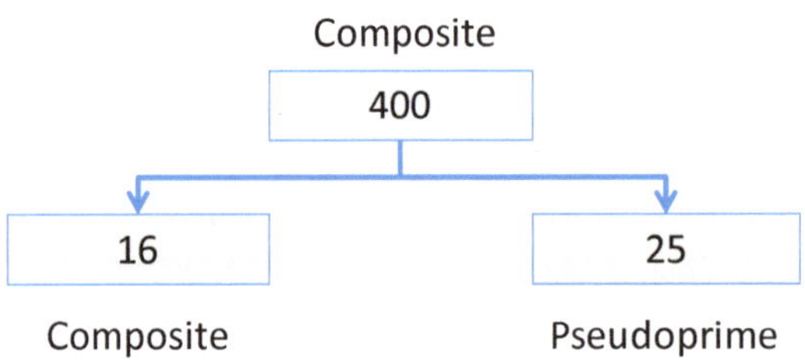

Figure 154

Furthermore, when $x_1 < 1$ when x_2=17, just a little lower than 22. Remember that this is an approximation, so the accurate value is 17. The reason why we have the last column as a check is because some products are valid but not their classification. In this case we get 400 = 8×50 but this would classify as class 2B which is incorrect. By noting the value of x_2 to be decimal, we know that this is not correct.

				X3	X2	X2
Factor-1	Factor-2	W	X1	67	17	Check
Class 1A	GB-2	Class 1A	1.145631	3.873786	103.2581	17.04302
Class 2A	GB-2	Class 2A	0.650485	3.902913	102.4876	16.91459
Class 1B	GB-2	Class 1B	0.980583	3.883495	103	17
Class 2B	GB-2	Class 2B	0.320388	3.92233	101.9802	16.83003
Class 1A	GB-4	Class 1A	1.158416	3.950495	101.2531	17.04219
Class 2A	GB-4	Class 2A	0.663366	3.980198	100.4975	16.91625
Class 2B	GB-4	Class 1B	0.326733	-0.0396	-10100	-1683.17
Class 1B	GB-4	Class 2B	1	8	50	8.5

Table 351

Consequently, this means no further factors besides 16 and 25 can be found to satisfy the condition of being composite and pseudoprime or prime. The reason is simple really, it because 16 is not a product of the form

$[C_0]\times[P]$, its an element of $[C_0]$ only. However, these can be factored further by transfer where 16 is factored using the technique for composites, while 25 would use the technique for the set [P].

We now consider the case $W = 3465$. This is a composite number. The number could be factored using the factorisation path approach. We are going to use this approach first and then compare it to the tree method. Whatever approach we use, the first step is to classify the set in which the number belongs. From the classification equations, we note that it belongs to Class 1A.

	Class 1A	Class 1B	GB-4	Class 2A	GB-2	Class 2B
x	6x-3	6x-2	6x-1	6x	6x+1	6x+2
W	(W+3)/6	(W+2)/6	(W+1)/6	W/6	(W-1)/6	(W-2)/6
3465	578	577.8333	577.6667	577.5	577.3333	577.1667

Table 352

We then input in Class 1A equations, starting with the control equations.

Input	54x-45	18x-3	18x+3	54x-27	54x-9
3465	65	192.6667	192.3333	64.66667	64.33333

Table 353

This tells us that we should use the equation $54x - 45$ for $x = 65$. This gives us the following factorization process.

X	54x-45	Factors
65	3465	3
	1155	3
	385	
Transfer	GB2	

Table 354

For the number 385, the transfer is to GB2.

	x_2	x_1	Factor-1	Factor-2
W=385	1	13	77	5
	1	9	55	7
	2	6	35	11
	6	2	11	35
	9	1	7	55
	13	1	5	77

Table 355

Hence the result is $3^2 \times 5 \times 7 \times 11 = 3465$.

Now we do the same process but using the SFM tree approach. Since the number is a composite, the first step in this approach is to find out if there are any factors of [P] that can be identified before doing a composite only factorization. For W = 3465, then $x_3 = 578$. We input this and let $x_2 = 1$.

X3	X2
578	1

Factor-1	Factor-2	X3	X1	Factor-1	Factor-2	X2 Check
Class 1A	GB-2	Class 1A	83	495	7	1
Class 2A	GB-2	Class 2A	82.57143	495.4286	6.993945	0.998991
Class 1B	GB-2	Class 1B	82.85714	495.1429	6.99798	0.999663
Class 2B	GB-2	Class 2B	82.28571	495.7143	6.989914	0.998319
Class 1A	GB-4	Class 1A	116	693	5	1
Class 2A	GB-4	Class 2A	115.6	693.6	4.995675	0.999279
Class 2B	GB-4	Class 1B	115.2	689.2	5.027568	1.004595
Class 1B	GB-4	Class 2B	116	698	4.964183	0.994031

Table 356

The matrix shows us that there are two ways of getting this result. That is;

- 3465 = 7×495
- 3465 = 5×693

Using $x_2 = (x_3 - 1)/3$, then we can find all the factors when $x_2 \leq 193$. However, we find all factors by $x_2 = 64$, but this is not very useful for deciding which factor pairs to use. Hence we try the iteration technique to see if it provides a better approach.

W= 3465	x_1	x_2	Factor-1	Factor-2
	116	1	693	5
	83	1	495	7
	53	2	315	11
	17	6	99	35
	11	9	63	55
	8	13	45	77
	2	64	9	385

Table 357

This gives us the following result.

Option A	x_1	x_2	Factor-1	Factor-2
W=3465	116	1	693	5
W=693	17	1	99	7
W=99	2	2	9	11

Option B	x_1	x_2	Factor-1	Factor-2
W=3465	1	1	495	7
W=495	17	1	99	5
W=99	2	2	9	11

Table 358

The 9 will be factored by transfer to give 3^2. This gives us:

X	54x-45	Power
1	9	3
	3	3
	1	
	GB2	

Table 359

This is much faster, and gives us the required format. When we enter 9 into the structure, it becomes decimal showing that 9 belongs to a composite construction only, that is class 1A. It does not matter whether we use Option A or B, we get the same result. So again we have $3465 = 3^2 \times 5 \times 7 \times 11$.

Let's pick up the number $W = 9020$, then this gives us the following using the SFM approach. The number is in class 2B.

W= 9020	x1	x2	Factor-1	Factor-2
	9	27	164	55
	34	7	44	205
	75	3	20	451

Table 360

We can use any of the three results, and if we choose 164, then we have;

x3	Equations	Factors	
7	24x-4	164	Power
	12x-2	82	2
	6x-1	41	2
		GB4	

Table 361

This transfers to GB2, but we know 41 is prime. When we test 55, its not prime, and comes to 55 = 5×11, hence we have for our result 9020 = 2^2×5×11×41.

For another example, consider the number W = 89756 picked up at random.

W= 89756	x1	x2	Factor-1	Factor-2
	787	3	4724	19

Table 362

When we test using the SFM tree, this shows that the number is a pure composite, hence we resort to the composite test. The classifying equation shows that it belongs to class 2B.

	Class 1A	Class 1B	GB-4	Class 2A	GB-2	Class 2B
x	6x-3	6x-2	6x-1	6x	6x+1	6x+2
W	(W+3)/6	(W+2)/6	(W+1)/6	W/6	(W-1)/6	(W-2)/6
4724	787.8333	787.6667	787.5	787.3333	787.1667	787

Table 363

The control equation identifies $24x - 2$.

12x-10	48x-40	24x-4	48x-16
394.5	99.25	197	98.75

Table 364

This then derives the following result.

x_3	Equations	Factors	
197	24x-4	4724	Power
	12x-2	2362	2
	6x-1	1181	2
		GB4	

Table 365

The transfer to GB4 procedure shows that 1181 is a prime number. Therefore we have the conclusion that 89756 = $2^2 \times 19 \times 1181$. The example uses all the techniques developed.

13.1 Small-Big Factorization Pattern

The pattern of a small prime number and big prime number as factors of a number is a distinctive behavior in factorization, especially of big integers. The following is noted about the model.

a) The pattern model developed here is based on the premise of the algebraic sieve under UNDT since pseudoprimes are given by $f(x_a) \cdot f(x_b)$.

b) The minimum number of factors is two, and this gives us a first order pattern. The next pattern consists of three factors and so on. If the order of factorisation is given by n, then the number of possibilities at a given order is $(n+2)$, and the number of factors by $(n+1)$.

c) If $n = 0$, then we have a prime number, if we consider 1 as a factor that is a trivial result.

d) The prime factors are consider as being either small or big. Therefore there is no medium prime sized factor. This is given by the following definition, constructed on a heuristic basis.

Definition

a) Let p_1 be the smallest prime factor. Then if $p_i > 10p_1$, then p_1 is a small prime factor and p_i is a big prime factor.

b) If $p_1 > 100\ 000$, then p_1 is classified a big prime.

Part b) of the definition is fixed and covers the situation where both prime factors are big, since part a) of the definition is based on a comparative scale. The factors are of type S for small, B for big and M for mixed, that is having both small and big.

This then gives us a way of classifying prime factors of a given number. Since the parameters of the model are arbitrary, it is very open to improvement.

Prime Factor Pattern

				Small	
2				Big	zero order pattern
			Small	Small	
3			Small	Big	1st order pattern
			Big	Big	
		Small	Small	Small	
		Small	Small	Big	
4		Small	Big	Big	2nd order pattern
		Big	Big	Big	
	Small	Small	Small	Small	
	Small	Small	Small	Big	
5	Small	Small	Big	Big	3rd order pattern
	Small	Big	Big	Big	
	Big	Big	Big	Big	

Figure 155

An example is given by the table below.

Number	Factors	Order	Type
26,161,157	11, 19, 41, 43, 71	4	S
7,007	7, 7, 11, 13	3	S
831,517	29, 53, 541	2	M
67,889,003	7, 13, 746033	2	M
78,902,135	5, 13, 1213879	2	M
90,007,123	19, 4737217	1	M
3 ,534,120,210,161	746033, 4737217	1	B

Table 366

Consider the number 831 517. Hence 541 is a big prime on a comparative scale since 29x10 = 290. The only number greater than 290 is 541, hence it is the only big prime. Then 29 and 53 are small prime factors. Similarly for the number 26,161,157, then 11x10 = 110, and all the prime factors are less than 110, hence these are small prime factors.

We can also conjecture as follows.

Conjecture

The population of prime factors in a given range [a,b] is such that $n_j < n_i$ *where* $j > i$*, and* n_j*,* n_i *are the number of factors.*

That is, we expect the prime factor occurrence to be less and less as the order increases in a specific given range on the number line. This is based on the premise that the next order is a repeat multiplication of the previous order, and this pushes the next order much higher in the number line. Consider the pseudoprime numbers from 26 161 091 to 26 161 325, then the following frequencies are recorded.

Number	Prime Factors					Number	Prime Factors			
26161091	11	293	8117			26161211	Prime			
26161097	23	1137439				26161217	Prime			
26161103	29	419	2153			26161223	47	53	105023	
26161109	13	37	137	397		26161229	73	358373		
26161115	5	5232223				26161235	5	23	227489	
26161121	7	101	37003			26161241	31	843911		
26161127	4817	5431				26161247	7	7	331	1633
26161133	641	40813				26161253	61	428873		
26161139	653	40063				26161259	Prime			
26161145	5	5232229				26161265	5	13	277	1453
26161151	Prime					26161271	19	149	9241	
26161157	11	19	41	43	71	26161277	29	53	17021	
26161163	7	757	4937			26161283	17	89	17291	
26161169	Prime					26161289	7	11	11	67
26161175	17	1538893				26161295	5	881	5939	
26161181	Prime					26161301	2459	10639		
26161187	13	47	47	911		26161307	Prime			
26161193	1223	21391				26161313	1237	21149		
26161199	109	240011				26161319	1621	16139		
26161205	5	7	747463			26161325	5	5	599	1747

Table 367

This is summarized as a frequency table.

Number of Prime Factors	Frequency
1	7
2	15
3	11
4	5
5	1

Table 368

One can explain it simply by saying small primes are also factors of big numbers, but it attracts interest because of its regular and particular characteristic. That is, we do not see a regular pattern of large integers having many small prime factors which should be the case as the integers get larger.

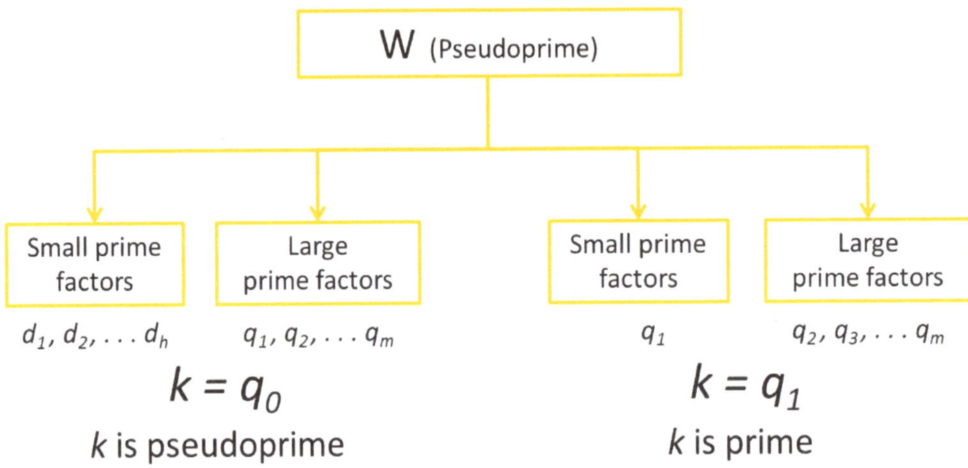

Figure 156

The intrigue that is provided by the UNDT framework is the fact that it is partly able to explain the pattern of small prime multiplying big primes. It gives us an understanding as to why it happens. Both $k = q_0$ and $k = q_1$ provide that possibility. The model does not include composite numbers, but this is easy to provide because all composites can be factored as 2^n, 3^m, $2^n 3^m$, where these are referred to as composite factors under UNDT. As can be observed, these are already small in size. We can take the total picture by including the composite group of integers.

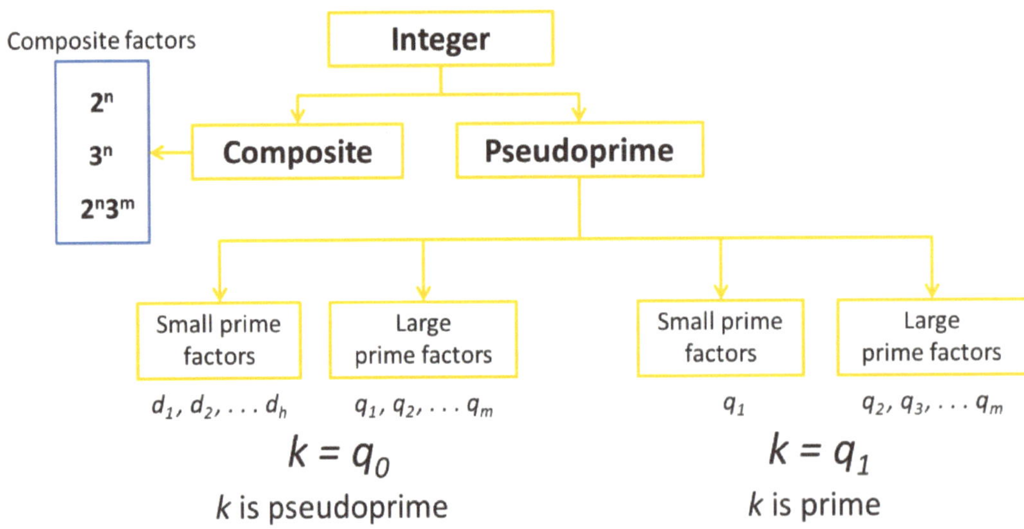

Figure 157

Basically what we note then is the fact that the Euclidean condition can be exploited to yield interesting insights into prime number behavior, as well as providing an anatomy of prime factors and composite factors of a given integer. We have an exact pattern of how we expect prime numbers to behave since they are doing so within a constraint called the algebraic sieve. This structure that has been derived is actually a consequence of the law of pseudoprimes $f(x_a) \cdot f(x_b)$ in action.

13.2 Merge Operator Arithmetic

Prime factors have also shown a surprising regularity of behavior when it comes to repeatable patterns. All sorts of descriptions have been found to put prime numbers in some form of behavior pattern by grouping them according to a particular trait. Prime numbers with repeating patterns are conventionally referred to as palindromes, as shown in the example below. Now palindromes are prime numbers, but the UNDT framework includes pseudoprimes in its repeat process. Therefore the numbers on the right are not necessarily prime numbers; in some cases they are pseudoprime. Our interest lies in the fact that they are not prime. It is also obvious that by not using the context of the UNDT framework, this aspect of analysis may not look relevant, and I picked up interest in the pattern because of the observation that the resulting numbers were pseudoprime.

Palindromes	Repeat Pattern	Type
11	1111	Pseudoprime
101	101101	Pseudoprime
10301	1030110301	Pseudoprime
1003001	10030011003001	Pseudoprime
100030001	100030001100030001	Pseudoprime

Table 369

The above result is not confined to palindromes. Any prime number produces the same result, either another prime or a pseudoprime. The operations occurs using elements of the set D_t, hence it is a closed operation.

Prime	Repeat Pattern	
61/	617617	Pseudoprime
727	727727	Pseudoprime
2609	26092609	Pseudoprime
3011	30113011	Pseudoprime
1201969	12019691201969	Pseudoprime
1205231	12052311205231	Pseudoprime

Table 370

If we take elements of the Class 1A, then we have a similar result. The operation is not particular to prime numbers.

Class 1A Composite	Class 1A Composite Repeat
6513	65136513
18879	1887918879
42003	4200342003
47691	4769147691
47997	4799747997

Table 371

There is one class that is an exception to being a closed set operation, and that is Class 1B. When we repeat a Class 1B element, this gives us Class 2B element. Similarly, for Class 2B in a repeat operation results in a Class 2A element. Hence only Class 1A, Class 2A and the set D_t have a closed operation.

	Composite	Composite Repeat	
Class 1B	17998	1799817998	Class 2B
	18088	18008818088	
Class 2A	20088	2008820088	Class 2A
	20412	2041220412	
Class 2B	21120	2112021120	Class 2A
	22092	2209222092	

Table 372

Since this is a systematic operation, it can have a specific symbol $\oplus$ to denote it, as well as a name. This actually leads to a new type of hybrid addition and product operation, where the result is always bigger.

Definition

Let there be an operator $\oplus$ *such that* $\alpha_1 \oplus \alpha_2 = \beta$ *denotes a merge operation, where* α_1*,* α_2 *are prime, pseudoprime or composite.*

Therefore when $\alpha_1 = \alpha_2$, then this is called a repeat. Consider a combination of a composite and a pseudoprime in the merge operation. In this case, there is no repeat but a merge with another number from a different class.

	Composite		**Prime/Pseudoprime**		**Merge**	**Result Type**
Class 1A	3045		3103		30453103	Pseudoprime
Class 1B	3070		3103		30703103	Prime
Class 2A	3102		3103		31023103	Prime
Class 2B	3002		3103		30023103	Class 1A
	Prime/Pseudoprime		Composite	Type	Merge	Result Type
	3103		3045	Class 1A	31033045	Pseudoprime
	3103		3070	Class 1B	31033070	Class 2B
	3103		3102	Class 2A	31033102	Class 1B
	3103		3002	Class 2B	31033002	Class 2A

Table 373

Therefore using the new notation, the following consistent results are observed in terms of class and type. Indeed this is strikingly similar in structure to the product rules between the classes, but the results do differ. The fact that the numbers used in the example are the same size is not responsible for the consistent result. Different values of digits still yield the same consistent outcome. For example 39$\oplus$40=3940, then 39 is Class 1A merges with 40 in Class 1B, and the result is a 1B result.

13.2.1 Merge Arithmetic Rules

Merge arithmetic is considered for $\alpha_1 = \alpha_2$, and for $\alpha_1 \neq \alpha_2$. The rules defined are heuristic and are constructed on the basis of the cyclic transform properties of the prime numbers. The rule is valid on the basis of the defined result. For example consider the following table for 4097731 a GB2 number, we have the following. The second result which does not conform to the expected outcome is referred to as an anomaly.

α_1		α_2		
409	GB2	7731	1A	GB2⊕ CLASS1A= GB2
4097	GB4	731	GB4	GB4⊕GB4 = GB2
40977	1A	31	GB2	Class 1A⊕ GB2= GB2

Table 374

1. Class 1A⊕Class 1A = Class 1A
2. Class 1B⊕Class 1B = Class 2B
3. Class 2A⊕Class 2A = Class 2A
4. Class 2B⊕Class 2B = Class 1B

5. Class 1A⊕Class 1B = Class 1B
6. Class 1B⊕Class 1A = GB2
7. Class 1A⊕Class 2A = Class 2A
8. Class 1A⊕Class 2B = Class 2B
9. Class 1B⊕Class 2A = 1B
10. Class 1B⊕Class 2B = 2A
11. Class 2A⊕Class 1A = Class 1A
12. Class 2A⊕Class 1B = 1B
13. Class 2B⊕Class 1A = GB4
14. Class 2B⊕Class 1B = 2A

15. Class 2B⊕Class 2A = Class 2B
16. Class 2A⊕Class 2B = Class 2B

17. Class 1A⊕GB2 = GB2
18. GB2⊕Class 1A = GB2
19. Class 1A⊕GB4 = GB4
20. GB4⊕Class 1A = GB4

21. Class 1B⊕ GB2= GB4
22. GB2⊕Class 1B = Class 2B
23. Class 1B⊕ GB4= Class 1A
24. GB4⊕Class 1B = Class 2A

25. Class 2A⊕ GB2= GB2
26. GB2⊕Class 2A = Class 1B
27. Class 2A⊕ GB4 = GB4
28. GB4⊕Class 2A = Class 2B

29. Class 2B⊕ GB2 = Class 1A
30. GB2⊕Class 2B = Class 1A
31. Class 2B⊕ GB4 = GB2
32. GB4⊕Class 2B = GB2

33. GB2⊕GB2= GB4
34. GB4⊕GB4 = GB2
35. GB2⊕GB4 = 1A
36. GB4⊕GB2 = 1A

Clearly ⊕ is a hybrid operator having properties of addition since you are joining two numbers, but also of a product because such a number becomes bigger. As a result of the consistency, it defines a specific type of mathematical outcome, but two or more operations are necessary to confirm a result if $\alpha_1 \neq \alpha_2$ to resolve the anomaly.

Unlike addition and multiplication, the operator is not commutative in terms of the outcome. However, it does have an identity operation as follows. Let $\alpha_1 \oplus \alpha_2 = \beta$, then;

a) $\emptyset \oplus \alpha_1 = \alpha_1 \oplus \emptyset = \alpha_1$, where $\emptyset$ is a null element
b) $0 \oplus \alpha_2 = \alpha_2$
c) $\alpha_2 \oplus 0 = \beta \in Class\ 1B, 2B, 2A$
d) $\alpha_1 \oplus \alpha_2 = \alpha_2 \oplus \alpha_1\ if\ and\ only\ if\ \alpha_1 = \alpha_2$, or $\alpha_1 and\ \alpha_2$ have the same digits

Note that zero is an element of 2A, and therefore will generate elements in Class 1B, 2B and 2A according to the merge operational rules. A combination of operations is also possible, for example;

$$(\alpha_1 \oplus \alpha_2) \oplus \alpha_3$$

This implies the inside bracket would be worked out first in order to arrive at the final result. Infact other possibilities with the usual arithmetic operations can also be established, for example;

$$(\alpha_1 \oplus \alpha_2) \oplus \alpha_3 + \alpha_1 \oplus (\alpha_2 \oplus \alpha_3)$$

defines a proper operation. There are rules that are defined for addition between the different classes, so the rules would follow for each operator type.

An opposite process can be defined that splits a number into two parts.

Definition

Let there be an operator ⊖ such that $\ominus\, \beta(n) = (\alpha_1, \alpha_2)$ denotes a truncation at point n, where α_1, α_2 are prime, pseudoprime or composite, and n is less than the total number of digits of β.

Therefore consider the following example of the number 5728190032447. Not only do we now know that it consists of the prime factors 34487 and 166097081, but it has other decomposition attributes as well that we can use to describe it.

Pseudoprime	α_1	Type	α_2	Type
5728190032447	57281	1A	90032447	Pseudoprime
5728190032447	572811900	2B	32447	Pseudoprime
5728190032447	572	2B	8190032447	Pseudoprime
5728190032447	572819003	Prime	2447	Prime
5728190032447	572819003244	Class 2A	7	Prime

Table 375

13.3 Alpha Factorization

A concept of factorization is developed that is particular to a given set of numbers. That is, not all number are described through this type of factorization. Any number that merges with itself defines a repeat merge according to the following definition.

Definition

Alpha numbers constitute an $\alpha_1$number and an α_2 number such that they are commutative in the $\oplus$ operation.

This means such numbers satisfy the condition that $\alpha_1 \oplus \alpha_2 = \alpha_2 \oplus \alpha_1$. The implication then is $\alpha_1 = \alpha_2$, or $\alpha_1 and\ \alpha_2$ have the same digit structure. For example 11 and 1111 will be commutative. Our interest here focusses on factorization of outcomes from the merge operation. The observations we make are:

a) If for example 657595 has 2 prime factors, then the merge operation results in a number that has more factors.
b) The prime factors of α_1 are repeated as prime factors of β. Since there is such a repetition of prime factors, this implies that the prime factors of β will always be more as a certainty.

Pseudoprime	Factors	No. of Factors	Alpha Number	Factors	No. of Factors
657595	5, 131519	2	657595657595	5, 101, 9901, 131519	4
443813	347, 1279	2	443813443813	101, 347, 1279, 9901	4
170093	7, 11, 47, 47	4	170093170093	7, 11, 47, 47, 101, 9901	6
777775	5, 5, 53, 587	4	777775777775	5, 5, 53, 101, 587, 9901	6
537907	167, 3221	2	537907537907	101, 167, 3221, 9901	4

Table 376

This is the first discovery that drew me to this pattern, it's intriguing because the number used is any random number. Secondly, it is intriguing because it is deterministic, since to predict that another number will have less prime factors of another number is unusual and unheard of in the current paradigm of mathematical framework. This is so because the bigger the prime number does not necessarily translate to a larger number of prime factors.

The condition $\alpha_1 = \alpha_2$ produces a very interesting result, expressed as a theorem.

Alpha Factor Theorem

Let α_1 have n prime factors, and $\alpha_1 \oplus \alpha_1$ have m prime factors, then for any α_1, $n < m$ such that all factors of $\alpha_1$are factors of β.

The theorem is also referred to as the factorization root law (FRL) because some numbers appear as common in the factorization process such as 101 and 9901.

In order to explore the concept of roots further, the alpha numbers are defined to have a particular characteristic that is consistent to them. We refer α_1 number as an alpha-1 number, and similarly an α_2 number as an alpha-2 number.

Definition

An α_1*number and an* α_2 *number consists of the following structures:*

a) *Uniform digit formation, where all the digits are the same.*
b) *Non-uniform digit formation, where a group of digits repeat and are distinct*
c) *Irregular digit formation, where there is no particular pattern to the digits.*

For example 444444 is a uniform digit formation, 57575757 is a non-uniform digit formation and 7789231 is an irregular digit formation. These numbers can be derived through a merge operation. The conditions for our example are as follows;

a) $\alpha_1 = \alpha_2$
b) α_1 is a prime number, pseudoprime or compoiste
c) α_1 is of uniform digits

The smallest alpha-1 or alpha-2 number is zero, but $0 \oplus 0 = 0$, that is, it is of no consequence. We can consider the next smallest alpha-1 number, that is 1, and $1 \oplus 1 = 11$. Therefore 11 is defined as the first alpha number, hence the terminology alpha number describes β. The operation continues as $11 \oplus 1 = 111$, and then we consider $11 \oplus 11$ and $111 \oplus 111$, and so on. Since 1 is the smallest alpha-1 or alpha-2 number that forms an alpha number, it is defined to form the set of fundamental alpha numbers, and the fundamental alpha numbers derive the fundamental alpha roots of factorization.

n	Type	Fundamental Alpha Number	Alpha Roots
1	GB2	1	1, 1
2	GB4	11	1, 11
3	Class 1A	111	3, 37
4	GB2	1111	11, 101
5	GB4	11111	41, 271
6	Class 1A	111111	3, 7, 11, 13, 37
7	GB2	1111111	239, 4649
8	GB4	11111111	11, 73, 101, 137
9	Class 1A	111111111	3, 3, 37, 333667
10	GB2	1111111111	11, 41, 271, 9091
11	GB4	11111111111	21649, 513239
12	Class 1A	111111111111	3, 7, 11, 13, 37, 101, 9901
13	GB2	1111111111111	53, 79, 265371653
14	GB4	11111111111111	11, 239, 4649, 909091
15	Class 1A	111111111111111	3, 31, 37, 41, 271, 2906161
16	GB2	1111111111111111	11, 17, 73, 101, 137, 5882353
17	GB4	11111111111111111	2071723, 5363222357
18	Class 1A	111111111111111111	3, 3, 7, 11, 13, 19, 37, 52579, 333667
19	GB2	1111111111111111111	Prime

20	GB4	11111111111111111111	11, 41, 101, 271, 3541, 9091, 27961

Table 377

This is the alpha set, or an alpha factorization table. Here are some of the properties of this set.

a) The roots do not appear to be related to any specific gap base.
b) It is not clear whether the prime numbers follow any sequence. For example 101 appears almost at the beginning. Others like 43 are not there and it is not clear if they will surface at higher values of the apha number.
c) At $n = 19$, we have an alpha number that is prime, which means the number is its own alpha root.
d) For n is even, then 11 will always occur as a factor.
e) If n is a multiple of 4, then 11 and 101 will always occur as prime factors.
f) For $n = 3$, we have factor 37 and the composite 3. All multiples of 3 in n will have these two factors.
g) Whenever 3 occurs, it implies the number is a composite.
h) The factor 7 occurs at $n = 6$, then at $n = 12$, and at $n = 18$. The pattern occurs in multiples of 6.
i) For 271, it appears at $n = 5$, then 10, 15, and 20.
j) We observe that the Factor Number Rule is obeyed. For example take $n = 4$, and keep on doubling it, the number of factor increase progressively.
k) If n is a prime number, then the prime factors are large numbers, hence they are fewer. The prime numbers are completely random, we cannot predict what prime number will be a factor.
l) Consider 11 and 101, then this will occur regularly too. That is at $n = 4, 8, 12, 16, and\ 20$. A similar observation can be made with other prime factors such as 3, 7, 11, 13, 37 occur as a group at multiples of 6 for n.
m) The absence of 2 and 3 in the factors implies that the number is a pseudoprime or a prime. These also occur in a regular manner as shown by the shading. The shading represents a gap base 2 or 4 number and the unshaded a Class 1A number.
n) For a particular n, the number of prime factors increases in the next multiple. The exception is when n is a prime number, or when the roots are in a composite band. Looking at $n = 5$, we have 41, 271, then at $n = 10$ we have 11, 41, 271, 9091, and $n = 20$ the result is 11, 41, 101, 271, 3541, 9091, 27961. The coloring shows the origination of the use of the word "roots" since they repeat in a fundamental manner. The roots indicate a condition of relationship between values of n.

Type	n	Alpha Roots	No of Roots	New Roots
Prime/Pseudoprime	5	41, 271	2	41, 271
Prime/Pseudoprime	10	11, 41, 271, 9091	4	9091
Composite	15	3, 31, 37, 41, 271, 2906161	6	31, 2906161
Prime/Pseudoprime	20	11, 41, 101, 271, 3541, 9091, 27961	7	3541, 27961

Table 378

The outcome $n = 1$ to $n = 5$ are called unique alpha roots. The property of such roots is that they repeat in a regular manner across the alpha table. However number like 7 also repeat in a regular manner, but it is not a unique root. Its definition will be clarified in due course. The unique alpha roots are not related to n as being a prime number. For example 1 and 5 are prime numbers, but 2, 3 and 4 are not.

n	Unique Alpha Roots	Type
1	1	P/Ps
2	11	P/Ps
3	3, 37	Composite

4	11, 101	P/Ps
5	41, 271	P/Ps

Table 379

It is possible, but it is not clear, that when n is a prime number, and it has a pseudoprime as the alpha number, then the factors will form unique roots. In this way the system keeps on generating new unique roots. The assumption made here is that these are the only unique roots by definition, and that the numbers corresponding to n is a prime define a class of their own. Whilst the argument is emphasized for n being prime, it is also possible that pseudoprimes produce large factors that are in this class, but this has not been attended to since 25 is the first pseudoprime, and the factorization program does not reach 25 digits, it was limited only to 20 digits. Hence these are areas for further research.

Unique roots from the prime/pseudoprime (P/Ps) band can appear alongside composite roots, but roots that are in the composite sets do not appear in the P/Ps group. It is an exclusive relationship. This is evident for $n = 12$ and $n = 15$, they have roots 3, 37 inclusive of roots from the P/Ps group. However, none of the P/Ps groups have roots 3, 37 or newly generated roots that appear in the group. For example, the newly generated factors 7 and 13 in the composite group will not appear as factors in the P/Ps group. The table below shows the permutation pattern, with prime numbers for n-values colored yellow.

	Permutation-1						Permutation-2						Perm-3
n	Cycle 1	n	Cycle 2	n	Cycle 3	n	Cycle 1	n	Cycle 2	n	Cycle 3	n	Cycle 1
1	P/Ps	6	Comp	11	P/Ps	16	P/Ps	21	Comp	26	P/Ps	31	P/Ps
2	P/Ps	7	P/Ps	12	Comp	17	P/Ps	22	P/Ps	27	Comp	32	P/Ps
3	Comp	8	P/Ps	13	P/Ps	18	Compo	23	P/Ps	28	P/Ps	33	Comp
4	P/Ps	9	Comp	14	P/Ps	19	P/Ps	24	Comp	29	P/Ps	34	P/Ps
5	P/Ps	10	P/Ps	15	Comp	20	P/Ps	25	P/Ps	30	Comp	35	P/Ps

Table 380

The influence between the three types of groups is referred to as a permutation effect. This probably adds to the random behavior of factorization patterns, since we cannot generally predict how numbers will behave when we factor them.

The way roots occur they can also intersect between themselves. Let us consider the group from $n = 4$, then we note that for example $n = 20$ represents an intersection between the group with $n = 5$. The relationship between values of n is determined from the operation $\alpha_1 \oplus \alpha_2$ and $\alpha_1 \neq \alpha_2$. Note that the composite band breaks the trend of increasing number of roots. The composite roots occur at $n = 12$, and $n = 15$.

Type	n	**Alpha Roots**	**No of Roots**	**New Roots**
Prime/Pseudoprime	4	11, 101	2	11, 101
Prime/Pseudoprime	8	11, 73, 101, 137	4	73, 137
Composite	12	3, 7, 11, 13, 37, 101, 9901	7	13, 37
Prime/Pseudoprime	16	11, 17, 73, 101, 137, 5882353	6	17, 5882353
Prime/Pseudoprime	20	11, 41, 101, 271, 3541, 9091, 27961	7	3541, 27961

Table 381

For $n = 12$ in the $n = 3$ group, what is interesting is that there are no new roots, all of them have already been generated. Clearly from this table, the composites as a group do not exhibit consistently increasing number of roots, though generally there is an upward trend. In contrast, the P/Ps group is consistent. It is possible that going beyond

$n = 20$ may reveal a different case for the P/Ps group, but the current assumption is that the next group will always be larger.

Then there is a second condition that describes relation between given values of n. Within this group we have other sub-groups as well that satisfy the condition $\alpha_1 \oplus \alpha_1$, that is;

a) for $n = 3$ and $n = 6$, all elements of $n = 3$ are in $n = 6$
b) for $n = 6$ and $n = 12$, all elements in $n = 6$ are in $n = 12$
c) for $n = 9$ and $n = 18$, all elements in $n = 9$ are in $n = 18$

Type	n	Alpha Roots	No of Roots	New Roots
Composite	3	3, 37	2	3, 37
Composite	6	3, 7, 11, 13, 37	5	7, 13
Composite	9	3, 3, 37, 333667	4	333667
Composite	12	3, 7, 11, 13, 37, 101, 9901	7	None
Composite	15	3, 31, 37, 41, 271, 2906161	6	31, 2906161
Composite	18	3, 3, 7, 11, 13, 19, 37, 52579, 333667	8	19, 52579

Table 382

When these underlying sets are mixed up in their distribution, it gives the condition of randomness in the behavior of the factorization process. All of these are amazing set properties showing that factorization is actually a structured process. The process is structured in a simple manner, but the complexity is created through the permutation of position and root intersection between sets.

However, the use of the $\oplus$ now helps to give some understanding of behavior when looking at particular numbers. If the number of digits in a "1" construction are even, then we know that we have a case $\alpha_1 \oplus \alpha_1$, and therefore the elements of the previous set will be a subset. This can be expressed as a theorem.

Beta Theorem

Let $\beta = \alpha_1 \oplus \alpha_1$ with n digits, then the alpha roots of $n/2$ are a subset of β.

This is another way of expressing the alpha factor theorem, the difference being that the alpha factor theorem specifically states that β will always have more roots. Hence the beta theorem suggests predictive factorization provided we know the elements contained in the subset. The theorem defines alpha factorization. The operation creates a foundation for predictive factorization because we can tell what group of prime factors will occur for a given number of digits. The graphic illustrates for the alpha number where $n = 18$, that is there are 18 digits.

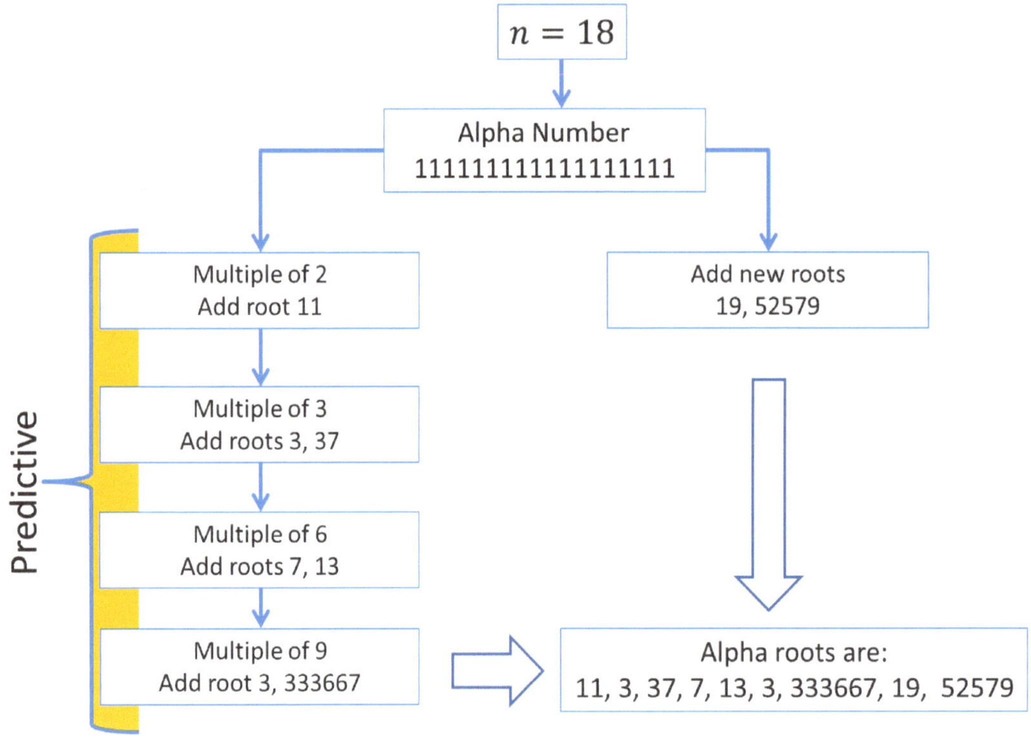

Figure 158: Deriving prime factors of a number

Let us consider the number for $n = 25$ and $n = 31$, and predict our expectation.

n	Number	Predictive Characteristics
31	1111111111111111111111111 111111	This number will have large prime factors because n is a prime number. The number is on the third permutation cycle.
25	1111111111111111111111111	The permutation number is 2 in cycle 2. Since $n = 24$ is a multiple of 3, hence the number at $n = 24$ will be a composite, which implies at $n = 25$ we have a pseudoprime or a prime. If the number is a pseudoprime, then the group of numbers that will be factors are *41, 271 as 5 is the only multiple of 25.Therefore we would expect two or more large numbers to be added here.*

Table 383

Hence we divide by 101 and 271 to derive our remaining prime factors. Now for any number k a prime, pseudoprime or composite, then;

$$y = k\beta$$

where β is of fundamental alpha numbers, that is consisting of 1 as a digit, then k is a factor that will be factorized further if it is pseudoprime or composite. Hence the roots of y will be determined by the value of n, and with the factorization of k, this will derive all the prime factors of y. The number y is referred to as the general alpha number.

n	y	k = 5 and Alpha roots
1	5	5
2	55	5,11
4	5555	5,11 , 101
8	55555555	5, 11, 73, 101, 137
16	5555555555555555	5, 11, 17, 73, 101, 5882353

Table 384

Hence for example 5555= 5(1111), therefore the roots of 5555 are 5, 11, 101. Similarly, 5555 is an alpha number while 1111 is a fundamental alpha number. Consider the following values of k for 111111111.

k-values	Alpha Number	Factors
$k = 1$	111111111	3, 3, 37, 333667
$k = 5$	555555555	3, 3, 5, 37, 333667
$k = 5 \cdot 7 = 35$	3888888885	3, 3, 5, 7, 37, 333667
$k = 5 \cdot 8 = 40$	4444444440	2, 2, 2, 3, 3, 5, 37, 333667

Table 385

Since these are multiples, it merely adds the factors of k itself. Note that the number 2 is not an alpha root, but it is a composite root. However in the set of prime factors 3, 3, 37, 333667 for $n = 9$, the number 3 is defined to be an alpha root.

Identify 3 as composite or alpha		Alpha root	Composite Root
111111111	3, 3, 37, 333667	3, 3	None
333333333	3, 3, 3, 37, 333667	3, 3	3

Table 386

Therefore since 3 can be either a composite root or an alpha root, it becomes a bit tricky to identify its particular characteristic unless one has the particular series for the alpha roots. In the case above the particular series is 3, 3, 37, and 333667.

13.4 Beta Factorization

The fundamental alpha numbers are developed on the basis of looking for factors of numbers whose digits are 1 only. Since there is infinity of such numbers, we also assume an infinitely long table describing the alpha roots. The previous section looked at alpha roots in terms of the classification of the fundamental alpha number, where this demonstrates a cycle of GB4-GB2-Class 1A.

Type for n	n	**Type**	**Fundamental Alpha Number**	**Alpha Roots**
Class 2B	2	GB4	11	1, 11
Class 1B	4	GB2	1111	11, 101
Class 2A	6	Class 1A	111111	3, 7, 11, 13, 37
Class 2B	8	GB4	11111111	11, 73, 101, 137
Class 1B	10	GB2	1111111111	11, 41, 271, 9091
Class 2A	12	Class 1A	111111111111	3, 7, 11, 13, 37, 101, 9901
Class 2B	14	GB4	11111111111111	11, 239, 4649, 909091
Class 1B	16	GB2	1111111111111111	11, 17, 73, 101, 137, 5882353
Class 2A	18	Class 1A	111111111111111111	3, 3, 7, 11, 13, 19, 37, 52579, 333667
Class 2B	20	GB4	11111111111111111111	11, 41, 101, 271, 3541, 9091, 27961

Table 387

The first column of the table above shows the systematic relationship between n and the fundamental number. This relationship is extended to the production of alpha roots also in a systematic manner. So its three constructs related to each other, the number n, which represents the number of digits of the alpha root, the alpha root which must consist of digits of 1, and the consequent prime factors and composite factors. Once more for clarity, in the UNDT framework, 2 and 3 are defined as composite factors, not prime factors.

The merge operation focused on a particular group of alpha numbers called the fundamental alpha numbers. These derived the unique alpha roots, as well as the general alpha roots. Since an alpha number is any number that satisfies the condition that $\alpha_1 \oplus \alpha_2 = \alpha_2 \oplus \alpha_1$, where $\alpha_1 = \alpha_2$, or $\alpha_1 and\ \alpha_2$ have the same digits. We now consider the case where;

a) $\alpha_1 = \alpha_2$
b) α_1 is a prime number
c) α_1 can be of non-uniform digits

Now a particular focus is given to prime numbers, since these cannot be simplified any further. An alpha number is any number that subscribes to the operation $\beta = \alpha_1 \oplus \alpha_2$, and according to the arithmetic rules for prime numbers in this case, then $D_t \oplus D_t = D_t$. That is, it is a closed operation, and this is true also for pseudoprimes.

Definition

A beta number is a merge operation of the same prime number.

Therefore a beta number is an alpha number, the emphasis is the fact that a prime number is used throughout in the merge operation. The property of this merge operation is illustrated by the table below. The major difference with the previous approach is that the pattern does not involve n, rather it depends on the structure of the number defined on the basis of a digit count.

Structure	Alpha-1	Alpha-2	Beta Number	Prime Factors	Beta Roots
One Digit	1	1	11	11	**11**
Two Digit	11	11	1111	11, 101	**101**
	23	23	2323	23, 101	
	31	31	3131	31, 101	
Three Digit	307	307	307307	7, 11, 13, 307	**7, 11, 13**
	401	401	401401	7, 11, 13, 401	
	877	877	877877	7, 11, 13, 877	
Four Digit	6451	6451	64516451	73, 137, 6451	**73, 137**
	8971	8971	89718971	73, 137, 8971	
	9547	9547	95479547	73, 137, 9547	
Five Digit	12007	12007	1200712007	11, 9091, 12007	**11, 9091**
	13441	13441	1344113441	11, 9091, 13441	
	17599	17599	1759917599	11, 9091, 17599	

Table 388

The operation $\oplus$ derives numbers that give us beta roots. Beta roots have a positional characteristic. However, pseudoprimes are not Beta numbers. It will be noticed that beta roots are also alpha numbers.

The following table gives the beta roots for up to ten digits of a prime number. Hence beta roots are tied to the power of ten in terms of range, that is, these are alpha roots that occur for a specific range. They occur as a pair or more in even numbers of n in alpha roots. Now 11 is the first alpha root, which explains why we have the single alpha root 101 occurring as the second beta root. While 11 and 1111 are beta numbers, they are also alpha numbers. The calculation is limited to ten digit numbers because of resource constraints.

Number of Digits	Alpha value of n	Beta Roots	Range of Numbers
1	2	11	1×10^0 *to* 9×10^0
2	4	101	1×10^1 *to* 9.9×10^1
3	6	7, 11, 13	1×10^2 *to* 9.99×10^2
4	8	73, 137	1×10^3 *to* 9.999×10^3
5	10	11, 9091	1×10^4 *to* 9.9999×10^4
6	12	101, 9901	1×10^5 *to* 9.99999×10^5
7	14	11, 909091	1×10^6 *to* 9.999999×10^6
8	16	17, 5882353	1×10^7 *to* 9.9999999×10^7
9	18	7, 11, 13, 19, 52579	1×10^8 *to* 9.99999999×10^8
10	20	101, 3541, 27961	1×10^9 *to* 9.99999999×10^9

Table 389

Hence for example, if we have a prime number where $\alpha_1 = 1200000611$, then the beta number is 12000006111200000611 and we know without any further factorization that its prime roots are 1200000611 because that repeats itself, and $101, 3541, 27961$. Hence we can find the prime factors of a range of number simply by compiling a table of beta roots. Since all beta roots are also alpha roots, this implies that they also repeat as indicated in the table below.

For example, if we consider the beta root 101, and we want to know the value of n at the tenth occurrence, then we have;

$$\frac{n}{2} = \{8(10) - 4\}/2 = 38$$

This means we look for any prime number with 38 digits, then the operation $\beta = \alpha_1 \oplus \alpha_1$ will yield a number that has the same prime number as a factor times 101. The number β will have 76 digits. Therefore this root will occur for all prime numbers that exist at the beginning of 1×10^{37} to $9.t \times 10^{37}$ where t represents 36 nines.

Beta Root	Equation for n
7	$6m - 3$
11	$4m - 2$
13	$6m - 3$
101	$8m - 4$

Table 390

The table can be extended for larger values as the number of digits of the alpha-1 number is increased.

The beta roots were considered for when n is even, hence a similar exercise can be done for n is odd. The table for the values is as presented below. This time the cycle is defined by *GB4-GB2-Class 1A*, and corresponds exactly to the same structure for the fundamental alpha number. The major difference with the even values of n is that an odd value cannot be twice another odd value. Consequently this affects the observed systematic relationship between the fundamental alpha numbers and the alpha roots. For example, 10 is twice 20, so we expect the relationship of alpha roots between these the values of n, where all the roots of $n = 10$ are in $n = 20$.

10	11, 41, 271, 9091
20	11, 41, 101, 271, 3541, 9091, 27961

Table 391

Such a relationship cannot occur for the odd values of n. However, Class 1A values have 3 as a common factor, therefore the alpha roots 3, 37 will be found in every Class 1A fundamental alpha number. The next number to be a common factor is 9, and the series 27, 45, 63, 81, ... will represent the values of n that have alpha roots of $n = 9$.

Type for n	n	**Type**	**Fundamental Alpha Number**	**Alpha Roots**
GB2	1	GB2	1	1, 1
Class 1A	3	Class 1A	111	3, 37
GB4	5	GB4	11111	41, 271
GB2	7	GB2	1111111	239, 4649
Class 1A	9	Class 1A	111111111	3, 3, 37, 333667
GB4	11	GB4	11111111111	21649, 513239
GB2	13	GB2	1111111111111	53, 79, 265371653
Class 1A	15	Class 1A	111111111111111	3, 31, 37, 41, 271, 2906161
GB4	17	GB4	11111111111111111	2071723, 5363222357
GB2	19	GB2	1111111111111111111	Prime

Table 392

The further implication is that the *GB4-GB2* cycle on the values of n will generate unique values of alpha roots. There is no repeatable pattern here.

This has effect on how factorization takes place in regard to alpha roots as shown in the table below. This is the expectation we have in factorization, appearing as a random operation rather than a systematic structure. The source of contribution towards this behavior is evidently from this relationship.

Alpha-1	Alpha-2	Alpha Number	Alpha Roots
23	5	235	4, 47
79	1	791	7, 113
419	29	41929	23, 1823
997	71	99771	3, 7, 4751
641	1693	6411693	3, 344483
1033	449	1033449	3, 37, 47, 1229

Table 393

Whilst odd values of n do not have predictive relationship except for those under Class 1A, the same behavior is observed when the pattern for the digits is not repeatable.

Observation	Alpha-1	Alpha-2	Factors
The alpha number is even, but factors not repeatable	547	409	19, 47, 613
	661	227	3, 7, 23, 37, 37
	401	409	3, 3, 3, 14867
Alpha-1 prime is repeated as a root, and alpha number is even.	823	823	7, 11, 13, 823
	1117	1117	73, 137, 1117
Alpha-1 and alpha-2 are not the same, but alpha number is even. There is no repeatable pattern.	823	937	179, 4603
	1117	1459	13, 401, 2143

Table 394

Hence predictive factorization is confined to even values provided $\alpha_1 = \alpha_2$, and Class 1A values of n. However, then if we perform the merge operation say on 823 and 937, and repeat merge to get 823937823937, then the roots 101, 179, 4603, 9901 are derived. Consequently this process of finding number with alpha roots is called derivative factorization, an alternative description to predictive factorization.

A similar observation is made with non-uniform alpha numbers, that is, for example 555555555⊕7777777.

Description	n	Irregular Alpha Number	Prime Factors
Even values of n	16	5555555557777777	59, 509, 112111, 1650097
	16	7777777775555555	5, 816911, 1904192201
	18	777777777555555555	3, 3, 3, 5, 37, 79, 293, 20161, 333667
	18	111111111555555555	3, 3, 3, 5, 37, 333667, 66666667
	18	111111115555555555	5, 11, 5519, 366044953979
Odd values of n	17	55555555557777777	3, 3, 6172839506419753
	17	55555555777777777	191, 367, 1697, 467031553
	19	5555555551111111111	73, 76103500700152207
	19	7777777775555555555	5, 1555555555111111111
	19	5555555777777777777	19, 67, 106189, 151937, 270493

Table 395

No particular patterns of interest are picked up here, except the occurrence of particularly large prime numbers comparatively.

13.5 Alpha-Beta Factorization

Generally we can consider the merge operation as constructing a number. Consider for example the number 15187, and we construct it through such an operation. We can then observe the prime factors and composite roots.

Alpha-1	Alpha-2	Beta Number	Alpha Roots	Composite Roots	Other roots
8	∅	8	None	2	None
8	7	87	3	None	29
151	∅	151	151	None	None
151	87	15187	15187	None	None

Table 396

The resulting number is a prime. This is not the only way to define the construction of the number, but what matters is the result as well as whether one can take advantage of the properties of alpha and beta factorization. Consider the examples given in the table below that gives the different combinations.

Class	**Alpha-1**	**Alpha-2**	**Beta**	**Prime Factors**	**Alpha Roots**	**Beta Roots**
1B⊕ D_t = D_t	718	997	718997	29, 24793	None	None
D_t ⊕ D_t = D_t	718997	718997	718997718997	29, 101, 9901, 24793	101, 9901	101, 9901
1B⊕ 2A = 1B	10	6	106	2, 53	None	None
1B⊕ D_t = D_t	106	79	10679	59, 181	None	None
D_t ⊕ D_t = D_t	10679	10679	1067910679	11, 59, 181, 9091	11, 9091	11, 9091

$D_t \oplus$ 1B = 2B	79	106	79106	2, 37, 1069	37	None
2B$\oplus$ 2B = 1B	79106	79106	7910679106	2, 11, 37, 1069, 9091	11, 37, 9091	11, 9091
$D_t \oplus D_t = D_t$	19	11	1911	3, 7, 7, 13	7, 13	7, 13
$D_t \oplus D_t = D_t$	1911	1911	19111911	3, 7, 7, 13, 73, 137	7, 13, 73	7, 13, 73
$D_t \oplus 1A = D_t$	59	111111	59111111	43, 1374677	None	None
$D_t \oplus D_t = D_t$	59111111	59111111	5911111159111111	17,43,1374677, 5882353	17, 5882353	17, 5882353
2A$\oplus D_t = D_t$	24	139	24139	101, 239	239	None
$D_t \oplus D_t = D_t$	24139	24139	2413924139	11, 101, 239, 9091	11, 101, 239, 9091	101, 9091

Table 397

For example the number 718997 has the factors 29 and 24793, and performing a merge operation gives the number 718997718997 which has factors 29, 101, 9901 and 24793. Now previously there was no trace as to where the additional factors besides the roots come from, but now this gives us the general picture. The factors 29 and 24793 are called determinant roots, or *D*-roots for short. Therefore 718997 is called the determinant number and 718997718997 is the resultant number.

718997718997	29, 24793	*D*-roots
	101, 9901	Alpha Roots

Table 398

If the alpha root is a beta root, then it denotes where in the range of powers of ten the number is, so then the *D*-root is taken to be like a characteristic signature of that particular number in that range. The *D*-roots are a complementary set of prime factors in a given number. Therefore we find them by elimination because when the other factors have been classified, what remains will be a *D*-root. But this is not always the case, that is when we know them we do not treat them as the remaining classification.

This means we can also do the same operation to look for such numbers in the following manner:

- We assume the numbers 29 and 1374677 as *D*-roots. Until proved otherwise by inclusion in the alpha roots table, such numbers are assumed not to be alpha roots.
- The product gives us the number 39865633
- Then we do a $\oplus$ operation on this number to derive 3986563339865633, and we know it will be with factors 29, 1374677.

Now because of the theory of factorization based on the number of digits, we know that 16 digits corresponds to the beta roots 17 and 5882353. Hence we can know the factors of the number without any due process simply by referring to the table that gives us the beta roots. Hence we have the prime factors as 17, 29, 1374677, and 5882353.

3986563339865633	29, 1374677	*D*-roots
	17, 5882353	Beta Roots

Table 399

Now consider the number 24139, this is a pseudoprime with alpha roots 101 and 239. The 101 is a fundamental alpha root, and 239 occurs for n is prime, that is $n = 7$. Then the operation $\oplus$ derives 2413924139, and because it has 10 digits, it will have beta roots 101, 9091. Hence the factors 101, 239 and 9091 are known, and from that the remaining factor 11 is determined.

2413924139	239	*D*-roots
	101, 9091	Beta Roots
	11	Fundamental alpha root

Table 400

Therefore we make an observation here in terms of pattern methodology and deductive approach that 239 is actually a *D*-root, though it is also an alpha root. If we follow that classification, this then implies we have two types of *D*-roots, those that are alpha roots and those that are not. A table is constructed of some of the determinant roots that have been derived.

Determinant Roots (*D*-Roots)	
Class 1	Class 2
Alpha Roots for n is a prime number	Non-alpha roots
41, 53, 79, 239, 271, 4649, 21649, 513239, 2071723, 265371653, 5363222357	23, 29, 59, 181, 241, 401, 601, 607, 691, 1069, 4831, 24793, 1374677

Table 401

The difference between the two classes is that the first class is tied to values of n being prime, whilst the other is not dependent on the value of n. The class 1 *D*-roots repeat at regular intervals determined by the value of n whilst the class 2 roots are irregular. All of the above shows the complexity of how an integer can be factorized in terms of roots or the way it is constructed and relates to the merge operation.

i. Composite root, that is 2 or 3
ii. Alpha root; This has four categories
    a. The fundamental alpha root; this predicts what factors we expect in terms of structure
    b. The Beta root; this predicts the factors in terms of position or range
    c. The *D*-root, Class 1; this is determined from n is prime
    d. The general alpha root; this occurs for $k\alpha_1$, where k defines the general root.
iii. The *D*-root, Class 2; this Is the signaure prime factors of the number, and accompany the position roots.

A number can be a product of all the listed categories, that is, $W = p \cdot q \cdot r \cdot s \cdot t \cdot k$, where $W = \alpha_1 \oplus \alpha_1$.

Number: W	p	Class 1	*D*-Roots
	q	Class 2	
	r	Beta Root	Positional
	s	Alpha Root-Fundamental	From digits of 1
	t	Composite Root	Roots 2 & 3. If $k \neq 1$, then it replaces t
	k	General alpha root	Can combine composite, prime and pseudoprime

Table 402

The general alpha root has a composite construct in it, hence this would be the same thing as defining t.

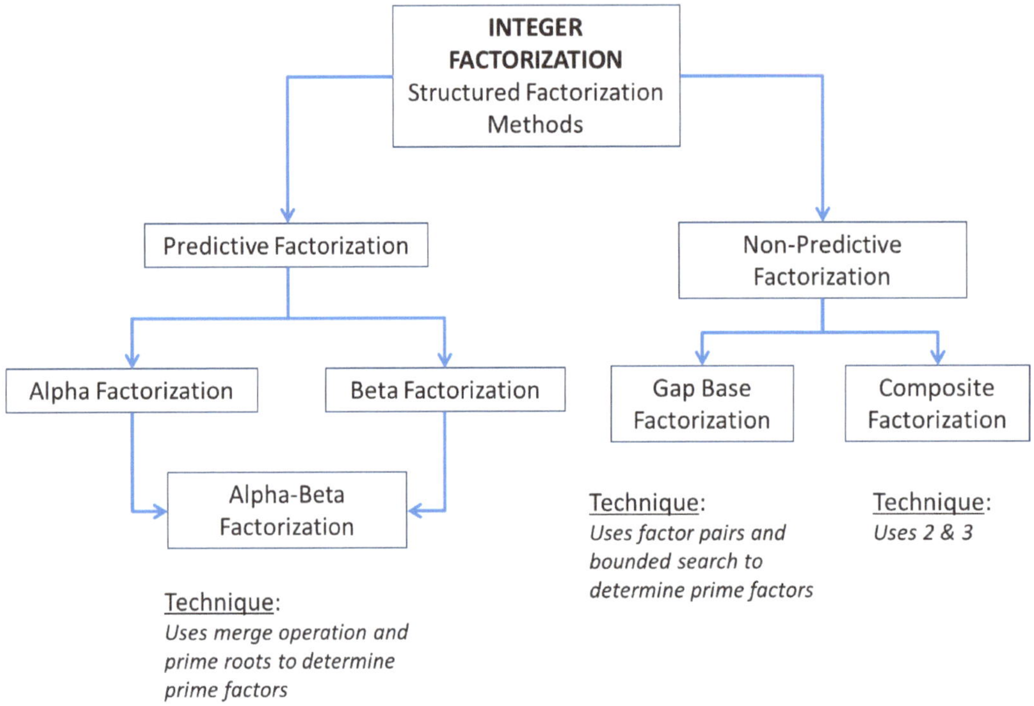

Figure 159

13.6 Number DNA

The merge operations can be used to define a concept of number DNA. This is based on the arithmetic rules that define the merge operations and from the realization that merge operations can actually be used to classify numbers.

Consider for example the number *W= 45670907* which is a prime in gap base 4. Now we can perform the following merge operations indicated in the table. This derives different combinations but the result is always the same classification, where DT stands for ditation set for gap base 2 or 4. Note that the lack of zero on 7090 does not change the classification as demonstrated in the last column.

Number	45670907			Outcome
Number breakdown	45	670	907	DT
Classification	1A	1B	GB2	
Number breakdown	456	7090	7	DT
Classification	2A	1B	GB2	
Number breakdown	4	5670	907	DT
Classification	1B	2A	GB2	
Number breakdown	456	709	7	DT
Classification	2A	GB2	GB2	

Table 403

Our standard approach in classification has been to divide by the 6 factor, but this approach is useful in the sense that all that is needed is to classify the number. Now DNA involves breaking up into constituent parts, so to do the DNA coding, we may break the number into single units, or into two digits or three digits. The parts are the classified to derive the coding. The breaking up may be done from the right or left of the number, and it is assumed that the zero in a number is always taken to the left; that is we accept 7090 rather than 709.

DNA Number Framework

Let there be a DNA number framework defined as follows;

Code	Category	DNA Coding
C1	*First Degree Numbers*	*1 DIGIT*
C2	*Second Degree Numbers*	*2 DIGIT*
C3	*Third Degree Numbers*	*3 DIGIT*
C4	*Fourth Degree Numbers*	*9 DIGIT*

Table 404

For example if we continue using the number 45670907, then this is broken down into its DNA coding as shown.

DNA Coding	LC3	2A	GB2	GB2	F3-GB4
Number		456	709	07	45670907

Figure 160

This can be represented in the following manner. LC3 is the category number of the classification where L indicates we start from the left, F3 is the family. Hence we break up the number from the left using three digits, but the last part will consist of two digits. Therefore the coding is;

LC3;2A:GB2:GB2; F3-GB4

The merge operation then tells us that when we combine the coding classification, the last part will be the result. The classification coding is not unique to a particular number, uniqueness is defined by the number itself. Infact, one may state the DNA coding and the range in which it is defined to generate numbers with the same code, for example **LC3[600, 900];2A:GB2:GB2; F3-GB4**. The square bracket is the usual definition of the prime space, though we now consider all the three types of numbers for DNA profiling. The answer here is that no such numbers exist because *W* will have three digits only, and the other classifications GB2;GB2 will not be defined.

Similarly however, one may pose a problem for example like;

Problem

*Look for a number that has the DNA coding **C3;2A:GB2:GB2; F3-GB4** and begins with 7 but ends with 1, where the number consists of the smallest for a given classification code. Let the middle GB2 number be between 200 and 300 and is divisible by 19.*

The solution can be constructed in the following manner. Note left or right operation is not defined in the code.

Solution

The smallest number that begins with 7 in class 2A is three digits is 702.The smallest three digit number that ends with 1 in gap base 2 is 121. Then we can consider the smallest GB2 number that has three digits divisible by 19, this being 247. Hence our number is 702-247-121, and if we test we find that it is a gap base 4 number, and it is prime.

This is a unique solution the way it is constructed. A second example is considered. Unless otherwise stated, it is always assumed that we are looking for the smallest number when investigating a solution as a general rule.

Problem

Look for a number that has the DNA coding **1A:GB4:GB2,** *it begins with 61 and* $x^2 - 1 = 6x - 1$ *for the GB4 number. The GB2 number is a square of the GB4 number. Determine its class using merge arithmetic.*

Solution

If it starts with 61, then 615 is the smallest number in Class 1A. For the GB2 number, solving gives us *x=6*, hence the GB2 number is 35. The square of 35 is 1225. Using the merge operation, the number required is 615351225. Now Class 1A⊕GB4 = GB4, and GB4⊕GB2 = 1A therefore this is a class 1A number using merge arithmetic.

Again this is a unique solution. Note that this time the category was not stated, and that this has no impact on the process of looking for a solution. Therefore we can define the classification of the number as **RC4;1A;F1-1A** or just **RC4;1A**, hence the construction is different from the final classification.

A final example is given as follows. Construct the number **C4:C1:C3:C2; GB4**. Note how the DNA code is written this time. The determining factor is the GB2 parameter. We can use the following merge operations; Class 1A⊕GB2 = GB2 and GB2⊕Class 1A = GB2 to define the number since GB2⊕GB2= GB4. Therefore a possible solution is 615351225⊕7 and 139⊕51 to derive 615351225713951 a GB4 number.

13.7 Merge Classification

The number DNA concept can be applied in reverse to determine the classification of a number, whether the number is gap base 2 or 4. Consider for example the number 67738901231, and we want to find out if the base of the number. Being more than nine digits implies that the computer program will not be able to handle such a number due to the memory limitation. The merge operator comes in handy through its arithmetic laws.

Now using the general form that GB2 is $f(x) = 6x + 1$, and GB4 is $f(x) = 6x - 1$, we can determine the gap base as follows.

67738901231	11289816871.67	GB2
67738901231	11289816872.00	GB4

Table 405

From the whole number, this is a gap base 4 number. Using the merge operation, we have;

Part 1		Part 2		Merge Result
67738	1B	901231	GB2	Class 1B⊕ GB2= GB4
6773	GB4	8901231	1A	GB4⊕Class 1A = GB4
677389012	1B	31	GB2	Class 1B⊕ GB2= GB4
677	GB4	38901231	1A	GB4⊕Class 1A = GB4
6773890	1B	1231	GB2 (PRIME)	Class 1B⊕ GB2= GB4
67	GB2 (PRIME)	738901231	GB2	GB2⊕GB2= GB4

Table 406

Hence any number, and in particular very long numbers can be classified by breaking into component parts and using the rules for merge arithmetic to determine the class. For example consider the following number which will be too large to process in an ordinary computer.

789900541289753188975009123215467980768572971397525750310229845601221980081261201243217863908703657843589532981178329079473829175386732109611

This can be broken down and classified as shown in the table, where $\alpha_1 \oplus \alpha_2 = (\alpha_1, \alpha_1)$. The last result gives us the classification.

Component Number	Component Classification	Merge Classification
789900541	GB2	
289753188	2A	(GB2,2A) = 1B
975009123	1A	(1B,1A) = GB2
215467980	2A	(GB2,2A) = 1B
768572971	GB2	(1B,GB2) = GB4
397525750	1B	(GB4,1B) = 2A
310229845	GB2	(2A,GB2) = GB2
6012219	1A	(GB2,1A) = GB2
80081261	GB4	(GB2,GB4) = 1A
201243217	GB2	(1A,GB2) = GB2
863908703	GB4	(GB2,GB4) = 1A
657843589	GB2	(1A,GB2) = GB2
532981178	2B	(GB2,2B) = 1A
329079473	GB4	(1A,GB4) = GB4
829175386	1B	(GB4,1B) = 2A
732109611	1A	(2A,1A) = 1A

Table 407

Hence the number is class 1A.

Integer Distribution and Infinity

The section is investigative, looking at Euclid's proof in the context of UNDT, and possible consequences as a result.

Outcome	Comment
Hypothesis *There exists a last prime number* $l_p \in E_3$.	This is derived from the distributive behavior of prime numbers.
The Last prime Concept $k = l_p$ will be generated by some value of p_n, where this is defined as p_t. There exists a value of $W = W_t$ that defines p_t. Hence there are three special numbers: $p_t,\ l_p, and\ W_t$	The Euclidean proof provides a framework for define the last prime number. W_t is the pseudoprime that defines the last prime from p_t.
Distributive Autonomous Law *The distributive behavior of prime numbers is independent of composite numbers.*	Composite numbers have "nothing" to do with prime number distribution, they don't influence it in any way.
Distributive Hypothesis of Prime Numbers *Then* $n\{f(x_i) \cdot f(x_j)\} = n_2\ increases\ as\ x_b \to \infty$, *consequently* $n_1 \to 0$.	The distribution of pseudoprimes has a direct consequence on the distribution of the prime numbers.
Law of Mutually Exclusive Infinity *For any infinitely large prime space* $[a, b]$, *It is impossible for prime numbers and pseudoprimes to be both infinite simultaneously in the same space.*	This is a consequence of the relationship between the two numbers. One of them has to be true.

Table 408

14 Are Prime Numbers Infinite?

The proof on number of primes has a focus on the counting aspect of infinity. The credibility of Euclid's proof lies in the fact that for any p_n, then $p_1, p_2, \cdots p_n < q_1,\ q_2, \cdots q_m$ where p_n is finite. The conversion of Euclid's argument to extend the proof to infinity is made from the assumption that not only is $p_n$finite, but it represents all the prime numbers possible. Therefore if $p_1, p_2, \cdots p_n < q_1, q_2, \cdots q_m$, then there are an infinite number of primes. That is, $q_1, q_2, \cdots q_m$ can exist at infinity as we will always factorize $p_n + 1$. In that case, the theorem then states its position as "There are infinitely many primes". This is the emphasis and context of the theorem. The question that can arise in one's mind is, *"is there a difference between infinitely many primes and q_i existing at infinity?"* This question is important in giving the proof a context of describing the behavior and distribution of primes at infinity or primes being infinite. The proof does not say q_i exists at infinity, but rather the number of primes is infinite.

The Euclidean proof is based on arithmetic properties rather than distributive and behavioral properties of the prime numbers. More so, it does not have the context of the UNDT framework. This brings about a limitation in scope in describing distributive properties of the prime numbers.

14.1 Axioms of Infinity

Infinity as a concept was not developed at the time the Greeks put forward the proof that prime numbers are many. This does not imply that a concept of infinity did not exist. The following passage from scripture shows that the notion has always existed in various manners.

Isaiah 40	Comment
12 Who has measured out the waters in the hollow of his hand, or carefully measured the sky, or carefully weighed the soil of the earth, or weighed the mountains in a balance, or the hills on scales?	The concept of infinity is expressed in comparative terms; that is, it is a comparative construct. This means one way of understanding infinity is through comparison (to weigh) using measure as the variable.
13 Who comprehends the mind of the LORD, or gives him instruction as his counselor? ; 14 From whom does he receive directions? Who teaches him the correct way to do things, or imparts knowledge to him, or instructs him in skillful design?	Whilst the above verse dealt with physical things in trying to illustrate the infinite, now a different approach is used. There is infinity in terms of character, of understanding; God has understanding without bound. Hence the construct is that of limitation and abstractness; implying that infinity is that which is limitless and abstract.
15 Look, the nations are like a drop in a bucket; they are regarded as dust on the scales. He lifts the coastlands as if they were dust.	Now infinity here is described through the definition of a construct, the "drop" and the "dust". The dust cannot be counted, but can be put on scales. That is infinity implies no counting can be made, but nevertheless comparison can be made. Comparison is determined by the constructs so developed.

Table 409

Three aspect are observed from the analysis and interpretation, that infinity can be understood in terms of;

i. Measure; this is a common platform for conceptualizing infinity, something can be infinitely small or infinetly big.

ii. Limit or boundary; this notion is also applied inmathematics
iii. Abstractness; The decimal numbers between 1 and 2 are infinite, yet the distance between 1 and 2 is finite. That is how abstract infinity can be.
iv. Constructionist approach; this is a technique of dealing with infinity, which mathematics applies
v. Counting; We cannot establish the concept of infinity without counting, yet counting is not possible at infinity.

We establish the notion of infinity through the following axioms. Some of the observations may appear to contradict each other because of the limitation of conceptual expression – using words created for a finite space to describe conditions for an infinite space.

Axioms of Infinity

a) **Infinity can be conceptualized.** This is a fundamental statement that allows us to form the axiomatic understanding of infinity.
b) **Axiom of Infinite Magnitude -** *Infinity is not a number*. Rather, it is about a number and its property. We consider infinity as an idea or a mathematical construct. Magnitude is both consequential and inconsequential. It is consequential because it describes infinity and inconsequential because it has no meaningful application at infinity.
c) **Infinity is an existence**. Therefore it is fully formed and understood through number behavior. The significant issue to look into is the behavior of prime numbers and pseudoprimes at infinity.
d) **Axiom of Infinite Order**. That is, if we choose an element of infinity called a, and another called b, we can still assume that $a < b$. Hence we expect order to exist at infinity, whilst at the same time we say infinity is not a number! The beauty of contradiction.
e) **Counting does not exist at infinity**. Generally, infinity plus one is infinity, which is a mathematical statement that implies there is no counting at infinity. Numbers therefore are understood in terms of characteristics, what they mean and stand for, rather than value.
f) **Infinity is irreducible.** This we can infer from the notion that infinity minus one infinity times is still infinity. That is, while we can think of reaching infinity, we cannot think of reducing infinity.
g) **Infinity is independent of the finite space.** This is a consequence of the fact that infinity is irreducible, and therefore cannot be directly related to the finite space.
h) **Infinity has no size**. The extent and depth of infinity cannot be specified. This we infer from the fact that infinity times infinity is infinity.
i) **Infinity is not comparable**. Therefore it cannot be subdivided and infinity divided by infinity is not defined.
j) **Converse axiom of infinity**. If a and b are finite, then $(a + b)\ or\ (ab)$ is finite, no matter how large. That is, we cannot define infinity in terms of finite operations. Hence these operations define the finite space.
k) **The finite can contain the infinite**. For example between 0 and 1, there are infinitely many decimals, but that does not imply that 1 is at infinity with respect to 0. This makes grasping infinity a challenge!
l) **The infinite can contain the finite**. That means a finite behavior can be established in an infinite space.

Whilst the above axioms may hold, it does not imply that we cannot describe what happens at infinity or within infinity. Therefore the axioms act as a guide in determining the type of descriptions that are acceptable.

In studying the prime numbers, the general argument is:

a) the gap between prime numbers is increasing but this does not translate to the fact that prime numbers are decreasing.

Comment: It is generally observed that prime numbers do decrease because of the gap in terms of the count function. In the current mathematical conclusion, no relation has been established between the gap and the count function, hence the assumption that there may be an infinite number of primes.

b) the assumption is that primes could actually be increasing but we meet less and less of them because of the ever increasing gap.

Comment: This is how we can relate the gap to an infinite number of primes.

c) The size of prime numbers is increasing even to millions of digits, and therefore a prime number with an infinite number of digits is a possibility. This being the case, then this creates a possibility of an infinite number of primes.

Comment: However, it may be noted that having an infinite number of digits does not translate to an infinite number of primes.

14.2 Euclid's Proof

One of the central questions in the study of prime numbers is whether they are infinite or not. The typical standard proof is due to Euclid. The observed assumptions behind the proof are as follows.

a) We have a finite set of prime numbers
b) There is a number V such that $V = p_1 \times p_2 \times \cdots p_n$ where p_i denotes a prime number.
c) There is a number $W = V + 1$, or $W = p_1 \times p_2 \times \cdots p_n + 1$. Hence $W > V$.
d) For the proof to be able to be conclusive, it must be assumed that V consists of all the possible set of prime numbers that define the finite set.
e) W may be a prime or it may not be prime.

Therefore Euclid's proof (Book IX) considers any finite set of primes, denoted p_1, p_2up to p_n. Suppose p_n is a prime number, and consider the number W defined as 1 plus the product of all prime numbers less than or equal to p_n. That is:

$$W = p_1 \times p_2 \times \cdots p_n + 1$$

For example;

$$W = 2 \times 3 \times 5 \times 7 \times 11 \times 13 \times \cdots p_n + 1$$

Like any natural number, W can be written as a product of prime numbers; where this is assured by the fundamental theorem of arithmetic. Therefore we have:

$$W = q_1 \times q_2 \times \cdots q_m$$

None of the primes denoted p_1, p_2 up to p_n can divide W, because dividing W by any of these leaves a remainder of 1. Therefore the primes $q_1, q_2, \cdots q_m$ are additional primes beyond the ones we started with. This is the key argument of the proof.

Alternatively we can reason as follows. If W itself is prime then already there is a prime greater than p_n. But suppose W is not prime, then there is a number m such that $m > 1$, where $m < W$ and m divides W exactly. Let q be the smallest such number, where $W \div m = q$. There are two possibilities for q, it may be prime or not prime.

Suppose q is not prime, then there would be a divisor $r > 1$ of q such that $r < W$. Hence $q \div r = s$, which implies r and s divide W exactly, so q would not be the smallest such number. Hence q is prime provided the condition is that it is the smallest such number dividing W.

Suppose $q \leq p_n$, then q is one of the primes in the definition of W above, so W/q would equal an integer plus $1/q$, which is impossible, since q divides W exactly. Therefore if W is not prime then there is a prime greater than p_n.

Hence we conclude that whether or not W is prime there is a prime greater than p_n. Consequently for any prime number there is a larger prime number, where this implies there are infinitely many prime numbers.

The weakness of the alternative argument is that we already assume rather than prove that q is the smallest number, or the fact that m is the biggest number dividing W. As a result it presents a circular argument. The fact that it is the smallest number already implies there cannot be r that is a divisor. Hence the consequential argument is based on whether $q \leq p_n$, but this is the same as the first argument presented above. Therefore Euclid's proof seems sufficient and strong enough to withstand scrutiny.

This does not mean there are no matters arising. We can also ask the question as to what is the weakness of Euclid's argument. This is found from the way W is defined. W is a new number that is derived from a product of prime factors, and there is no reason to confine it to +1. That is we can also define;

$$W = p_1 \times p_2 \times \cdots p_n - 1$$

We observe that the role of +1 is conceptually defined from the fact that we attempt to divide W by p_i and get a remainder that is 1, whereas we cannot do the same if we subtract 1. However, this does not invalidate the suggested approach in constructing W, it just presents a manipulation limitation. The resulting W may be a prime q, which means the condition $p_n < q$ is met, or it may have factors that are either smaller or greater than p_n, where $W = q_1 \times q_2 \times \cdots q_m$. The issue then is to validate if $q_i > p_n$ when we subtract 1.

14.3 Observations From UNDT Perspective

The first major difference is the definition of the prime number, because 2 and 3 are not prime under UNDT. The truth of the above proof hinges on 2 being a prime number. This is based on the fact that the product of any two odd numbers is odd, and an odd number plus 1 always yields an even number.

The next observation is the fact that we assume that V is a product of all prime numbers. It is obvious that V and W will have different prime factors, and W is not defined to be a product of all prime factors as V. The assumption is that $+1$ generates a set of entirely new prime factors outside V, where;

$$p_1, p_2, \cdots p_n < q_1, q_2, \cdots q_m$$

This is because $p_1, p_2, \cdots p_n$ are not factors of W since they should always result in a remainder of 1.

Now lets us define a factorial function for the prime and pseudoprime numbers as follows.

Definition

$$[P]! = 5 \times 7 \times 11 \times 13 \times 17 \times 19 \times 23 \times 25 \times 29 \ldots$$

The alternative expression for $W = V + 1$ is that V can also consist of pseudoprimes instead of prime numbers only as a way of defining a general approach.

n	Factors	x = [P]!	V = 6x	W = V+1		FACTORS
1	5,1	5	30	31	prime	none
2	5,7	35	210	211	prime	none
3	5,7,11	385	2,310	2,311	prime	none
4	5,7,11,13	5,005	30,030	30,031	pseudo	59, 509
5	5,7,11,13,17	85,085	510,510	510,511	pseudo	19, 97, 277
6	5,7,11,13,17,19	1,616,615	9,699,690	9,699,691	pseudo	347, 27953
7	5,7,11,13,17,19,23	37,182,145	223,092,870	223,092,871	pseudo	317, 703763
8	5,7,11,13,17,19,23,25	929,553,625	5,577,321,750	5,577,321,751	pseudo	432, 3016399
9	5,7,11,13,17,19,23,25,29	26,957,055,125	161,742,330,750	161,742,330,751	pseudo	89, 22811, 79669
10	5,7,11,13,17,19,23,25,29,31	835,668,708,875	5,014,012,253,250	5,014,012,253,251	pseudo	1019, 4 920 522 329
11	5,7,11,13,17,19,23,25,29,31,35	29,248,404,810,625	175,490,428,863,750	175,490,428,863,751	pseudo	101,547, 3 176 469 833

Table 410

Now consider the expression $2 \times 3 \times 5 \times 7 \times 11 \times 13 \times .... p_n$. This can be simplified in the following manner because $6 = 2 \times 3$.

$$2 \times 3 \times 5 \times 7 \times 11 \times 13 \times .... p_n = 6 \times (5 \times 7 \times 11 \times 13 \times .... p_n) = 6 \cdot [P]!$$

This then derives the above table, where the result is always an element of gap base 2, that is the result of $6x + 1$. The shaded lines indicate the fact that the same principle works even if the factors are not prime numbers, for example using the pseudoprimes 25 and 35. The observation that can be made is that:

a) The product $2 \times 3 = 6$ gives us the basic format for prime number generation, and this is the reason why the proof is reliable. It is an indicator that we can use both $6x + 1$ and $6x - 1$.
b) If we assume that $6x + 1$, then x can be any number; prime, pseudoprime or composite. For example; $(35 \times 7) = 245$, and $(6 \times 245 + 1)$ = 1471, which is a prime number.
c) V is a composite number in Class 2A, since $V = 6x$.
d) W is a prime or pseudoprime in gap base 2, since $W = 6x + 1$.
e) V can include elements of $[P]$ that are not prime.
f) The value of p_n does not have to be at infinity for the proof to hold, that is p_n can be finite. The credibility of the proof lies in the fact that for any p_n, then $p_1, p_2, \cdots p_n < q_1, q_2, \cdots q_m$.

The use of consecutive prime numbers to multiply each other defines a concept of product gap of prime numbers. Consider for example starting at 2 and 3, then the prime generated is 7.

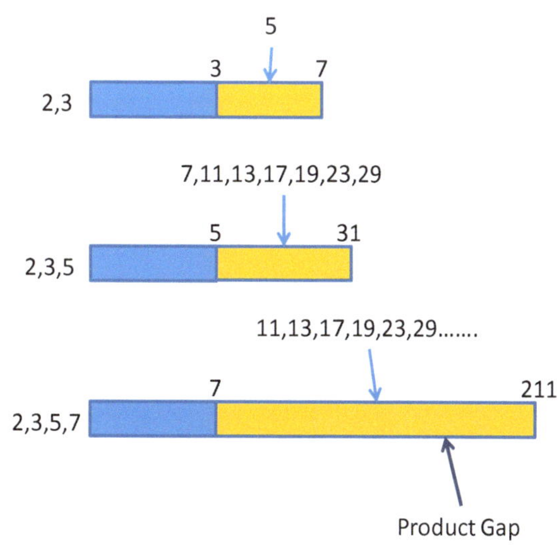

Figure 161

Using the Euclidean approach, the largest prime is $p = 3$, the generated prime is 7, so the product gap is 4. We define 3 as the lower prime and 7 as the upper number where this is prime or pseudoprime. Similarly $W = 31$ for the primes 2,3,5, and we have primes in between 5 and 31. Hence $p_1 \times p_2 \times \cdots p_n$ defines the value of the upper number, and p_n the lower prime. Therefore generally the product gap $PG = W - p_n$.

The assumption of the proof is that there will always be prime numbers bigger than the lower prime number p_n. That is, the factors of W cannot be greater than W but are greater than p_n, which means they lie within the product gap. The above graphic probably represents Euclid's concept of prime number distribution. It will be slightly different when 2 and 3 are not included. In his proposition 20, instead of saying that there are an infinity of primes as is the common modern trend, Euclid preferred to say that there are more prime numbers than contained in any given collection of prime numbers up to p_n. The two concepts are very different statements. Whilst Euclid confined himself to saying there are more primes than found in any finite list of primes, the modern notion involves infinity, hence both views will be explored in the context of UNDT.

Under UNDT, the following sets are defined with regard to the above graphic.

Definition

a) $E_1 = \{p_1, p_2, \cdots p_n\}$, where p_i is a prime number or a pseudoprime
b) $E_2 = \{q_0, q_1, q_2, \cdots q_m\}$, where these are called ditation factors provided q_0 is pseuodprime and q_i is a prime number, and
$$W = q_0 \cdot q_1 \cdot q_2 \cdot \ldots q_m$$
c) $E_3 = \{r_1, r_2, \cdots r_m\}$, where r_i is a prime number or a pseudoprime, and $p_n < r_i < W$, and
$$W \neq r_1 \cdot r_2 \cdot r_3 \cdot \ldots r_m$$

Hence these three sets are related to each other through the Euclidean condition $p_1 \times p_2 \times \cdots p_n + 1$, but without limiting the values to prime numbers only. The smallest element of p_1 is assumed to be 5 since 2 and 3 are not defined as prime numbers, and 1 is taken as an identity. For example if 5 is the only element, then $W = 6(5) + 1 = 31$, and we find elements between 5 and 31 that are prime only as;

i. $E_1 = \{5\}$
ii. $E_2 = \{31\}$
iii. $E_3 = \{7,11,13,17,19,23,29\}$

When $E_2 = \{q_1\}$, that is, it has one element only then E_2 must consist of one prime number and not pseudoprime. These define the first group of Euclidean sets for the construction, where $n(E_1) = n(E_2)$. In E_3, 29 is called the last prime, that is $l_p = 29$. For E_2, then $W = q_1 = 31$, and $q_1 > l_p$. To form a new set, E_1 takes an element of E_3 in order to derive higher value elements of E_2. For example, taking 7 into E_1, then $W = 6(5 \times 7) + 1 = 211$, and we have a range of primes between 7 and 211. Hence taking the smallest element of E_3 to E_1 will shift the value of q_1 in E_2.

i. $E_1 = \{5,7\}$
ii. $E_2 = \{211\}$
iii. $E_3 = \{11,13,17,19,23,29 \cdots 191, 193, 197, 199\}$

Therefore clearly $E_1 \not\subset E_3$ and $E_2 \not\subset E_1$ as long as $W = q_1$ and q_1 is prime. We can also call E_2 the observation set because it determines the outcomes in E_1 and E_3. In each of the above cases, we have $W = q_1$ and $n(E_2) = 1$ if q_1 is prime. We can give another example as follows where W is not prime and we restrict the results to prime numbers only. We have $W = 6(85085) + 1 = 510510 + 1 = 510511$. This is a pseudoprime with factors 19, 97, and 277.

i. $E_1 = \{5,7,11,13,17\}$
ii. $E_2 = \{19,97,277\}$
iii. $E_3 = \{19, 23,29,31,37, \ \cdots 97, \cdots 277, \cdots, 510449, \ 510451, \ 510457, \ 510481\}$

The Euclidean condition $p_1, p_2, \cdots p_n < q_1, q_2, \cdots q_m$ still holds. But we observe that if W is a pseudoprime, then $E_2 \subset E_3$ and $W > q_m$, but if W is prime then $E_2 \not\subset E_3$ and $W = q_1$.

14.4 The +1 Argument

The power of Euclid's proof is the fact that you have the product V of all known factors up to a number p_n, and if you find factors for the number $W = V + 1$, such factors are greater than p_n, where such factors are $q_1, q_2, \cdots q_m$. That is, you can keep on adding these new factors to make another finite set and get a new product such that $W = V + 1$. This implies that that are infinitely many factors since you can keep on growing the set infinitely many times.

Another proof by a mathematician Kummer (1878) makes the assumption that we subtract 1 instead of adding. He assumes that if we suppose the number of primes be finite, then we have the primes $p_1, p_2 \cdots p_n$. If V is the product of all such primes, then it implies $V - 1 > p_n$ and is not prime. In that case if it is not prime, then there exists a p_i from the finite list that divides $V - 1$. But p_i also divides V, which then implies p_i divides $V - (V - 1) = 1$, which is an absurdity. From this one makes the conclusion that primes is infinite. Therefore the major difference between the two approaches is that one uses a contradiction while the other demonstrates that there are other prime factors outside the finite set $p_1, p_2 \cdots p_n$.

Infact both methods make the same assumption that there are other factors more than p_n. Kummer's method has a different framework for the proof in respect to Euclid. But it is then noted that we can actually continue Euclid's approach. For example;

$$2 \times 3 \times 5 \times 7 \times 11 \times 13 = 30030$$

$$30030 - 1 = 30029$$

The number 30029 is a prime number. Therefore this argument can be extended if we consider the fact that from $W = V - 1$, we also get factors greater than p_n as shown by the example given below. If this is equally true for the subtraction process, then it questions the validity of the +1 process which is based on induction.

n	Factors for [P]	W=V-1	Factors	W = V+1	Factors
1	5,1	29	Prime	31	Prime
2	5,7	209	11,19	211	Prime
3	5,7,11	2309	Prime	2311	Prime
4	5,7,11,13	30029	Prime	30031	59, 509
5	5,7,11,13,17	510509	61, 8369	510511	19, 97, 277
6	5,7,11,13,17,19	9699689	53, 197, 929	9699691	347, 27953
7	5,7,11,13,17,19,23	223092869	37, 131, 46027	223092871	317, 703763
8	5,7,11,13,17,19,23,25	5577321749	1951, 285, 8699	5577321751	432, 3016399
9	5,7,11,13,17,19,23,25,29	161742330749	255133, 633953	161742330751	89, 22811, 79669
10	5,7,11,13,17,19,23,25,29,31	5014012253249	Prime	5014012253251	1019, 4 920 522 329
11	5,7,11,13,17,19,23,25,29,31,35	175490428863749	443, 396140922943	175490428863751	101,547, 3 176 469 833

Table 411

The strong implication is that the +1 method does not have anything to do with an infinite number of primes, because this is what the inductive methodology suggests. However, the fact that the -1 process derives a similar characteristic implies a structural argument rather than a proof for an inductive process.

What the UNDT framework has shown is that we are actually using the function $6x + 1$ for Euclid's approach, and producing primes or pseudoprimes in gap base 2. But we can also apply $6x - 1$ to derive a similar result to produce primes or pseudoprimes in gap base 4. That is, we have a condition $6x \pm 1$ as the framework for deriving factors of W.

14.5 The +*k* Argument

There is no specific rule or implication that grounds us to the use of 1 as the structural result is independent of whether we are factoring a number that will have factors less than or greater than p_n. In order to understand the process involved in general terms, consider the expression $W = V \pm k$ where $k \in D_t$. Then there are two possibilities.

a) Let $V = 6kf$, hence $W = 6kf \pm k = k(6f \pm 1)$ it then means $k = q_1 < p_n$
b) If $V/k \neq Int(V/k)$ then $k \notin E_1$, hence either:
   i. $q_1 < p_n$, where this implies k is pseudoprime
   ii. $q_1 > p_n$, where this implies k is a prime number

Hence if $k \in V$, then k will obviously be a factor of W, and $k < p_n$. Now $(6f \pm 1) \in D_t$, and it will either be a prime or a pseudoprime. Similarly, if $k = 1$, then $W = 6f \pm 1$, which implies that this is what Euclid has assumed in his proof. That is in general we are assuming $W = 6kf \pm k = k(6f \pm 1)$, and that $k = \pm 1$ is a specific consideration. In such a case then $k \in E_1$.

Let us start with a simple example to illustrate the use of $+k$ argument. Let $E_1 = \{5,7,11\}$, hence $V = 6 \cdot (385) = 2310$. Now:

a) Let $k = 7$. Then this gives $V + 7 = 2310 + 7 = 2317$, and the factors are 7, 331, and we have the condition $p_n > q_1$where $q_1 = 7$.
b) Let $k = 13$. Then this gives $V + 13 = 2310 + 13 = 2323$, and the factors are 23, 101. We have the condition $p_n < q_1$ where $q_1 = 13$.
c) Let $k = 17$. Then this gives $V + 17 = 2310 + 17 = 2327$, and the factors are 13, 179. Again we have the condition $p_n < q_1$ where $q_1 = 17$.

As a result, we observe that the condition of $6 \cdot (p_1 \times p_2 \times \cdots p_n) + 1$ is by no means unique in terms of generating values of prime factors that are greater than p_n. That is, if we have $k \in D_t$ and it is prime;

$$6 \cdot (p_1 \times p_2 \times \cdots p_n) \pm k, \text{ and } p_n < k, \text{ then } p_n < q_1$$

The prime factors generated will be greater than the ones of the set $E_1 = \{p_1, p_2, \cdots p_n\}$. If it is pseudoprime, then the factors of k being be smaller than p_n is a possibility, hence for certainty, it is assumed that k is prime.

It has been assumed that $E_1 = \{p_1, p_2, \cdots p_n\}$ where these are both prime and pseudoprime. That was the new introduction under UNDT. The other assumption to consider is whether elements should always start at the first prime number and whether they should be consecutive.

Definition

For the set $E_1 = \{p_1, p_2, \cdots p_n\}$, let $p_1$start at any prime or pseudoprime number and p_1 is not necessarily consecutive.

The table below gives an example of an application of the definition. Note that the example also includes a multiplication by a pseudoprime 25.

Product	Result	Factors
$6(17 \cdot 23 \cdot 31) + 1 = 72\,727$	Prime	None
$6(7 \cdot 23 \cdot 29 \cdot 31) + 1 = 868\,435$	Pseudoprime	5, 173687
$6(5 \cdot 7 \cdot 11 \cdot 17 \cdot 25) + 1 = 981\,751$	Pseudoprime	67, 14653

Table 412

The second example is of interest because we have a case where $q_1 < p_n$. That is, $q_1 = 5$ and $p_n = 31$. Hence if $k = \pm 1$ and elements of E_1 are not consecutive, then it is possible that $q_1 < p_n$. Adding a pseudoprime to the multiplication process should not change anything much since it consists of a product of prime numbers anyway. Now consider the addition of the pseudoprime $k = 667$. It is observed that also $q_1 < p_n$ in the third example.

Product	Result	Factors
$6(7 \cdot 11 \cdot 13) + 667 = 66\,673$	Prime	None
$6(7 \cdot 11 \cdot 13 \cdot 17) + 667 = 102\,769$	Prime	None
$6(7 \cdot 11 \cdot 13 \cdot 17 \cdot 19) + 667 = 1\,940\,605$	Pseudoprime	5, 137, 2833

Table 413

As an example, let us consider the set $E_1 = \{1, 7, 19, 23, 25, 37\}$, hence $7 \cdot 19 \cdot 23 \cdot 25 \cdot 37 = 282955$, $V = 16977450$ and $p_n = 37$. Note that the primes are not consecutive. For example, let $0 \leq n \leq 6$ then this gives us a space over which the input set can generate possible values of primes and pseudoprimes. As already observed, each value of k gives a pair of equations, but the value of k need not be equal, that is we can have k_1 and k_2, where $V + k_1$ and $V + k_2$. Now $n = 0$ gives us the Euclidean condition, therefore we define the general Euclidean condition to be applicable when $1 \leq n$.

Hence this is a linear ditation space since $V = 6x$ and $W = V \pm k$. In this example $0 \leq n \leq 6$ and we use have k_1 and k_2. This is called a Euclid factors group E_f, where E_1 is referred to as the base sequence, and V is the identifier of the group. In this case $V = 16977450$ is the group identifier. Hence k_1 and k_2 become the generators of the group, where n gives us the range where the group starts and ends. Hence prime factors found under this group E_f form a particular family because of the parameters that define the numbers being factored.

n	k_1	k_2	$+k_1$	Factors $+k_1$	$-k_2$	Factors $-k_2$
0	1	1	16,977,451	1, 137, 123923	16,977,449	1, 71, 239119
1	5	7	16,977,455	5, 11, 308681	16,977,443	7, 2425349
2	11	13	16,977,461	419, 40519	16,977,437	53, 320329
3	17	19	16,977,467	13, 1305959	16,977,431	19, 893549
4	23	25	16,977,473	23, 738151	16,977,425	5^2, 97, 7001
5	29	31	16,977,479	Prime	16,977,419	1289, 13171
6	35	37	16,977,485	5, 7, 41, 11831	16,977,413	37, 458849

Figure 162

Therefore we can say $738151 \in E_f$. And for this group, we can note that $E_f \cap E_1 \neq \emptyset$. The following is observed.

a) When E_1 has a pseudoprime for example 25, this then defines a ditation factor that can be simplifed further. For example $k_2 = 25, W = 16977425$, then the ditation factors are 25, 97, 7001 and the prime factors of the number are 5, 5, 97, 7001.
b) Now $E_1 = \{1, 7, 19, 23, 25, 37\}$ contains the the prime factor 5 indirectly through the number 25. In a simialr manner the set contains the pseudoprime 35 indirectly through the product $35 = 5 \cdot 7$. Hence the ditation factors of 16977485 are 35, 41 and 11831, hence the prime factors are 5, 7, 41 and 11831.
c) If $k \in E_1$, then $k \in E_2$ as shown by the purple coloured results.

The examples as well as the discussion have confined itself to the set D_t which consists of primes and pseudoprimes. The example from the tables below shows that adding one is not a magic approach; other numbers can also be used from other classes. The next two tables show an addition of 663 and 666 from Class 1A and Class 2A respectively.

Product	Result	Factors
$6(7 \cdot 11 \cdot 13) + 663 = 66\,669$	Composite – 1A	3, 313, 71
$6(7 \cdot 11 \cdot 13 \cdot 17) + 663 = 102\,765$	Composite – 1A	3, 5, 13, 31, 17
$6(7 \cdot 11 \cdot 13 \cdot 17 \cdot 19) + 663 = 1\,940\,601$	Composite – 1A	3, 13, 17, 2927

Table 414

Product	Result	Factors
$6(7 \cdot 11 \cdot 13) + 666 = 66\,672$	Composite – 2A	$2^4, 3^2$, 463
$6(7 \cdot 11 \cdot 13 \cdot 17) + 666 = 102\,768$	Composite – 2A	2^4, 3, 2141
$6(7 \cdot 11 \cdot 13 \cdot 17 \cdot 19) + 666 = 1\,940\,604$	Composite – 2A	2^2, 3, 161717

Table 415

The simple rule seems to be:

$$V \pm k \in Class\ X = Result\ in\ Class\ X$$

The factorization will contain elements from class X, for example 3 from Class 1A and 2, 3 from Class 2A. We can also consider k being consecutive and a pseudoprime. An example is given from the following case where $k = 91$. This time k is not a prime, and $k = 7 \times 13$.

n	Factors	W = V-91	Factors
1	5,7	119	7, 17
2	5,7,11	2219	7, 317
3	5,7,11,13	29939	7, 7, 13, 47
4	5,7,11,13,17	510419	7, 13, 71, 79
5	5,7,11,13,17,19	9699599	7, 13, 15227
6	5,7,11,13,17,19,23	223092779	7, 13, 419, 5851
7	5,7,11,13,17,19,23,25	5577321659	7, 7, 7, 13, 1250801
8	5,7,11,13,17,19,23,25,29	161742330659	7, 7, 13, 13, 19531739
9	5,7,11,13,17,19,23,25,29,31	5014012253159	7, 13, 27799, 1982051
10	5,7,11,13,17,19,23,25,29,31,35	175490428863659	7, 13, 398459, 4839811

Table 416

At first we started with $k = 1$ and saw its particularity, then k was a prime greater than 1, but now k is a pseudoprime. Now we consider the example where E_1= {7, 19, 23, 25, 37}, hence $V = 16977450$ and $p_n = 37$ but 1 is not an element, $k_1 > p_n$ and $k_2 > p_n$ are not elements of E_1. This derives the following Euclidean factorization space.

n	k_1	k_2	$+k_1$	Factors $+k_1$	$-k_2$	Factors $-k_2$
7	41	43	16,977,491	31,547,661	16,977,407	17, 307, 3253
8	47	49	16,977,497	433, 39209	16,977,401	7,83,29221
9	53	55	16,977,503	937, 18119	16,977,395	5, 3395479
10	59	61	16,977,509	17, 157, 6361	16,977,389	11,11,13,43,251
11	65	67	16,977,515	5, 3395503	16,977,383	29, 151, 3877
12	71	73	16,977,521	11, 431, 3581	16,977,377	Prime
13	77	79	16,977,527	7, 2425361	16,977,371	Prime
14	83	85	16,977,533	Prime	16,977,365	5,239, 14207
15	89	91	16,977,539	1933, 8783	16,977,359	7, 2425337
16	95	97	16,977,545	5, 13, 19, 59, 233	16,977,353	Prime
17	101	103	16,977,551	89, 190759	16,977,347	653, 25999
18	107	109	16,977,557	29, 227, 2579	16,977,341	Prime

Table 417

For n = 7, 10, 12, and 18, we observe that $q_1 < p_n$ and they are not elements of the set E_1 whereas for n = 8, 9, 15 and 17 then $q_1 > p_n$. However, elements that are in E_1 and contained in a pseudoprime appear as factors less than p_n as expected.

The above leads to the observation that the $+k$ operation can actually be used to define the next prime number after p_n because where $q_1 < p_n$ and not in E_1, this generates prime numbers less than p_n. The condition for generating the next prime number after p_n is that E_1 must consist of elements that are consecutive, and that start from the prime number 5.

A simple example is made to derive the next prime number after 11 and finding out what value of $+k$ is necessary to achieve that. Hence $5 \cdot 7 \cdot 11 = 385$ and therefore $V = 6 \cdot 385 = 2310$. Consequently, to find the value of k that we must add, we add $k_1, and\ k_2$, and factorize these.

Therefore, when $n = 3$, we derive the next prime number to $p_n = 11$ which is 13. This occurs for the number 2327 when $k_1 = 17$.

385	5,7,11						
V	n	k_1 in GB4	k_2 in GB2	k_1	Factors	k_2	Factors
2310	0	-1	1	2309	Prime	2311	Prime
2310	1	5	7	2315	5, 463	2317	7, 331
2310	2	11	13	2321	11, 221	2323	23, 101
2310	3	17	19	2327	13, 179	2329	17, 137

Table 418

Note that the process found 23 before it found 17 because as we saw from the example, this is defined by 13. In a similar manner, we can find out when the next prime number occurs after 17. Interestingly, it happens at $n = 0$. This implies that $k = \pm 1$ has the possibility of generating the next prime number after p_n. Observe also in this example that the pseudoprime can also generate the next prime, where this occurs at $n = 13$.

85085	5,7,11,13,17						
V	n	k_1 in GB4	k_2 in GB2	k_1	Factors	k_2	Factors
510510	0	-1	1	510509	61, 8369	510511	19, 97, 277
510510	1	5	7	510515	5, 102103	510517	7, 72931
510510	2	11	13	510521	11, 46411	510523	13, 173, 227
510510	3	17	19	510527	17, 59, 509	510529	prime
510510	4	23	25	510533	17, 17, 83, 383	510535	5, 5, 102107
510510	5	29	31	510539	31, 43, 383	510541	307, 1663
510510	6	35	37	510545	5, 7, 29, 503	510547	139, 3673
510510	7	41	43	510551	prime	510553	prime
510510	8	47	49	510557	439, 1163	510559	7, 72937
510510	9	53	55	510563	37, 13799	510565	5, 11, 9283
510510	10	59	61	510569	prime	510571	103, 4957
510510	11	65	67	510575	5, 5, 13, 1571	510577	23, 79, 281
510510	12	71	73	510581	prime	510583	prime
510510	13	77	79	510587	7, 11, 19, 349	510589	prime

Table 419

1616615	5,7,11,13,17,19						
V	n	k_1 in GB4	k_2 in GB2	k_1	Factors	k_2	Factors
9699690	0	-1	1	9699689	53, 197, 929	9699691	347, 27953
9699690	1	5	7	9699695	5, 1939939	9699697	7, 28279
9699690	2	11	13	9699701	11, 593, 1487	9699703	13, 263, 2837
9699690	3	17	19	9699707	17, 17, 33563	9699709	19, 19, 97, 277
9699690	4	23	25	9699713	prime	9699715	5, 389, 4987
9699690	5	29	31	9699719	47, 47, 4391	9699721	23, 421727

Table 420

The next example is finding when the prime number 23 occurs, and this happens at $n = 5$, and the number is 9699721. Therefore the method has the approach that after choosing a position p_n:

a) finding whether all the primes before p_n exist,
b) what the next prime after p_n is, defined as p_{n+1}
c) or the set of prime numbers $(q_1, q_2 \cdots q_m)$ after p_n, where this is the Euclidean approach.

The following observation can be derived by observing the patterns already illustrated. We can define k in several ways, that is;

a) $k \in E_1$, where $k \leq p_n$
b) $k \notin E_1$, where $k \leq p_n$ or $k > p_n$
c) $k = 1$, which gives us the original Euclidean condition.
d) $1 < k, and\ k = q_1$, where $q_1 < p_n$ and q_1 is prime.
e) $1 < k, and\ k = q_0$, where q_0 is a pseudoprime

Furthermore we can say;

f) $k = q_1$ provided $k \leq p_n$. This is the same as $k \in E_1$.
g) $k = q_0$ provided $d_i \leq p_n$. This is because $q_0 = d_1 \times d_2 \cdots d_l$ and $d_i \in E_1$.

The above conditions define a criterion within which the observation operates. The case where $k = q_1$ is called predictive factorization since there is at least one prime factor that we already know if k is given.

Euclidean Factorization Criteria

Given that $W = V \pm k$, where $k \in D_t$ and $V = 6(p_1 \times p_2 \times \cdots \times p_n)$ for $p_i \in D_t$, therefore the factors of W are $(q_0, q_1, q_2 \cdots q_m)$, where $(q_1, q_2 \cdots q_m)$ are prime and q_0 is pseudoprime. Then factorization criteria are defined as follows:

<table>
<tr><th>Category of k</th><th>$k \in E_1$</th><th colspan="3">$k \notin E_1$</th></tr>
<tr><td>Standard Euclidean Function
$k = 1$
k is prime</td><td>$k = q_1$
$q_2 \notin E_1$
$p_n < q_2$</td><td colspan="3">Not defined</td></tr>
<tr><td>Semi-Advanced Euclidean Function
$1 < k$
k is prime</td><td>$k = q_1$
$q_1 \in E_1$
$q_1 \leq p_n$</td><td><u>Class 1</u>
$k \neq q_1$
$q_i \notin E_1$
$q_1 < p_n$</td><td colspan="2"><u>Class 2</u>
$k \neq q_1$
q_1=1
$q_i \notin E_1$
$q_2 > p_n$</td></tr>
<tr><td>Advanced Euclidean Function
$1 < k$
k is pseudoprime</td><td>$k = q_0$
$q_0 = d_1 \times d_2 \times \cdots d_m$
$q_0 \in E_1$
$d_i \in E_1$
$d_i \leq p_n$</td><td>$k = q_0$
$q_0 \notin E_1$
$d_i \in E_1$
$d_i \leq p_n$</td><td>$k = q_0$
$q_0 \notin E_1$
$q_i \notin E_1$
$q_1 < p_n$</td><td>$k = q_0$
$q_0 \notin E_1$
q_1=1
$q_i \notin E_1$
$q_2 > p_n$</td></tr>
</table>

Figure 163

The examples are given as follows, where $p_n = 37$ and E_1= {7, 19, 23, 25, 37}. We start with $k \in E_1$.

<table>
<tr><th>Value of k</th><th>Criteria</th><th>Example</th></tr>
<tr><td>$k = 1$
k is prime</td><td>$k = q_1$
$q_2 \notin E_1$
$p_n < q_2$</td><td>a) 16,977,451; Factors 137, 123923 and $q_2 = 137$
b) 16,977,449; Factors 71, 239119 and $q_2 = 71$
c) 71 and 137 $> p_n = 37$</td></tr>
<tr><td>$k > 1$
k is prime</td><td>$k = q_1$
$q_1 \in E_1$
$q_1 \leq p_n$</td><td>a) 16,977,455; Factors 5, 11, 308681 and $q_1 = k = 5$
b) 16,977,473; Factors 23, 738151 and $q_1 = k = 23$
c) 5 and 23 < $p_n = 37$</td></tr>
<tr><td>$k > 1$
k is pseudoprime</td><td>$k = q_0$
$q_0 = d_1 \times d_2 \times \cdots d_m$
$q_0 \in E_1$
$d_i \in E_1$
$d_i \leq p_n$</td><td>a) 16,977,425; Factors 5^2, 97, 7001 and $k = q_0 = 25, d_1 = 5, d_2 = 5$
b) 16,977,485; Factors 5, 7, 41, 11831 and $k = q_0 = 35, d_1 = 5, d_2 = 7$
c) 5 and 7 $< p_n = 37$, and in each case $q_0 \in E_1$</td></tr>
</table>

Table 421

We now consider $k \notin E_1$.

Value of k	Criteria	Example
$k = 1$ k is prime	Not defined	Not defined
$k > 1$ k is prime	***Class 1*** $k \neq q_1$ $q_i \notin E_1$ $q_1 < p_n$	a) **16,977,509**; Factors 17, 157, 6361 and $q_1 = 17, k = 59$ b) **16,977,389**; Factors 11,11,13,43,251 and $q_1 = 11, k = 61$. c) Similarly, 11 and 17 < $p_n = 37$ d) In each case $q_i \notin E_1$
$k > 1$ k is prime	***Class 2*** $k \neq q_1$ $q_1 = 1$ $q_i \notin E_1$ $q_2 > p_n$	a) **16,977,551**; Factors 89, 190759 and $q_2 = 89, k = 101$ a) **16,977,347**; Factors 653, 25999 and $q_2 = 653, k = 103$ b) Implied assumption q_1=1 b) Also 89 and 653 $> p_n = 37$ c) In each case $q_i \notin E_1$
$k > 1$ k is pseudoprime	***Class 1*** $k = q_0$ $d_i \in E_1$ $d_i < p_n$	a) **16,977,359**; Factors are 7, 2425337 and $d_1 = 7, k = q_0 = 91$ b) **16,977,545**; Factors are 5, 13, 19, 59, 233 and $d_1 = 5,\ d_2 = 19, k = q_0 = 95$ c) Also 5, 7 and 19 $< p_n = 37$ d) In each case $d_i \in E_1$
$k > 1$ k is pseudoprime	***Class 2*** $k = q_0$ $q_i \notin E_1$ $q_1 \leq p_n$	a) **16,976,999**; Factors are 13, 17 , 76819 and $q_1 = 13,\ q_2 = 17,\ \ k = q_0 = 451$ b) **16,979,471**; Factors are 29, 317, 1847 and $q_1 = 29, k = q_0 = 2021$ c) Also 13, 17, and 29 $< p_n = 37$ d) In each case $q_i \notin E_1$
$k > 1$ k is pseudoprime	***Class 2*** $k = q_0$ q_1=1 $q_i \notin E_1$ $q_2 > p_n$	c) **16,977,593**; Factors are 103, 164831 and $q_2 = 103, k = q_0 = 143$ d) **16,975,687**; Factors are 197, 86171 and $q_2 = 197$, $k = q_0 = 1763$ e) Implied assumption q_1=1 f) Also 103, and 197 $> p_n = 37$ g) In each case $q_i \notin E_1$

Table 422

14.5.1 The P_{gap} and Q_{gap} Populations

It is observed that if W is a pseudoprime, then $E_2 \subset E_3$ and $W > q_m$, but if W is prime then $E_2 \not\subset E_3$ and $W = q_1$. Infact we can assume that $q_1 = q_m$. Therefore in terms of the construct, the observation is that for every p_n there exists a number W and q_m. The three numbers related to create the three sets that have been defined. Consequently we can also define a the gaps;

$$P_{gap} = W - p_n$$

$$Q_{gap} = W - q_m$$

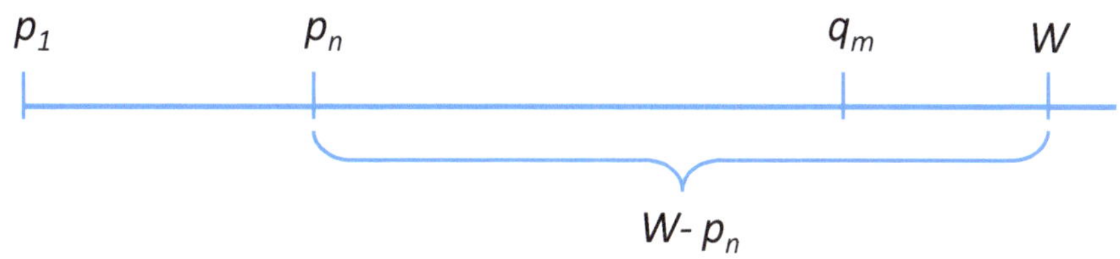

Figure 164

It is obvious that $P_{gap} > Q_{gap}$ since $p_n < q_m$. Now here are the interesting statistics. As usual we note the fact that the number of primes or pseudoprimes for gap base 2 and gap base 4, the number is almost exactly the same. The space being considered here is [0,45] and [0,85079] using $f(x) = 6x + 7$. It's a technicality that means 1 and 5 are excluded from the count. Note again the fact that the distribution occurs in a fixed space, hence the total must be the same both sides like an accounting system of credit and debit.

	Gap Base 2	Gap Base 4		
Gap 2 primes	17	16		E_1 ={5,7}
Gap 6 primes	6	11		
Over gap 6 primes	4	3		W=211
TOTAL	27	30	62%	
Pseudoprimes	19	16	38%	
[0,45] space	46	46	100%	
Gap 2 primes	4,635	4,629		E_1 ={5,7,11,13,17}
Gap 6 primes	3,768	3,836		
Over gap 6 primes	12,731	12,729		W=510511
TOTAL	21,134	21,194		
Pseudoprimes	63,946	63,886		
[0,85079] space	85,080	85,080		

Table 423

Our main interest lies in the fact that $E_1 = \{5,7\}$ and $E_2 = 211$, the number of primes from 7 to 211 is (27+30)= 57. Similarly, if $E_1 = \{5,7,11,13,17\}$ and $E_2 = \{19,97,277\}$ the number of primes between $q_m = 277$ and $W = 510511$ is given by the table as (21108+21165)= 42273.

	Gap Base 2	Gap Base 4	
Gap 2 primes	4,618	4,614	
Gap 6 primes	3,763	3,825	
Over gap 6 primes	12,727	12,726	
TOTAL	21,108	21,165	25%
Pseudoprimes	63,927	63,870	75%
[45,85079] space	85,035	85,035	100%

Table 424

Therefore whilst we expect more prime numbers between $q_m = 277$ and $W = 510511$, it is noted that there is also a dramatic increase in the number of pseudoprimes. The Q_{gap} population has (63927+63870)= 127797 pseudoprimes whilst the other had only (19+16)=35 pseudoprimes. The point is to illustrate that the Q_{gap} can be fairly large, and that it will consist mostly of pseudoprimes. Secondly, it actually defines a prime space $[a, b]$ =

$[q_m, W]$ where the number of pseudoprimes is expected to increase as p_n increases. This is an average distributive behavior since we know that $q_m = W$ is a possible occurrence.

n	Base sequence	$W = V + 1$	q_m	$SQRT(W)$
1	5,1	31	prime	6
2	5,7	211	prime	15
3	5,7,11	2311	prime	48
4	5,7,11,13	30031	509	173
5	5,7,11,13,17	510511	277	715
6	5,7,11,13,17,19	9699691	27953	3114
7	5,7,11,13,17,19,23	223092871	703763	14936
8	5,7,11,13,17,19,23,25	5577321751	3016399	74681
9	5,7,11,13,17,19,23,25,29	161742330751	79669	402172
10	5,7,11,13,17,19,23,25,29,31	5014012253251	4920522329	2239199
11	5,7,11,13,17,19,23,25,29,31,35	175490428863751	3176469833	13247280

Table 425

The next observation is that given the two possibilities that W is either a prime and equals q_m or it is a pseudoprime greater than q_m, then q_m must at least be greater than $SQRT(W)$ on average. Therefore $q_m > SQRT(W)$ implies that as W increases, so does q_m. This observation is similar to the fact that we have a twin prime gap, but at the same time we can calculate and estimate the average gap between primes and predict that it is generally on an increase. The same result is used in finding prime factors of a number, that is, one of them being less than $\sqrt{W}$.

This observation can be taken to support Euclid's proof because as W increases, so does q_m, which implies that as W goes to infinity, the same is true for q_m. We can even put the expression in a more interesting manner;

$$(q_m)^2 > W$$

Another set defined as E_4 can also be defined since we have to consider the number of pseudoprimes.

Definition

Let the set of all pseudoprimes $E_4 = \{u_1, u_2, u_3, \cdots u_n\}$, *where* $u_1 > p_n$ *and* $u_n < W$.

Generally it appears that as in the space $[q_m, W]$ we have $n(E_4)$ increasing due to the pseudoprime law, as W increases in size. Note that q_m is controlled by the value of W .

14.6 Concept Of The Last Prime

Euclid's proof led to the conclusion that there are an infinite number of primes on the assumption that there is always a prime that can be generated that is outside the finite set consisting of all prime numbers. The proof did not take into consideration the fact that there are pseudoprimes, and the relationship between prime numbers and pseudoprimes. Consequently the proof did not consider what was creating the gap between the prime numbers, and how this relates to distribution. For example we can consider the question; what if Euclid's assumption is true to a certain extent, in the sense that there exists a last prime? Therefore, this section considers the concept of a last prime, and whether this is a reasonable assumption to make.

A hypothesis may be made that there is a last prime number on the basis of the following.

a) The number of pseduoprimes in $[a, b]$ = $[q_m, W]$ is increasing compared to the number of primes. Therefore, for a given space, the pseudprimes are taking more and more positions in the sieve.

b) The pseuodprimes are self-generating according to $n\{f(x_i) \cdot f(x_j)\}$.

$f(x_i)$	$f(x_j)$	**Product Result**	**Comment**
Prime	Prime	Pseudoprime	Prime Dependent
Prime	Pseudoprime	Pseudoprime	Prime Dependent
Pseudoprime	Prime	Pseudoprime	Prime Dependent
Pseudoprime	Pseudoprime	Pseudoprime	Self-generating
Self-generating	Pseudoprime	Pseudoprime	Self-generating
Pseudoprime	Self-generating	Pseudoprime	Self-generating
Self-generating	Self-generating	Pseudoprime	Self-generating

Table 426

Of interest is the fact that one or two pseudoprimes that are the result of self-generation continue the process of self-generation of pseudoprimes, where this is independent of the formation of a new prime number.

c) It is also important to consider if there is a general increase in the number of pseudoprimes between W and q_m. That is, does Q_{gap} keep on increasing? The expectation is that this should be the case, given that pseudoprimes are self-generating. This may have the implication that the probablity of W being pseudoprime becomes 1 either as we approach infinity or at infinity.

d) The set E_3 consists of prime numbers such that $q_m \in E_3$. Since this set consists of all prime numbers less than W, it then implies that the last prime number if it does exist will be found here.

e) Now $n(E_1) \leq n(E_1 \cup E_2)$ since $E_2 \neq \emptyset$ as W will always have factors and $n(E_1 \cup E_2) < n(E_3)$ since primes in E_3 are determined by the value of W and P_{gap}, where $P_{gap} > Q_{gap}$. Hence $n(E_2) < n(E_1) < n(E_3)$. This explains that while the Euclidean condition holds, therefore E_1 will always be a finite set. The fact that E_1 will always be a finite set is significant because it answers the question as to whether we can interpret Euclid's proof to imply that there are an infinite number of primes. Since E_2 is an observation set, and E_3 is determined from E_1, then E_3 will consequently be a finite set because it is not produced independently of E_1. Hence $n(E_2) < n(E_1) < n(E_3)$ defines an Euclidean space in which a prime number may be found.

Therefore, tentatively from this relationship $n(E_2) < n(E_1) < n(E_3)$ we can suggest the following.

Hypothesis

There exists a last prime number $l_p \in E_3$.

Let $E_u = E_1 \cup E_2 \cup E_3$. There are two possibilities for W as already indicated. That is, $W = q_m$ because this will give the biggest prime on the basis of the Euclidean condition. However for $W \neq q_m$, it implies W is pseudoprime, and $E_2 \subset E_3$. Hence $l_p < W$, where l_p is the last prime number in E_3. The implication is that E_u is also a finite set. Now $q_m \in E_3$ if W is pseudoprime. If the last prime number in E_3 is l_p, it is noted that $q_m < l_p$ but $W > l_p$ provided W is pseudoprime. This is called the critical condition of the Euclidean premise. That is, E_3 will always have a last prime l_p such that $q_m < l_p$. Since $W > l_p$, then $W \in \infty$ is valid even if l_p is finite. Therefore, according to the hypothesis;

$$q_m < l_p < W$$

But if we assume this to be correct, it actually implies that the set E_3 is continuously having new prime numbers such that we are able to say $q_m < l_p$. But if there is a last prime indeed, then there is a probability of one that $q_m = l_p$. Therefore an accurate statement is;

$$q_m \leq l_p < W$$

If there is a last prime, then it also implies that there exists a last value of the product of all primes defined as p_t. Therefore:

$$p_n \leq p_t$$

Furthermore, if we assume a last prime number, then the critical implication is that W is always a pseudoprime since $q_m = l_p$.

Now we know that if $p_n < k$, then k generates new prime numbers $p_n < q_i$. Then it means $k = l_p$ will be generated by some value of p_n, where this is defined as p_t. And similarly, there exists a value of $W = W_t$ that defines p_t or the last prime. Hence there are three special numbers:

$$p_t,\ l_p, and\ W_t$$

We then define pseudoprimes greater than W_t as W_g, hence $W_t < W_g$. Since there are no new primes after l_p, pseudoprimes after W_g define a region consisting mostly of self-generating pseudoprimes. That is;

$$k > l_p$$

generates only pseudoprimes and no new prime numbers. All of the above then define a concept of the last prime.

14.6.1 Law of Mutually Exclusive Infinity

Euclid's assumption had not taken into context the fact that there are pseudoprimes, and that the position that a prime number occupies is governed by a specific law, the law of pseudoprimes. If there exists a prime number l_p, and W is a pseudoprime at infinity, then pseudoprimes are being consistently generated independently of prime numbers since no new prime numbers are being generated. That is, W at infinity implies pseudoprimes are infinitely many and self-generating.

The hypothesis then leads to a revision in the way we conceptualize or think about the prime number by posing the following possibilities.

I. Case 1. Prime numbers are finite in number, and finite in size.
II. Case 2. Prime numbers are finite in number, and infinite in size.
III. Case 3. Prime numbers are infinite in number, but finite in size.
IV. Case 4. Prime numbers are infinite in number, and consequently infinitely large in size.

There are two possibilities in regard to q_m when W is at infinity. The first is shown by graphic A, where q_m is an element of the finite space. The second is q_m is an element of the infinite space as indicated in graphic B. In other words, problem question is:

Can q_m and W be both elements of the infinite space?

The standard argument assumes that q_m can indeed be at infinity, which is why we can further deduce that there are an infinite number of primes from Euclid's proof. But in the approach here, that assumption has not been taken for granted. The question is relevant in the light of the inequality $q_m \leq l_p < W$. It is noted that E_u is a finite set, therefore q_m can only be finite. Hence q_m can only be an element of the infinite space provided W is a pseudoprime. The following conditions are postulated. Graphic A and B illustrate the idea.

1. That q_m is finite but W is infinite
2. That q_m is infinite and W is infinite

There is nothing to suggest that the second condition is not a possibility, and the fact that W can be infinite is not really questionable. This condition is then considered in the context of the last prime l_p. This condition is the key determinant.

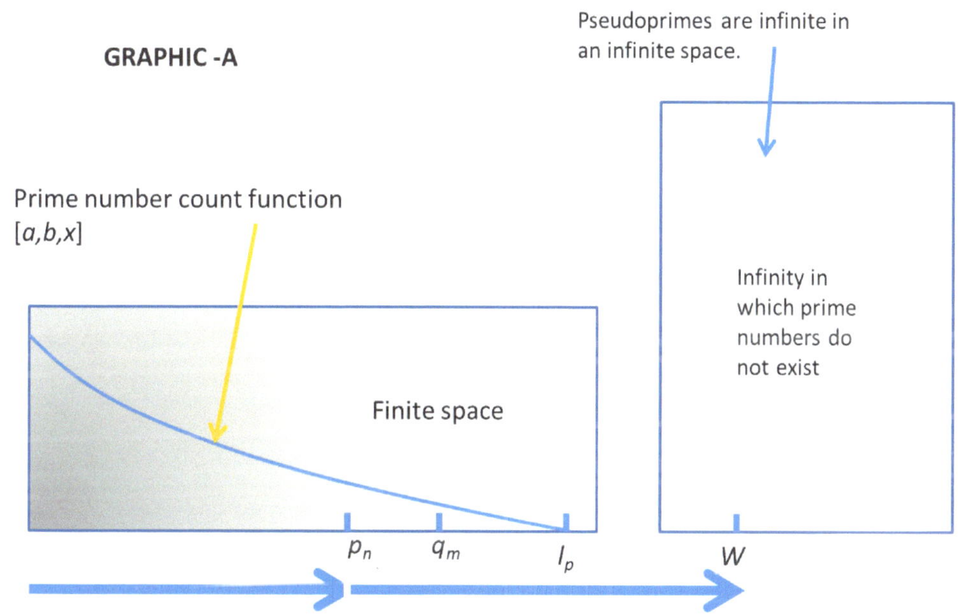

Figure 165

Hence it is further hypothesized that the following are strong possibilities for consideration:

1. W as a pseudoprime can be infinitely large.
2. The number of pseudoprimes W are infinitely many.
3. Prime numbers exist in a finite space, consequently l_p is finite, but W can be simultaneously infinite.
4. Prime numbers exist in an infinite space, that is q_m is at infinity, but are finite in that space as a consequence of the assumption of the last prime l_p.
5. If the number of pseudoprimes W are infinite, then pseudoprimes in general are infinite.
6. Pseudoprimes exist in an infinite space where there are no prime numbers, since there exists a last prime l_p.
7. Composite numbers exist both in the finite and infinite space occupied by prime numbers and pseudoprimes.

Therefore we derive the law of mutually exclusive infinities with regard to prime numbers and pseudoprimes.

Law of Mutually Exclusive Infinity

For any infinitely large prime space $[a,b]$, It is impossible for prime numbers and pseudoprimes to be both infinite simultaneously in the same space.

The two are not friends, they are in constant confrontation for a specific 8-position framework in the context of the relationship *Probability(Prime Occurrence)* $\leq 8.33\%$. Hence this probability tends to zero towards infinity.

GRAPHIC -B

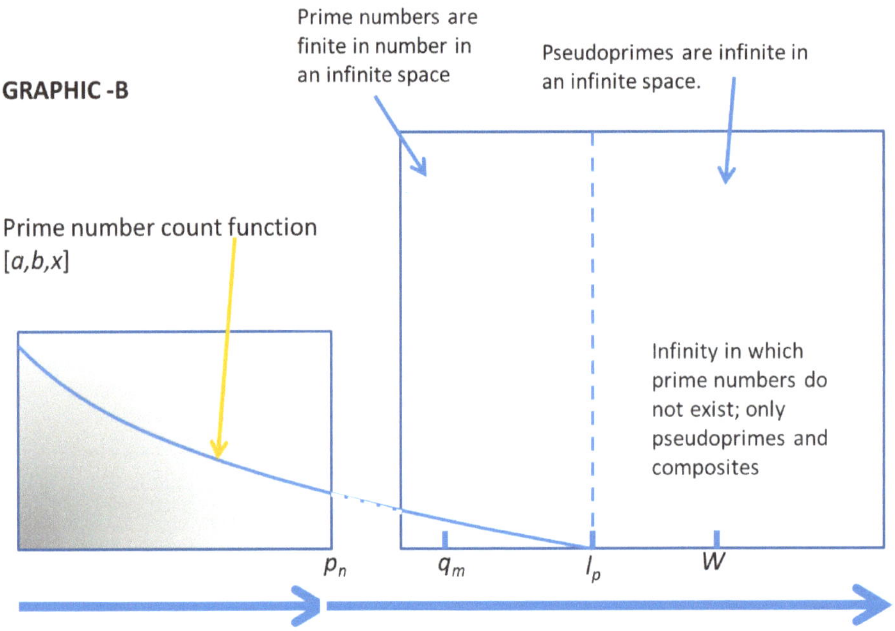

According to these observations, we therefore have two main possibilities.

a) We have a finite space and an infinite space in which q_m and l_p can exist.

b) We have a finite space p_n, an infinite space with finite prime numbers space in which q_m and l_p can exist, and an infinite space with no prime numbers but only pseudoprimes.

All these define a number spectrum in that sequence as shown by the two graphics. If the law of mutually exclusive infinity holds, then option B is our answer. The interesting assumption about graphic B is the fact that there exists a situation at infinity where we only have composite numbers and pseudoprimes, and prime numbers do not occur. This may look an unusual assumption to make until we note that such a pattern exists on a small scale at low values as shown in the table below. Therefore we assume it is amplified at infinity.

X	Family-1	Family-2			Family-3		
	Class1A	Class1B	Class2A	Class2B	GB-4	GB-2	
1082	6489	6490	6492	6494	6491	6493	
1083	6495	6496	6498	6500	6497	6499	Continuous band of composite numbers with no prime occurrence
1084	6501	6502	6504	6506	6503	6505	
1085	6507	6508	6510	6512	6509	6511	
1086	6513	6514	6516	6518	6515	6517	
1087	6519	6520	6522	6524	6521	6523	

Table 427

The numbers 6491 and 6521 are prime, therefore our continuous band consisting of composites and pseudoprimes only is from $x = 1083$ to 1086. The colored band of 6495 to 6518 shows a spectrum within the number distribution where there is no prime number occurrence, only composites and pseudoprimes. This implies that the model of number distribution and interpretation of infinity with respect to the type of distribution in numbers is then a possibility. The context of the argument is that such a band will have to be infinitely large and without an end prime.

This concept is the same as saying the prime gap on average is increasingly getting bigger. An infinitely large prime gap can only make sense provided there is a last prime l_p. If that it is the case, then as a consequence there is no twin prime gap or other smaller gaps between primes.

14.6.2 Proof of Law of Mutually Exclusive Infinity

We begin by defining the following framework.

Let $f(x) = 6x \pm 1$, where $f(x) \in D_t$, and $D_t = f_1(x) \cup f_2(x)$ provided $f_1(x)$ represents a prime number and $f_2(x)$ is defined as a pseudoprime. Therefore;

a) $f_1(x) = 6x + 1$ defines a prime number, and $f_2(x) = 6x + 1$ defines a pseudoprime and this is referred to as gap base 2.
b) $f_1(x) = 6x - 1$ defines a prime number, and $f_2(x) = 6x - 1$ defines a pseudoprime and this is referred to as gap base 4.
c) $f(x) \neq 6x \pm 1$ defines a composite.

The above forms the fundamental gap theorem of numbers. It results in a construct called the UNDT framework, or an algebraic sieve. Now let $W \in D_t$ where;

$$W = f(x_i) \cdot f(x_j)$$

then W is not a prime number, hence it is a pseudoprime. Consequently, W defines a pseudoprime generation. Similarly, $f(x_i)$ and $f(x_j)$ are called ditation factors of W, hence such factors are either prime or pseudoprime. For example $W = 125$ has ditation factors 5 and 25 where both satisfy the definition $f(x) = 6x \pm 1$. The prime factors of W are 5, 5, and 5.

For $x_1 = a$ and $x_2 = b$ that are constant, then $f(x_1) = 6a - 1$ and $f(x_2) = 6b + 1$. If $a < b$, then $(6b + 1) - (6a - 1) = 6(b - a) - 2$. Now since the difference $(b - a)$ is constant, then this defines a prime space $[a, b]$ such that;

$$f(b) - f(a) = n_1 + n_2 + n_3$$

where;

- n_1 number of primes
- n_2 number of pseudoprimes
- n_3 number of composites, provided a composite is any number z in the same range such that $z \notin D_t$.

This is referred to as the law of integer distribution. Therefore a prime space consists of all the three types of numbers, and provides a relationship between them. This then defines the P-index that shows the dynamic relationship between the prime numbers, the pseudoprimes and the composites. That is;

$$P(prime) = \frac{n_1}{f(b) - f(a)}$$

$$P(pseudoprime) = \frac{n_2}{f(b) - f(a)}$$

$$P(composite) = \frac{n_3}{f(b) - f(a)}$$

Since $z \notin D_t$, then $n_1 + n_2 = c$, a constant for a given $f(b) - f(a)$. That is, we can focus on the algebraic sieve only in studying the distributive behavior. The implication of this is that the composites do not influence the distribution of prime numbers, resulting in the following law. They are a distinct autonomous class of numbers.

Distributive Autonomous Law

The distributive behavior of prime numbers and pseudoprimes is independent of composite numbers.

From this rule, we can focus on $n_1 + n_2 = c$ as sufficient to consider prime number behavior. The rule also means the composite numbers do not influence the prime number gap.

If the difference $(b - a)$ is as large as we want, then $1 \leq c \leq \infty$. Now consider sliding the difference $(b - a)$ then $a = x_a$ or $b = x_b$ defines a position of the space $[a, b]$ on the algebraic sieve. Hence for a given space $[a, b]$ then we derive $2(x_b - x_a + 1) = c$. Consequently, we have;

$$2(x_b - x_a + 1) = n_1 + n_2$$

This indicates a bounded region for primes and pseudoprimes. Similarly, if x_b locates the space $[a, b]$ in the set D_t, then $1 < x_b < \infty$, and the location is defined as $[a, b, x_b]$. Now, because of the generative property of pseudoprimes, then for a given $[a, b, x_b]$ and the number of pseudoprimes n;

$$n\{f(x_i) \cdot f(x_j)\} \ increases \ as \ x_b \to \infty$$

Hence the distributive hypothesis is defined as follows for prime numbers.

Distributive Hypothesis of Prime Numbers

Given the following conditions of distribution for prime numbers:

a) $f(x) = 6x \pm 1$, where $-\infty < Int(x) < \infty$
b) $f(x_i) \cdot f(x_j) \in D_t$
c) $[a, b]$, where a and b are fixed as the initial variables of the prime space
d) $n_1 + n_2 = c$, and c is constant
e) $1 \leq c \leq \infty$
f) $1 < x_b < \infty$, where x_b locates $[a, b]$

Then $n\{f(x_i) \cdot f(x_j)\} = n_2$ $increases \ as \ x_b \to \infty$, *consequently* $n_1 \to 0$.

Hence for *P* the probability, if $n_1 \to 0$, then $P(Prime) \to 0$ which means the P-index will decrease towards zero. The hypothesis explains the formation of the prime number gap, explains why prime numbers are closer at the beginning, and demonstrates the concept of the last prime number.

The distributive hypothesis of prime numbers can be given a context, which in this case is the Fixed Population Theory of the Prime Number (FPT-PN). According to the FPT-PN, $[a, b]$ can be precisely defined as a population space for prime numbers. Therefore applying the theory, assume that;

$$[a, b] = [a, a + 18] \text{ for a cycle } 1 - 7 - 3 - 9$$

Then we are certain that $P(Prime) \to 0$ in the population space $[a, b]$. But we have to also ascertain where and how in terms of the context of distribution.

Therefore consider a Euclidean distribution, where this is defined as a system that consists of an input set and an output set that relates primes numbers and pseudoprimes in a particular manner. Let there be a finite set $p_1, p_2, \cdots p_n$consisting of primes in sequence. This is called the initial set E_1. Then define the Euclidean prime number distribution conditions as follows.

1) $W = V \pm k$, where $k \in D_t$ and is a prime number, but W is either prime or pseudoprime.
2) $V = 6x$, where $x = p_1 \times p_2 \times \cdots \times p_n$ where p_i is prime.
3) For W a pseudoprime, an output set $E_2 = \{q_1, q_2, q_3 \cdots q_m\}$ is defined to consist of factors of W.
4) Two conditions will exist, either for $W = 6kf \pm k = k(6f \pm 1)$ then f is integer or decimal since k is an integer. For f is an integer, the following applies.
    a. If $k \in V$, then k will obviously be a factor of W, and $k < p_n$. Then the condition:
    $$p_1, p_2, \cdots p_n < q_1, q_2, q_3 \cdots q_m$$
    holds and $k = q_1$ where $q_1 \neq 1$. Therefore for $k < p_n$, we have $q_1 < q_2$, so k will always be the first prime factor.
    b. But if $k = 1$, then $W = (6f \pm 1) \in D_t$. This is the Euclidian assumption, and therefore this will derive $p_1, p_2, \cdots p_n < q_1, q_2, q_3 \cdots q_m$ if W is pseudoprime. This explains why the result is either prime or pseuodprime.
    c. If $k = 1$ and W is a prime number, then $p_1, p_2, \cdots p_n < q_2$ and $W = q_2$, because it is assumed that $k = q_1 = 1$
5) Now let $k > p_n$, then $k \notin V$. Hence for $W = k(6f \pm 1)$ then $V = 6kf$, and k is an integer. Hence if $k \notin V$, then f is a decimal, the second condition.
    a. For W is a pseudoprime, then $p_1, p_2, \cdots p_n < q_1, q_2, q_3 \cdots q_m$, where $k \neq q_i$ since $k \notin V$. Therefore the prime number k generates a new set of prime numbers. That is, new prime numbers greater than p_n are generated on the basis of two conditions; by $k = \pm 1$ or $k > p_n$.
    b. If $k < q_1$ and $k \notin V$ then k is the next prime number after p_n. This provides a test for identifying the next prime number of p_n.
    c. Therefore for $k > p_n$ then $k < q_1$or $k > q_1$.
    d. If $k > q_1$, then $q_1 < k < q_m$ since $k \neq q_i$ provided $k < V$. This is derived from the observation that in general $(q_m)^2 > W$.
    e. If $k > V$ then $k > q_m$.
6) Therefore we have three ways of determining the next value of p_n.
    a. For $W = V \pm k$, W can either be prime or pseudoprime. If prime, then it determines a new value of p_n.
    b. For $k < q_m$, it implies we can define p_n up to q_m, giving us a new value of p_n.
    c. If $k > q_m$,then it defines a new value of p_n.

The above gives a firm relationship between p_i, q_i and k. Now $E_1 = \{p_1, p_2, \cdots p_n\}, p_i \in D_t$, where this is the set of prime numbers and E_1 is a finite set. Assuming that W is a pseudoprime and $k > p_n$, then there exists a set

$E_2 = \{q_1, q_2, \cdots q_m\}, q_i \in D_t$ where these are prime factors of W, and it has been established that $k \notin V$ if $k > p_n$. Hence:

Postulate

For every set E_1 and E_2, there exists a set E_3 consisting of prime numbers $\{r_1, r_2, \cdots r_m\}$ such that;

a) $p_n < r_1$
b) $q_m \leq r_m < W$
c) $p_n < k \leq r_m$
d) Either $n(E_1) < n(E_2)$ or $n(E_1) > n(E_2)$

Therefore E_3 consist of prime number less than W but greater then p_n. Consequently we can define the following.

Definition

There exists a set of all pseudoprimes $E_4 = \{u_1, u_2, u_3, \cdots u_n\}$, where $u_1 > p_n$ and $u_n < W$.

As a result, $E_3 \cup E_4$ defines the prime space $[p_n, W]$. Consequently;

a) E_3 results in a prime space $[p_n, W]$ if we include the pseudoprimes.
b) $n(E_4)$ increases as W increases in $[p_n, W]$.
c) $p_n < r_1 < W$, hence $E_1 \not\subset E_3$.
d) If $p_n < r_1$ then $p_n < r_m < W$, hence it is possible that $r_i \in E_2$. That is, if W is a pseudoprime, then $E_2 \subset E_3$.
e) $E_2 = E_3$ if $W = q_2$ since W is a prime number. That is $k = 1$.

In $W = V \pm k$, let $k > p_n$, which means we are constantly generating new prime numbers. If the last prime number in E_3 is $r_m = l_p$, then;

$$q_m \leq l_p < W$$

Let us assume that W is a pseudoprime and the distributive hypothesis applies. Now if W is at infinity, still we have $q_m < l_p = r_m < W$. Since $x_b \to \infty$, consequently $n_1 \to 0$, and the above is an algebraic prime space, then it implies $n_1 \to 0$ for W at inifnity. The center of the argument is W, that it is a pseudoprime.

Now either;

$$n(E_1 \cup E_2) > n(E_3) \text{ as } W \to \infty \text{ or } n(E_1 \cup E_2) < n(E_3) \text{ as } W \to \infty$$

On the basis of the the algebraic space $[q_m, W]$, and its distributive property as determined by the distributive hypothesis of prime numbers, then;

a) $n(E_1 \cup E_2) < n(E_3)$ if W is finite
b) $n(E_1 \cup E_2) > n(E_3)$ if W is infinite since $n_1 \to 0$ as $x_b \to \infty$ where E_3 defines a space $[p_n, W]$.

Therefore from $q_m \leq l_p < W$ as $n_1 \to 0$ means we seek l_p, the last prime being a constant. This is refered to as the compression behaviour of prime numbers due to the increase of pseudoprimes that are self-generating. Hence the deterministic conditions are:

a) $W = V \pm k$, where $k > p_n$
b) Let $x_b = \infty$, hence $W = W_\infty$ is at at infinity
c) Therefore the distributive hypothesis applies; $n_1 = 0$ such that $q_m \leq l_p < W_\infty$, then the condition implies that the P-Index = 0.

d) Whereupon we derive $n(E_1 \cup E_2) > n(E_3)$, and $l_p \in E_3$

e) Consequently $k > l_p$ and no new prime numbers are generated.

f) Hence W_∞ is a self-generating pseudoprime.

Now E_3 is finite and generally $q_m > SQRT(W)$, that is $(q_m)^2 > W$. In that case W is finite, since such an operation is not defined at infinity. That being the case, then q_m is a finite quantity, indicating that there exists a last prime number l_p in a finite space. The number l_p is extremely large but still finite. Similarly, when $W = W_\infty$ at infinity and l_p constant, then as any quantity in a finite space $(q_m)^2$ is less than a quantity W_∞ in an infinite space, it implies;

$$(q_m)^2 < W_\infty$$

If q_m is always finite, then the prime number space is finite with respect to the pseudoprime space of W_∞. The relationship is not absolute but relative, hence they are mutually exclusive in terms of their distribution in space with respect to infinity. They cannot both occupy the same infinity at the same time.

Hence when $n_2 = W_\infty$ then $P(Prime) = 0$, and there exists the last prime l_p such that the law of mutually exclusive infinity holds.

■

What remains is to find l_p through some super computer.

14.6.3 Concluding Argument of Proof

Note that the argument has not involved the set of composite numbers. This does not mean they are absent, but it has been established that prime number behavior is independent of the set of composite numbers.

The proof has the following results.

1. It indicates that only two types of numbers exist at infinity where we have the pseudoprime W_∞, the pseudoprimes and the composites, there are no prime numbers.
2. That being the case, then there exist a last twin prime pair, which means twin primes are not infinite.
3. Similarly, there exists a last quad prime pair, and such pairs also are not infinite.

Between the twin pair and the quad pair, it will be interesting to guess which one comes first.

The proof provides a platform for hypothesizing how the integers distribute. The following laws contribute towards development of a model on how integers distribute.

1. The Prime Gap Law
2. The Law of Pseudoprimes
3. Gap Population Law
4. Pseudoprime Gap Law
5. Prime Gap Structure Law
6. The Co-existence Theorem
7. Fixed Space Distribution Law
8. Law of Integer Distribution

They are supported by the following laws as well.

Law	Consequence and attribute of law
Distributive Autonomy Law	*The distributive behavior of prime numbers is independent of composite numbers.* Consequence and attribute This arises from $n_1 + n_2 = c$, where c is constant in a space $[a, b]$. The primes and pseudoprimes do not involve the composites in their fight for position.
Law of Mutually Exclusive Infinity	*For any infinitely large prime space $[a, b]$, It is impossible for prime numbers and pseudoprimes to be both infinite simultaneously in the same space.* Consequence and attribute There is only one winner in this game of probability, and that winner is the pseudoprimes.

Table 428

The fixed space distributive law is almost similar to the distributive autonomy law. The fixed space law states that the space consisting of primes and pseudoprimes is an independent space whereas the other law emphasizes the relationship between primes and composites. The fixed population theory relates the same issues in terms of probability and distribution.

One can then consider all these laws and come up with a concept of how the number space looks like in the context of UNDT. It must be remembered that the concept of such a space began with the conceptualization of the natural numbers to the real numbers which also had its own geometric foundation of the number line. Infact infinity has had the influence of the real numbers in the way it was developed. This was also supported by the concept of sets, where these could be finite or infinite. Hence we have had five constructs with firm conceptual boundaries;

a) The positive numbers
b) The negative numbers
c) Zero
d) Finite space
e) Infinite space

The finite and infinite space have been understood and interpreted in the context of arithmetic operations rather than in the context of integer properties of distribution and behavior. As a result, it becomes natural to consider a revised approach that accommodates distributive properties of integers. The positive numbers in particular are defined to distribute according to the law of integer distribution that expresses a precise relationship between the three type of numbers. However the real number concept, the decimal presentation of numbers, and the real number line have provided a limited scope of development in studying how integers actually distribute and behave. There has been a lack of paradigm shift in this area, where this has also had an impact on the conceptualization of infinity.

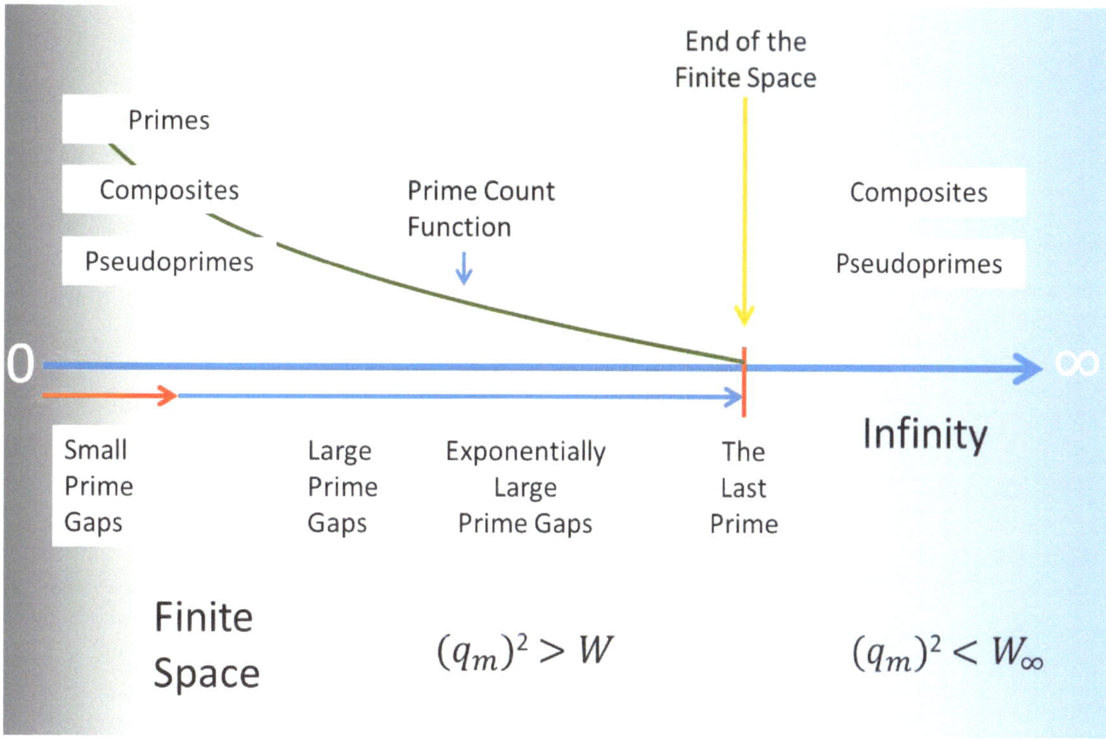

Figure 166

This then is the possible linear and relational model for the distribution of integers, where we assume positive numbers. It is also assumed that it is practical to consider the last prime as an appropriate definition of the end of the finite space.

It can be further noted that the limitation of our conceptualization is the fact that we assume that there is a fixed finite space and a firm boundary where the infinite space begins. Normally in mathematics we express the concept of x approaching infinity, which means in a way we are thinking of a transition between the finite space and the infinite space but we are really not at infinity.

This can be taken a step further by assuming a space that is both finite and infinite in behavior through the following definition.

Definition

A space that exhibits attributes of both the finite and infinite is called a transition space of infinity.

Consequently we expect the last prime to exist in this space, and we also assume that the space consists of prime numbers. In a purely infinite space, we then expect only composites and pseudoprimes. Whilst this has been discussed for positive numbers, the same applies for negative numbers. Informally, we may even consider p_n as the start of the transition space though the conceptual diagram does not suggest this.

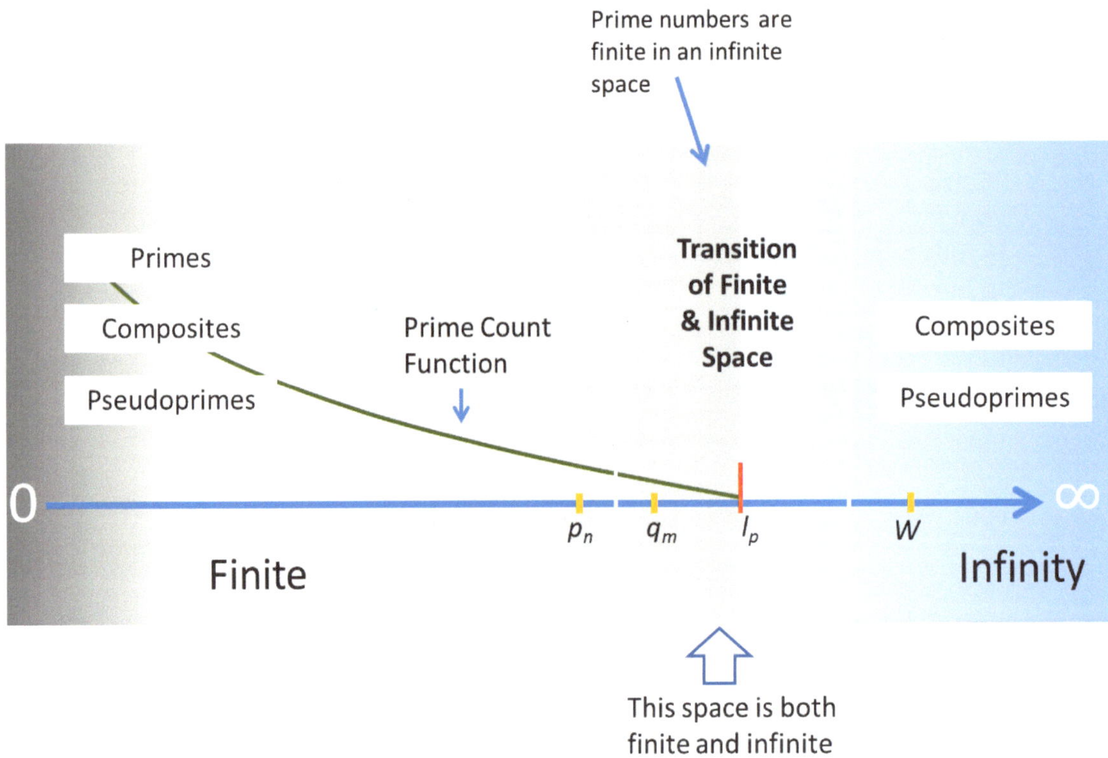

Figure 167

The finite space and the infinite space relationships are therefore determined as a consequence of the distributive property of numbers rather than just as a concept where there are infinitely many numbers. It is also important to note the fact that there is a transition from the finite to the infinite.

Infinity Number Theorem

The infinite number space consists of composites and pseudoprimes only.

The conclusion is a consequence of the mutually exclusive infinity describing the exact type of numbers that are to be found at what is defined as infinity. The infinity number theorem is a far cry from the simplistic real number line interpretation of the infinite.

Index

A

B

C

D

E

F

G

H

I

L

M

N

O

P

R

S

T

U

www.ingramcontent.com/pod-product-compliance
Ingram Content Group UK Ltd.
Pitfield, Milton Keynes, MK11 3LW, UK
UKHW061028310726
14090UKWH00026B/255
9781483630861